ANALYSIS OF NOBLE METALS
Overview and Selected Methods

Analysis of Noble Metals

OVERVIEW AND SELECTED METHODS

F . E . BEAMISH

Department of Chemistry
University of Toronto
Toronto, Ontario, Canada

J . C . VAN LOON

Department of Geology
University of Toronto
Toronto, Ontario, Canada

With the assistance of CLYDE L. LEWIS
Falconbridge Nickel Mines Ltd.
Toronto, Ontario, Canada

ACADEMIC PRESS   New York   San Francisco   London   1977

A Subsidiary of Harcourt Brace Jovanovich, Publishers

6414 - 1925

Chemistry Lib.

ACADEMIC PRESS, INC.
111 Fifth Avenue, New York, New York 10003

United Kingdom Edition published by
ACADEMIC PRESS, INC. (LONDON) LTD.
24/28 Oval Road, London NW1

Library of Congress Cataloging in Publication Data

Beamish, Fred Earl,   Date
        Analysis of noble metals.

        Includes bibliographical references and index.
        1.   Precious metals–Analysis.   I.   Van Loon, Jon
Clement, Date   joint author.   II.   Lewis, Clyde
L., joint author.   III.   Title.
QD132.B43   546'.63   76-27433
ISBN 0–12–083950–4

PRINTED IN THE UNITED STATES OF AMERICA

To our wives
Dorothy and Maureen

CONTENTS

Chapter 3   Spectrophotometry   **77**

Chapter 4   Gravimetric Determinations   **126**

Chapter 5   Volumetric Determinations   **155**

Chapter 6   Fire Assay   **177**

PART II

Chapter 7 **Methods of Separation**   215

References   311

PREFACE

It is anticipated that this book will fulfill the needs of practicing analytical chemists. It is not intended to be a substitute for the two monographs, "The Analytical Chemistry of the Noble Metals" and "Recent Advances in the Analytical Chemistry of the Noble Metals," which have different orientations.

The number of spectrophotometric methods for the determination of noble metals published during the past three decades almost equals the total number published prior to this period. Up to 1970 there were published 160 spectrophotometric procedures for palladium, 68 for gold, and 33 for platinum. For gravimetric methods there were, respectively, 68, 38, and 26.

Faced with these facts, even the experienced analytical chemist will find much uncertainty in choosing the most suitable procedure for separation and determination of noble metals in a material of unknown composition. Unfortunately, there is some tendency even for analytical chemists to accept the more recently published procedures as somehow superior to those well tried and documented.

In a previous monograph "Recent Advances in the Analytical Chemistry of the Noble Metals," the authors have included procedures and discussions of X-ray fluorescence, neutron activation, and electrometric analytical techniques. During the preparation of that monograph it became clear that X-ray fluorescence and associated analytical methods for the noble metals were used quite sparingly in practice. This was true also of electrometric methods with the exception of potentiometric techniques. In the case of neutron activation and associated methods, the published procedures were recorded largely by specialists in these techniques with a relatively small number of analytical chemists contributing to the literature. To the best of the authors' knowledge, few, if any, of the industrial noble metal companies of all countries employ neutron activation methods for routine metal determinations. These methods are very often used to determine traces of impurities of the order of nanogram and microgram quantities for which the method has no peer. During the past two years some 700 papers dealing with the analytical chemistry of the noble metals were recorded in chemical abstracts; approximately 10% of these were concerned with neutron activation and some 5% or more of the latter dealt with nanogram and microgram amounts of metals. Some 2% or less were concerned with X-ray methods and some 5% with electrometric noble metal methods, and these were comprised almost entirely of potentiometric or amperometric procedures with a few polarographic methods. Furthermore, the published neutron activation methods involve relatively simple analytical techniques, e.g., they use chemical yield determinations which almost always need no quantitative recoveries. On the other hand, there are notable research organizations which use the neutron activity more or less regularly. Among these are the Institute of Ghent, Belgium, and the National Bureau of Standards, Washington, D.C.

With this information in mind, it was decided to restrict analytical material to atomic absorption and spectrochemical, spectrophotometric, gravimetric, and volumetric determinations, the latter however including potentiometric.

Therefore, the presentation was designed to provide methods which have been generally useful, involving simple techniques, useful sensitivity, and providing accuracy and precision limited only by the characteristics of measuring instruments. Cases in point are the dimethylgyloxime–palladium reaction and the hydrolytic separation of platinum from other platinum metals, the applications of which remain at least as generally profitable as any of the host of methods now available. To a good degree the separational and determinative methods included in this volume are those which continue to give satisfaction in the authors' laboratory. Some are used extensively in large noble metal analytical laboratories in Canada, the United States, Africa, and Britain.

The authors wish to avoid the impression that the analytical methods chosen here are unequivocally the superior procedures. The majority of the methods recorded in detail in the two earlier monographs as well as those recorded in abstract form in these monographs are, by and large, acceptable methods, some of which, however, have not been reported as having extensive laboratory application. In addition to the detailed procedures, this volume contains a few restricted references in abstract form, which relate to specific analytical problems.

In the authors' opinion, future attempts to review the status of the analytical chemistry of the noble metals will meet a discouraging situation arising from difficulties in obtaining pertinent Russian literature. During the past two decades there has been a rapidly growing volume of research reports dealing with the analysis of noble metals. A recent survey of the number and origin of such reports indicated the preponderance of Russian contributions. In 1971 and 1972 Russian chemists contributed, respectively, 40 and 50% of the total emanating from all sources. To this may be added the number from politically associated countries. One answer to this situation lies in a cooperative effort. The senior author has made some considerable effort to secure this cooperation but without success. This problem must be solved if wasteful and inconsequential researches are to be avoided.

J.C.V.L. must sadly record the death of Professor Frederick E. Beamish, senior author of this book, on April 8, 1976. His passing leaves a vacuum in the field of noble metal analytical chemistry which can never be filled.

ACKNOWLEDGMENTS

The authors are indebted to researchers who graciously gave permission to use their recorded procedures. In particular the National Bureau of Standards, Washington, D.C. and the American Chemical Society allowed the inclusion of the classical work of the late Raleigh Gilchrist and co-workers. In the authors' opinion, Gilchrist's report is the most outstanding single contribution yet recorded in the analytical field of the noble metals. We were fortunate, too, in receiving permission from the National Institute for Metallurgy, Johannesburg, to publish a selection of some of their excellent researches on the analysis of the noble metals; most of these have appeared in the analytical literature, subsequent to the publication of the authors' monographs. The authors are greatly indebted to the Department of Chemistry, University of Toronto, for providing office space and numerous related services.

We wish to acknowledge specifically the following publishers for permission to use the material in the procedures.

The American Chemical Society: Procedures 13, 14, 26, 27, 33, 34, 35, 37, 43, 44, 45, 53, 54, 58, 61, 65, 66, 70, 72, 78, 96, 98, 100, 120, and 122 from *Analytical Chemistry;* Procedures 50, 63, 68, 90, 91, 94, and 104 from *Industrial and Engineering Chemistry, Analytical Edition;* Procedure 30 from *The Journal of the American Chemical Society*. Copyright by the American Chemical Society.

American Institute of Physics: Procedure 7 from *Applied Spectroscopy*.

American Journal of Science: Procedure 67.

American Society for Testing Materials: Procedures 17, 18, and 19 from Methods for Emission Spectrochemical Analysis.

Canadian Institute of Mining and Metallurgy: Procedure 12 from *Canadian Mining and Metallurgical Bulletin;* Procedure 16 from *Transactions of the Canadian Institute of Mining and Metallurgy.*

The Chemical Society (London): Procedures 69, 106, 107, 111, 118, and 119 from *Analyst.*

Economic Geology Publishing Company: Procedure 1 from *Economic Geology.*

Elsevier Scientific Publishing Company: Procedures 3, 4, 21, 39, 48, 57, 75, and 103 from *Analytica Chimica Acta.*

Engelhard Industries: Procedure 15 from *Engelhard Industries Technical Bulletin.*

Institution of Mining and Metallurgy (London): Procedure 76 from the *Bulletin;* Procedures 73 and 74 from the *Transactions.*

McGraw-Hill Book Company: Procedure 95 from *Engineering and Mining Journal.*

Mineralogical Society of America: Procedure 36 from *The American Mineralogist.*

National Business Publications, Ltd.: Procedure 87 from *Canadian Mining Journal.*

National Research Council of Canada: Procedures 32, 40, 41, and 55 from *Canadian Journal of Chemistry;* Procedure 59 from *Canadian Journal of Research.*

Nature (London): Procedure 109.

Pergamon Publishing Company: Procedures 80, 81, and 86 from "Analytical Chemistry of the Noble Metals"; Procedures 5, 6, 25, 79, and 85 from *Talanta.*

South African Institute of Mining and Metallurgy: Procedures 88, 89, and 93 from *Journal of the Chemical, Metallurgical and Mining Society of South Africa.*

South African National Institute for Metallurgy: Procedures 9, 10, 11, 42, 83, 84, and 101 from Reports 970, 1086, 1179, 1232, 1273, 1341, and 1432.

Springer-Verlag: Procedures 8, 24, and 38 from *Fresenius Zeitschrift für Analytische Chemie;* Procedures 108 and 110 from *Mikrochimica Acta.* Copyright by Springer-Verlag.

Thomas Newspapers: Procedure 102 from *South African Industrial Chemist.*

United States Bureau of Standards: Procedures 60, 64, 99, 115, and 117 from *Journal of Research;* Procedure 97 from *Scientific Papers.*

United States Geological Survey: Procedure 2 from Circular 544.

John Wiley and Sons (Interscience): Procedures 20, 22, 28, 29, 31, 46, and 47 from "Colorimetric Determination of Traces of Metals."

PART I

CHAPTER 1

ATOMIC ABSORPTION

INTRODUCTION

Atomic absorption spectroscopy (A.A.) methods were applied to the determination of noble metals immediately following its establishment as a useful analytical technique some ten years ago. Rapid acceleration in A.A. methods has occurred within the past two or three years. For this reason it is not possible to produce a report which will remain up to date for any appreciable time.

The work to the present has shown that platinum metal determinations with the commonly employed, lower temperature flames, (e.g., air–acetylene, air–propane) are fraught with a complex sequence of chemical cationic and anionic interferences. Gold determinations appear to be less complex. Recent work with the hotter nitrous oxide flame gives hope for a much simpler interference pattern for all the noble metals. In any case, a final assessment of best A.A. operating parameters awaits further research.

Because of the almost exclusive use of low temperature flames in early work and the recognition of interference complexity for which no easy solution existed, early publications frequently recommended matrix standards as a means of circumventing problems. Another popular early approach was the use of solvent extraction as a means of removing the noble metals from the complex sample solutions and hence potential interferences. Recent work, while frequently employing a separational technique, has also outlined one or a combination of several releasing and buffering reagents which are capable of overcoming interferences.

Most papers contain a discussion of potential interference problems* in which data pertaining to chemical interference of associated cations and anions is presented (usually for ion concentrations lower than 1000 ppm). As a result this type of interference problem is now fairly well documented for the analytical applications covered to date in the literature. Interference in high salt content solutions ($\sim$10,000 ppm range) due to light scattering and/or molecular absorption phenomena have received little attention. This is surprising since solutions high in salts and low in noble metal content are commonly encountered. Carlson and Van Loon (1a) showed that soil sample solutions containing 10,000 ppm total cations (Na, Ca, Mg, Al, Fe) mainly as chlorides gave absorption signals at 242.8 nm corresponding to 4 ppm gold even when gold was not present. In some types of high salt content gold is sufficiently concentrated that dilutions can be made to remove this interference (e.g., Mikhailova et al. (2) electrolytes and plant liquors Mikhailova et al. (3)).

A variety of releasing agents and buffers have been suggested for suppressing interference in the determination of noble metals using low temperature flames. Mallett et al. (4, 5) used vanadium as a suppressor in a perchloric acid base metal solution and uranium as a suppressor in silver and gold prill sample solutions respectively in the analysis of platinum, palladium, rhodium, ruthenium, and gold. Uranium was also used by Jansen and Umland (6) to overcome interference in the determination of rhodium, palladium, iridium, and platinum. Pitts et al. (7) and Schnepfe and Grimaldi (8) used lanthanum to eliminate interferences due to basic metals and other cationic interferences in the determination of platinum and rhodium respectively. Interference in the determination of iridium was suppressed by adding mixtures of sodium and copper in procedures described by Van Loon (9) and Grimaldi and Schnepfe (10). Mallett and Breckenridge (11) found that either vanadium or uranium were effective in suppressing interferences in the determination of iridium and osmium. The combination copper and cadmium was used by Rowston and Ottaway (12) to remove interferences in ruthenium determinations.

A variety of good separational procedures developed for noble metal analysis, are applicable to A.A. methods. Of these, solvent extraction and fire assay are

---

*See article on interferences and their elimination in the determination of the noble metals by atomic absorption spectrometry, Mallett et al. (1).

most commonly employed. Fire assay techniques provide a means of concentration as well as yielding a product for dissolution which has a relatively simple matrix. In addition to providing a means of concentrating small quantities of noble metals, solvent extraction yields an organic medium for aspiration which frequently results in substantially enhanced signals over those obtained for a similar concentration of metals in aqueous media.

Of the available fire assay procedures the classical lead assay followed by a cupellation giving a silver or gold bead (prill) is most commonly utilized. It must be borne in mind that silver beads are only widely applicable to the determination of gold, platinum, or palladium. The other noble metals are relatively insoluble in a silver matrix.

Van Loon (13) collected gold, platinum, and palladium in a silver bead, dissolved the bead in aqua regia, and performed the A.A. determination on a chloride sample solution. The silver bead was also used by Huffman et al. (14) to collect gold. After an aqua regia parting and treatment of the sample with hydrobromic acid and bromoaurate complex obtained was extracted into methyl isobutyl ketone (MIBK). In the determination of rhodium (8) and iridium (10) Grimaldi and Schnepfe employed a gold bead collection. Analyses were performed on chloride sample solutions. Mallett et al. (5) performed a five element analysis (platinum, palladium, rhodium, ruthenium, and gold) on silver or gold prills dissolved in aqua regia.

Solvent extraction is commonly employed to separate noble metals from wet chemical leach solutions of ores and rocks, industrial plant wastes, and cyanide plating baths. Pitts and Beamish (15) utilized an ethyl acetate extraction of the stannous chloride platinum complex leaving a wide range of potentially interfering cations in the aqueous phase. Gold was extracted from prill sample solutions using di-Bu carbitol by Fowler et al. (16). Following a tellurium coprecipitation of gold, Hildon and Sully (17) extracted the chloroaurate complex into MIBK. Boncetta and Fritshe (18) and Ichinose (19) also utilized the extraction of the chloroaurate complex into MIBK as a method of extracting gold from geological sample solutions. Gold bromide extraction into MIBK is the basis of a method proposed by Thompson et al. (20) for the analysis of geological sample solutions. Chao (21) combined an ion exchange concentration of gold in river and waste waters with a solvent extraction of the bromoaurate complex into MIBK.

The atomic absorption determination of gold has been extensively investigated with the result that a large variety of procedures have been recorded for its determination in numerous sample types. Methods for the other noble metals are not as readily available. Table 1.1 gives a representative list of A.A. methods.

Recent trends in A.A. noble metal methods development center around research into improved atom reservoirs. There is good evidence that interferences with noble metals determinations are greatly diminished and frequently eliminated in the hotter nitrous oxide flames. Pitts et al. (7) found that all cationic

TABLE 1.1
Methods for the Atomic Absorption Determination of Noble Metals

Metal	Absorption (nm)	Flame	Metal content (ppm)	Sample type	Ref.
Au	242.8	C_2H_2–air	0.05–50 WR[a]	Plant liquors	2
	242.8	C_2H_2–air	2 DL[b]	High salt content solutions	1
	242.8	C_2H_2–air	0.01–2 WR	Electrolytes	3
	242.8	C_2H_2–air	0.02–0.2 WR	Rocks	18
	242.8	Organic solvent–air	0.002–2S	Rocks	19
	242.8	C_2H_2–air	0.2 DL	Ores	10
	242.8	Flameless carbon rod	0.0004 DL	Geologic materials	22
	242.8	C_2H_2–air	0.01–0.5 WR	Cyanide solution	23
	242.8	C_2H_2–air	nanogram	River waters	21
	242.8	C_2H_2–air	0.02 DL	Rocks and soils	20
	242.8	C_2H_2–air	0.02 DL	Geologic materials	14
Pt	265.9	C_2H_2–air	0.5 DL	Base metal solutions	15
	265.9	C_2H_2–air	—	Precious metal concentrate	24
	265.9	C_2H_2–N_2O	—	Precious metal concentrate	7
Rh	343.5	C_2H_2–air	0.07 DL	Chromite ores	8
	349.9	C_2H_2–air	5–100 WR	—	12
Os	290.9	C_2H_2–N_2O	1 Sen[c]	Receiver solutions	35
Ir	264.0	C_2H_2–air N_2O–air	32 Sen	Mafic rocks Complexes	10 26
	285.0	C_2H_2–air	15 Sen	Precious metal concentrate	9
Au	242.8	C_2H_2–air	1 Sen	Ores (Ag bead)	13
Pd	247.6	C_2H_2–air	1 Sen	Ores (Ag bead)	13
Pt	265.9	C_2H_2–air	5 Sen	Ores (Ag bead)	13
Pt	265.9	C_2H_2–air	0.5 DL	Fire assay prills	5
Pd	244.9	C_2H_2–air	0.03 DL	Fire assay prills	5
Rh	343.5	C_2H_2–air	0.05 DL	Fire assay prills	5
Ru	349.9	C_2H_2–air	0.5 DL	Fire assay prills	5
Au	242.8	C_2H_2–air	0.05 DL	Fire assay prills	5
Pt	265.9	C_2H_2–air	20–100 WR	Base metal solutions	4
Pd	244.8	C_2H_2–air	2–10 WR	Base metal solutions	4
Rh	343.5	C_2H_2–air	2–10 WR	Base metal solutions	4
Ru	349.9	C_2H_2–air	10–50 WR	Base metal solutions	4
Au	242.8	C_2H_2–air	2–10 WR	Base metal solutions	4

[a]WR   Working Range.
[b]DL   Detection Limit.
[c]Sen   Sensitivity.

interferences, noted in air–acetylene flames, were eliminated using nitrous oxide flames. Unfortunately, the sensitivity was also somewhat less. Carlson and Van Loon (1a) found simpler interference patterns using nitrous oxide flames for gold determinations but increases in noise levels resulted in poorer detection limits. No chemical interferences were reported by Luecke and Zeike (26) using nitrous oxide flames in the determination of iridium in relatively simple matrix sample solutions. Of the noble metals, osmium is the only one definitely requiring nitrous oxide to obtain suitable absorption magnitudes.

The recent development of flameless absorption cells will undoubtedly have a great impact on A.A. methods for noble metals. Flameless cells show great potential for improving detection limits by at least one order of magnitude. In general a solvent extraction step is essential to provide a suitable sample solution for the cell. Bratzel *et al.* (22) extracted the chloroaurate complex into MIBK and added micro quantities to a carbon rod atomizer. The determination limit using a 2.5-μl sample was 6×10^{-13} g (absolute).

Atomic fluorescence (A.F.), a recently introduced closely related technique of A.A., may be applicable to the determination of noble metals. The main potential advantage of this technique over A.A. is its presently publicized adaptability to multielement simultaneous analysis. Matousek and Sychra (27) observed A.F. for gold at 242.8, 267.6, and 312.3 nm. Recently Aggett and West (29) investigated the A.F. of gold at 242.8 and 267.6 nm. They found serious interference with fluorescence signals when high salt content solutions were used. Much more research will be necessary to evaluate fully the potential of atomic fluorescence as a routine analytical tool.

GOLD

Gold in Waters

A method for the atomic absorption determination of gold in the nanogram range in river waters or waters discharged from gold mining operations is described by Chao (21). An anion exchange concentration procedure is used to bring gold levels up to determinable amounts.

PROCEDURE 1 (21)

Use an ion exchange apparatus as shown in Fig.1.1, which permits the automatic gravitational flow of water from a large container through a 2.5-cm resin column containing 3 g of AG 1-X8 anion resin of 100–200 mesh. The rate of flow must be adjustable by a Teflon stopcock. With a Perkin Elmer 303 atomic absorption spectrophotometer and a

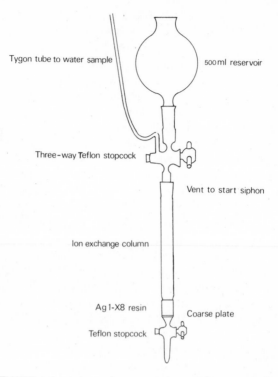

Tygon tube to water sample

500 ml reservoir

Three-way Teflon stopcock

Vent to start siphon

Ion exchange column

Ag 1-X8 resin

Coarse plate

Teflon stopcock

FIGURE 1.1.   Anion exchange apparatus used for concentration.

Boling burner use the following operating conditions: wavelength, 2428 Å; slit, 5 (3 mm, 20 Å); lamp current, 14 mA. With the burner lit, adjust the flow meters to approximately 2.8 on the acetylene and 3.0 on the air while aspirating water. Minor adjustments may be necessary to obtain a blue flame when the MIBK is aspirated.

To prepare a standard gold solution dissolve 0.200 g of gold wire in a 250-ml covered beaker with 30 ml aqua regia on a steam bath, remove cover, and evaporate to dryness. Add 100 ml (1 + 1) nitric acid to dissolve the gold salt. Transfer the solution to a 500-ml volumetric flask. Rinse the beaker with an additional 100 ml (1 + 1) nitric acid and pour the rinsing solution into the flask. Dilute to volume with demineralized water.

Collect 10 or more liters of water and filter through a 4-in. Gelman membrane filter (0.10 μ pore size). (A plastic pressure filtration unit described by Skougstad and Scarbro (28) may be used to aid the filtration of water under a nitrogen pressure of 40 psi.) The sample size of water depends on the concentration of gold in the water. Fifty to 1000 nanograms of gold per liter can be accurately determined with 10 l of water. With lower concentrations of gold, the volume of water is proportionally increased. The filtered water is immediately acidified to pH 1 with hydrochloric acid and brominated with 50 mg of bromine per liter. If gold associated with suspended materials is to be determined, save the filter for that purpose.

In the laboratory, pass the water through a 2.5-cm AG 1-X8 anion exchange resin column (3 g of the resin with 40% moisture) at a flow rate of 2.5 ml/min using an automatic siphon system. Rinse the resin column with 250 ml demineralized water and drain the water completely. By turning the upper 3-way Teflon stopcock (Fig.1.1), the column is opened to the 500-ml reservoir from which 250 ml of an acetone nitric acid (sp gr 1.48)–water mixture (volume ratio = 100:5:5) are allowed to elute the gold from the resin at a rate of 50 ml/hr into a 250-ml Teflon beaker. Evaporate the acetone at 50–55°C in a constant temperature water bath. Transfer the remaining solution to a screw-capped test tube and dilute to 50 ml with water. Add 0.25 ml (5 drops) of bromine water and mix. Using a pipet, measure 10.0 ml MIBK into the tube and shake vigorously for 1 min. Allow the MIBK layer to separate completely. The MIBK layer is then aspirated into the flame of the atomic absorption spectrophotometer for gold determination.

Prepare gold reference standard solutions containing 0.1, 0.2, 0.3, 0.4, 0.6, 0.8, and 1.0 μg gold per milliliter in the MIBK phase by adding appropriate aliquots of a gold standard solution to a series of tubes. Add 20 ml water and 10 ml nitric acid (sp gr 1.42), and dilute to 50 ml. Proceed with addition of bromine water and MIBK extraction as above. A demineralized water blank is treated in the same manner as the sample. Within the above range a linear relationship is observed between the scale reading and the gold concentration.

Gold in Geologic Materials*

A procedure for the determination of gold in geologic materials has been reported by Huffman *et al.* (14). The method, applicable to methyl isobutyl ketone extracts of sample solution, can be used after fire assay and bead parting or on cyanide leach solutions. The following procedure describes the method for fire assay beads.

PROCEDURE 2 (14)

To prepare a standard gold stock solution containing 400 ppm gold, dissolve 0.200 g of 99.999% gold metal in a covered beaker with 30 ml of aqua regia on a steam bath, remove cover, and evaporate to dryness. Heat on a hot plate to expel completely nitric acid fumes. Add 250 ml of (1 + 1) hydrobromic acid and digest on the steam bath. Transfer the solution to a 500-ml volumetric flask and dilute to volume with distilled water. Use best available reagent grades 48% hydrobromic acid and methyl isobutyl ketone.

Prepare reference standards containing 0.1, 0.2, 0.4, 0.8, 1.0, 2.0, 4.0, 8.0, and 10.0 ppm gold in the organic phase for working curves by adding appropriate aliquots of a gold standard solution to a series of culture tubes. Add 2 ml of 48% hydrobromic acid to each and adjust the volume to about 40 ml with water. Add 10.0 ml of MIBK and shake for 1 min to extract the gold into the organic phase. Measure the absorption of these standards

---

*Determination of gold and silver in parts per billion or lower levels in geological and metallurgical samples by atomic absorption spectrometry with a carbon rod atomizer (Bratzel *et al.* (29a)).

on the 1× or 10× scales of the atomic absorption spectrophotometer. Periodically, standards should be checked.

Transfer the silver gold bead obtained by a fire assay to a 50-ml beaker. Add 5 ml of (1:1) nitric acid, digest on a steam bath for 45 min, and evaporate to 2.5 ml. Add 6 ml of concentrated hydrochloric acid, cover, and let stand 20 min. Place the beaker on a steam bath and digest for 30 min. Evaporate to remove nitrogen oxides. Add 7 ml of concentrated hydrochloric acid and digest on the steam bath to dissolve salts. Add about 15 ml of water, and digest an additional 30 min. Cool and transfer the solution to a Pyrex culture tube by washing the beaker with about 10 ml of water. Add 2 ml of 48% hydrobromic acid, mix, add 10 ml of MIBK, and shake for 1 min. Determine the absorbance of the organic phase along with reference standards.

Gold can be determined down to 50 ppb with this method.

The following parameters may be used with a Perkin Elmer Model 303 atomic absorption spectrophotometer: wavelength, 2428 Å; slit, 3 mm; lamp current, 14 mA.

Gold in Cyanide Waste

The very important determination of gold in cyanide solution was done by Strelow *et al.* (27) using atomic absorption. This procedure dates back to 1966, the early days of atomic absorption. But to the authors' knowledge no method has been proposed in more recent times which has any distinct advantages. Gold is extracted from the complex matrix using methyl isobutyl ketone.

PROCEDURE 3 (27)

A Zeiss P.M.Q. II spectrophotometer with a gold hollow cathode lamp on an aluminum burner, similar to that described by Butler (30), fitted with a flat water-cooled burner plate and a standard Zeiss atomizer operated at 20 psi air pressure may be used. Use the 242.8-nm gold line. Separation funnels with Teflon stopcocks must be used. Because the ketone is flammable and causes the flame to become smokey and fuel rich a supplementary air supply may be necessary when the ketone is sprayed. This makes possible the slightly oxidizing flame most suitable for this work.

To prepare the standard curve add a standard solution of gold, containing 0.01–0.5 ppm Au in 0.1 *M* hydrochloric acid, to separating funnels. Dilute with distilled water to give a total volume of 100 ml. Then add 43.0 ml of concentrated hydrochloric acid, followed by 10.00 ml of methyl isobutyl ketone. Shake the funnels for about 5 min, after settling separate the organic layer. Aspirate the ketone into the flame.

Measure 100 ml of waste solution into a beaker together with 43 ml of concentrated hydrochloric acid and 1 g of solid potassium permanganate. Boil the solution for 10 min and, after cooling to room temperature, transfer into a 250-ml separatory funnel marked at 143 ml and dilute with water to this volume. Measure 10 ml of methyl isobutyl ketone accurately and add to the funnel. Shake the solution for 2 min. After the phases have settled, separate the ketone and then aspirate into the flame of the spectrophotometer. Prepare a standard curve, including the extraction step at the same time, and use this to read the gold concentration. Because absorption readings are fairly sensitive to slight changes in flame conditions such as fuel gas and air pressure or solvent concentration, readings for standards must always be taken together with those for the unknown.

The procedure is applicable to gold in the range 0.01 to 0.5 ppm in the extract.

PLATINUM

Platinum in a Sulfide Concentrate

The determination of platinum in a sulfide concentrate was studied in detail by Pitts *et al.* (7, 29). A complex variety of cationic and anionic interferences were outlined in air–acetylene flames. These could be overcome in all cases by the addition of lanthanum (0.2% in final solution). No interferences with the platinum absorbances were noted using the higher temperature nitrous oxide flame. A slight loss in sensitivity is however a negative feature of the nitrous oxide flame.

PROCEDURE 4 (7, 29)

Reagents and Equipment

A Techtron Model AA4 atomic absorption spectrophotometer fitted with a 5-cm Techtron AB-50 nitrous oxide–acetylene burner was used. Lanthanum chloride solution was prepared by dissolving high-purity lanthanum(III) oxide in a minimum amount of hydrochloric acid.

Procedure (Tested on a Precious Metal Concentrate)

Weigh a sample of appropriate size into a 25-ml nickel crucible (0.2 g with this ore). Add a few drops of water to moisten the sample and follow with 5 ml of hydrofluoric acid. Heat to expel all the silica and hydrofluoric acid. Mix an excess of sodium peroxide (2 g for 0.2 g of sample) intimately with the residue by means of a glass rod and brush the rod clean of the sample. Sinter the material over a flame for 30 min and then fuse at red heat for 5 min. Cool the crucible and contents and place in a 250-ml covered beaker. Add water to fill the crucible and after 5 min, quantitatively transfer the crucible contents to the beaker; finally rinse the crucible with a 50% solution of hydrochloric acid. Slowly acidify the solution in the beaker with 20 ml of concentrated hydrochloric acid and 5 ml of concentrated nitric acid, digest on a steam bath, and evaporate to 20 ml. Then destroy most of the nitrates by four successive additions of concentrated hydrochloric acid followed by evaporation after each addition. Dilute the sample to 75 ml with doubly distilled water and filter into a 100-ml volumetric flask; silver which is precipitated as silver chloride is thus eliminated from the sample solution. Depending on the platinum concentration expected, a dilution of the above may be necessary. The sample solutions can be run using air–acetylene or nitrous oxide. If air–acetylene is used the samples and standards should contain 0.5% lanthanum.

The following are the instrument parameters recommended: lamp current, 10 mA (Perkin Elmer Intensitron), burner height, air–C_2H_2 0.1 cm (burner to beam), N_2O–C_2H_2 1.0 cm; slit, 4.9 Å; gas flow rate, air/C_2H_2, 4/1; N_2O/C_2H_2, 4.5/3. The detection limits for platinum using the above dissolution procedure are 0.5 and 2.0 ppm for air–C_2H_2 and N_2O–C_2H_2 flames, respectively.

RHODIUM

Rhodium in a Chromite Concentrate

Schnepfe and Grimaldi (8) outlined a procedure for the determination of rhodium which was applied to chromite concentrates. Rhodium is concentrated by either a classical fire assay followed by a gold bead cupellation or by co-precipitation with tellurium. A large number of cationic interferences are overcome by the use of 1% lanthanum as the sulfate. Other platinum metals as well as substantial amounts of Ag, Al, Au, Bi, Ca, Cd, Co, Cr, Cu, Fe, Ho, Hg, K, Mg, Mn, Mo, Na, Ni, Pb, Te, Ti, U, Y, and Zn could be present under these circumstances without causing interference.

PROCEDURE 5 (8)

Reagents and Apparatus

Tellurium solution   (1 mg/ml in 10% (v/v) hydrochloric acid). Dissolve tellurium metal in aqua regia and remove nitrate by evaporation with hydrochloric acid.

Tin(II) chloride solution.   Dissolve 20 g of fresh tin(II) chloride dihydrate in 17 ml of hydrochloric acid. Dilute to 100 ml with water.

Gold wire for fire assay.   Wire is 99.999% pure and 0.1 mm in diameter. Cut into 2.5-mg segments.

Lanthanum sulfate solution.   Dissolve 14.66 g of lanthanum oxide in approximately 25 ml of hydrochloric acid and then add 15 ml of sulfuric acid (1 + 1). Evaporate the solution. Dissolve the residue in 125 ml of hydrochloric acid and dilute to 500 ml with water.

Standard solutions of rhodium.   Prepare from the ammonium chloro-salt a stock solution containing 1.00 mg of rhodium per ml in 2% (v/v) hydrochloric acid. Prepare other solutions by dilution by factors of 10 with 2% (v/v) hydrochloric acid.

Alumina crucibles for sodium peroxide fusions.   Coors Ad-999 Alumina ceramic, available from Coors Porcelain Company, Golden, Colorado.

Instrument parameters and settings. A Perkin Elmer Model 303 instrument was used with the following operating conditions:

Wavelength	343.5 nm	Air flow setting	6.8
Slit	0.3 mm	Flame	oxidizing
Hollow-cathode current	20 mA	Burner	standard head
Acetylene flow setting	6	Aspirator	adjusted for optimum uptake

Tellurium Precipitation Procedure

Fuse over a burner in an alumina crucible 3.0 g of chromite concentrate with 10 g of fresh sodium peroxide. Heat for approximately 15 min after the charge becomes molten. After cooling the melt, place the crucible in approximately 100 ml of water in a beaker and carefully add 60 ml of hydrochloric acid. Detach the melt and remove the crucible. Heat the solution to approximately 60°C and while stirring, very carefully add 5–7 ml of 30% hydrogen peroxide to reduce chromium. Heat this solution on the steam bath for 30 min or more to destroy peroxide and then filter the solution through Schleicher & Schüll 589 White Ribbon paper (or equivalent).

Add 2.5 ml of tellurium solution, then 15 ml of tin(II) chloride solution by pipet while stirring. Adjust solution volume to approximately 200 ml. Digest the tellurium precipitate on a steam bath for approximately 2 hr, filter it off on a Schleicher & Schüll 589 White Ribbon paper, and then wash with hot 10% (v/v) hydrochloric acid. Discard the filtrate.

Dissolve the precipitate off the paper by slowly adding 50 ml of hot aqua regia (8 volumes of hydrochloric acid, 2 of nitric acid, and 5 of water) collecting the filtrate in a 100-ml beaker. Wash finally with hot 10% (v/v) hydrochloric acid. Pass this filtrate through a 15-ml medium-porosity fritted glass Buchner-type filter funnel to remove paper fibers. Wash the filter with 10% (v/v) hydrochloric acid and then evaporate the solution to dryness. Remove nitrate by several evaporations with 3-ml portions of hydrochloric acid. Add 2 ml of lanthanum sulfate solution to the residue and warm briefly on the steam bath. Transfer the solution to a 5-ml volumetric flask and adjust to volume with water. Prepare rhodium standards and a blank containing 2 ml of the lanthanum solution in a 5-ml volume. Determine rhodium on all solutions by atomic absorption.

Fire Assay Procedure

Add 3.0 g of sample to a flux consisting of 35 g of sodium carbonate, 11 g of silica, 19 g of anhydrous sodium tetraborate, 50 g of lead oxide, and 4.2 g of flour contained in a fire assay crucible and mix thoroughly. Place in a furnace at 850°C and gradually raise the temperature to 925°C. Heat for 10 min at this temperature. Total heating time should be approximately 50 min. Pour the melt into an iron mold. Collect the lead button and shape it into a cube. Make a linear indentation on one surface of the cube by tapping a knife edge against this surface. Place a segment of gold wire in the indentation and then carefully hammer the cube to secure the wire in place. Cupel the lead button at approximately 950°C.

Transfer the gold bead to a 5-ml beaker. Add 2 ml of aqua regia and allow the mixture to stand overnight at room temperature. Heat the solution to ensure complete dissolution of the bead. Evaporate the solution, and if necessary repeat the treatment with hot aqua regia, evaporating to dryness on the steam bath each time. Add 1 ml of hydrochloric acid (1 + 1) and evaporate the solution to dryness. Repeat the treatment with hydrochloric acid (1 + 1).

Add 2 ml of lanthanum sulfate solution to the dry residue and continue as in the previous procedure. Using a 3-g sample as low as 0.07 ppm, rhodium can be determined.

IRIDIUM

Iridium in Mafic Rocks

Grimaldi and Schnepfe (10) have published a method for the atomic absorption determination of iridium in mafic rocks. A fire assay concentration of iridium into a gold bead is used. Interferences are overcome using a modification of a buffering technique developed by Van Loon (9) employing a mixture of copper and sodium sulfates.

PROCEDURE 6 (10)

Reagents and Equipment

Gold wire. This is 99.999% pure.

Copper sulfate solution (7% (w/v) copper). Dissolve 27.5 g of copper sulfate pentahydrate in water and dilute to 100 ml.

Mixed copper–sodium solution. Dissolve 13.75 g of copper sulfate pentahydrate and 4.64 g of sodium sulfate in 50 ml of hydrochloric acid and make up to 100 ml with water.

Standard solutions of iridium. Prepare from ammonium hexachloroiridate a stock solution containing 1.000 mg of iridium per ml in 2% (v/v) hydrochloric acid. Prepare other solutions by dilution with 2% (v/v) hydrochloric acid.

Cylindrical alumina crucibles, 2-ml capacity, for sodium peroxide fusions. Coors AD-999 alumina ceramic crucibles available from Coors Porcelain Company, Golden, Colorado.

Synthetic standards. Two synthetic standards were prepared, using U.S.G.S. dunite DTS-1 as a base. Analyses cited by Flanagan (31) show it to contain negligible amounts of platinum metals. The first standard was made to contain 10 ppm each of iridium and rhodium, and the second 10 ppm each of iridium, osmium, palladium, platinum, rhodium, and ruthenium. Iridium contents were verified by activation analysis while palladium, platinum, and rhodium contents were verified by fire assay/atomic absorption procedures (8). The fate of osmium and ruthenium was not ascertained. The standards are prepared as follows. To weighed samples of dunite finer than 200-mesh in porcelain casseroles, add water to form a slurry. Add the required amounts of standard solutions of the appropriate platinum metals and mix. Evaporate the solution on a steam bath, stirring the slurries frequently. Dry the residues in an oven at 110°C and then heat in a furnace at 450°C for approximately 30 min. Roll the samples in a ceramic jar mill with alumina balls for 3 hr. Mix and bottle.

Instrument parameters and settings. A Perkin Elmer Intensitron hollow-cathode tube with a Perkin Elmer Model 303 instrument was used in the experiments. The conditions were as follows: wavelength, 263.97 nm; slit, 0.3 mm; hollow-cathode current, 30 mA; fuel, acetylene pressure, 69 kN/m^2 (10 psi); flow-meter setting, 8.5; oxidizer, air pressure, 190 kN/m^2 (28 psi); flow-meter setting, 7.5; flame, oxidizing; burner, standard head; aspirator, adjusted for uptake of ~3 ml/min. Conditions for other instruments must be found by trial and error.

Procedure

The fire assay fusion and cupellation procedures used here follow generally accepted practices as described by Bugbee (32). A bisilicate slag composition is to be preferred. For a 20-g sample of dunite, a flux consisting of 50 g of lead(II) oxide, 35 g of sodium carbonate, 15 g of silica, 19 g of sodium tetraborate, and 4 g of flour will yield both a satisfactory lead button and a bisilicate slag. The fusion is made in the presence of 50 mg of added gold for the quantitative collection of iridium, palladium, platinum, and rhodium.

Transfer the gold bead obtained on cupellation to a small beaker. Add 5 ml of aqua regia, cover and allow the mixture to react at room temperature for approximately 1 hr. Heat on a steam bath for several hours more. Add 3 ml of water and a small amount of paper pulp, mix, and filter through a 42.5-mm medium porosity filter paper. Wash the iridium residue with water. Reserve the filtrate for the determination of palladium, platinum, and rhodium. Ignite the residue in a 2-ml alumina crucible at 600°C starting with a cold furnace. Add 100 ± 10 mg of sodium peroxide by calibrated dipper and carefully fuse the residue. Heat for 5 min more, maintaining in the molten state. Cool. Add 1.5 ml of water, cover and allow to stand at room temperature for approximately 15 min. Warm the mixture until the melt disintegrates, and transfer the solution to a 25-ml beaker by several alternating washes with 1-ml portions of concentrated hydrochloric acid, nitric acid, and water. Add 0.15 ml of sulfuric acid (1 + 1) by pipet to convert sodium salts into sulfate and evaporate the solution on a steam bath. Add 2 ml of hydrochloric acid (1 + 1), warm briefly to dissolve salts, and transfer the solution to a 10-ml volumetric flask with water. Dilute to volume with water and mix. Transfer a 5-ml aliquot of the solution to another 10-ml volumetric flask, and reserve the remainder for the determination of palladium, platinum, and rhodium. Add 1 ml of hydrochloric acid (1 + 1) and 1 ml of copper sulfate solution. Adjust to volume with water and mix. Prepare iridium standards and a blank, each containing 1 ml of mixed copper–sodium solution in a 5-ml volume. Determine iridium in all solutions by atomic absorption. Samples containing as low as 2.5 ppm iridium can be determined.

OSMIUM*

Osmium in Aqueous and Nonaqueous Solutions

Oslinski and Knight (25) used the 2909-Å line for the determination of osmium. A nitrous oxide flame is required. Osmium can be determined in aqueous and nonaqueous samples in any valence state. The method has not been proven over a wide range of conditions but should be useful for determinations following distillation separation. The sensitivity is 1 ppm and less than 5 min of operator time is required per sample.

PROCEDURE 7 (25)

Reagents and Equipment

A Jarrell–Ash atomic absorption spectrometer, Model 82–526, with 3000-Å grating, type R-106 photomultiplier tube, 5-cm laminar flow burner (Aztec Instruments Inc.), and a Westinghouse osmium hollow cathode lamp are used.

Osmium tetroxide is obtained from Mallinkrodt in 1-g ampoules.

Use the primary resonance line of 2909.0 Å. The experimental parameters established are the following: 8 mA current to the cathode, 1¼ psi acetylene, 32 psi nitrous oxide, and either 700 or 720 V applied to the photomultiplier tube. Use a single optical pass. At the fuel pressures cited, the reddish part of the flame is about ½ in. in height. Adjust the burner height so that the light emitted from the hollow cathode tube will pass directly over the thin white part of the flame, since the concentration of neutral dissociated atoms is highest in this region of the flame.

Standard osmium solutions. Prepare standards by dissolving osmium(VIII) oxide from the ampoules in chloroform or water. These may be standardized by treatment with thiourea, or by precipitation with thionalide (see Procedure 66). Prepare working solutions by dilution of the above.

Procedure

Determination of osmium in water or chloroform solutions. Adjust the equipment as indicated in the instrumental parameters and aspirate the solutions directly. Owing to incomplete combustion of chloroform, the burner slot must be cleaned after each analysis. The glass burner chamber and burner head are cleaned periodically of any solid residue on the inside surfaces. The combustion products must be vented into a hood.

---

*Determination of atomic absorption spectrophotometry of osmium and iridium in solution (Mallett *et al.* (32a)).

Determination of ruthenium and osmium in residues resulting from a leaching of mattes (Mallett *et al.* (32b)).

Determination of osmium in thiourea complexes (25). The thiourea complex is a familar occurrence for osmium during analytical manipulations. Before the atomic absorption determination can be performed the thiourea complex must be destroyed as follows. Add 5 ml of a solution of the complex to a 25-ml volumetric flask with 2 ml of 30% hydrogen peroxide. Dilute each flask to the mark with distilled water, stopper, and allow to stand at room temperature for 2½ hr for the reaction to proceed to completion. Then determine osmium in the flask by direct aspiration.

MULTIELEMENTS

Platinum, Palladium, and Gold in Ores and Concentrates*

Van Loon (13) recorded a simple atomic absorption method for platinum, palladium, and gold in ores and concentrates. The method requires a fire assay preconcentration into a silver bead. The silver bead is parted in aqua regia, nitrates are removed, and silver is stabilized in the sample solution by using sufficient hydrochloric acid to retain silver in its complexed form. One percent lanthanum is used to overcome interelement interferences.

PROCEDURE 8 (13)

Reagents and Equipment

A Perkin Elmer 303 Atomic Absorption spectrophotometer fitted with a Boling burner was used. The respective wavelengths, slit widths, and lamp currents were as follows: platinum 266 nm, 1.0 mm, 25 mA; palladium 248 nm, 1.0 nm, 30 mA; and gold 243 nm, 3.0 mm, 14 mA.

Standard solutions were prepared from pure metals of each material. Other reagents were Fisher Analyzed Grade and were checked for the absence of the precious metals.

Procedure

Add 5 ml of concentrated nitric acid to the bead in a 50-ml beaker and heat to leach the silver. Evaporate the solution to 0.5 ml and add several ml of concentrated hydrochloric acid. After all bubbling has ceased add additional portions of concentrated hydrochloric acid until no further gases are evolved. Add enough lanthanum to make the final concentration in the flask one percent. Wash the mixture into a volumetric flask and dilute to volume with 6 N hydrochloric acid. Determine the platinum, palladium, and gold absorbances versus standards containing 1% lanthanum and 6 N hydrochloric acid by running a lower concentration standard, the sample, and a higher concentration sample in quick

---

*Use of extraction concentration during platinum palladium and gold determination by A.A. flame photometry (Fishkova (32c)).

succession. The lower limits of detection are 1 ppm palladium, 1 ppm gold, and 5 ppm platinum.

Platinum, Palladium, Rhodium, Ruthenium, and Gold with Other Metals

The above elements can be determined in the presence of up to 5 g/l of Sn, Mg, Cr, Co, Ni, Al, Se, Te, Ca, Cu, or Na by atomic absorption spectroscopy using the method of Mallett *et al.* (4). Vanadium in perchloric acid solution is used as a releasing agent. No sample dissolution method is given but the authors recommend a chlorination.

PROCEDURE 9 (4)

Reagents and Equipment

Perchloric acid. Concentrated A.R.-grade as supplied by Baker Analyzed (70–72%).

Vanadyl chloride solution (1 ml≡ 100 mg of V). Transfer 69 ml of 50% $VOCl_2$ (w/v) density 1.3, as supplied by British Drug Houses, to a 100-ml volumetric flask, add 10 ml of hydrochloric acid, and dilute to the mark with water. Do not keep this solution for more than a month.

Lithium chloride solution (10 mg Li per milliliter). Dissolve 6.14 g of LiCl A.R.-grade in water and dilute to 100 ml.

Stock solutions. These solutions are made from pure metals or metal salts so that all the solutions, with the exception of that for platinum, contain 1 mg of the noble metals per milliliter of solution. The platinum concentration is 10 mg/ml.

Working solutions. Platinum. Dilute 10 ml of the stock solution to 100 ml with water. 1 ml ≡ 1 mg of platinum.
Palladium. Dilute 10 ml of the stock solution to 100 ml with water. 1 ml≡ 100 µg of palladium.
Rhodium. Dilute 10 ml of the stock solution to 100 ml with water. 1 ml≡ 100 µg of rhodium.
Ruthenium. Use the stock solution. 1 ml ≡ 1 mg of ruthenium.
Gold. Dilute 10 ml of the stock solution to 100 ml with water. 1 ml ≡ 100 µg of gold.

The amount of sample taken should be such that, after the material has been dissolved and diluted, the concentration of noble metals in the solution for measurement is in the ranges (ppm) platinum, 20–100; palladium, 2–10; rhodium, 2–10; ruthenium, 10–50; gold, 2–10.

Instrument Parameters (See Table 1.2.)

The Techtron AA3, AA4, and AA5 Atomic Absorption Spectrophotometers were used.

TABLE 1.2
Instrument Parameters[a]

	Pt	Pd	Rh	Ru	Au
Wavelength (Å)	2659.5	2447.9	3434.9	3498.9	2428.0
Width of slit (μm)	50	50	100	50	100
Lamp current (mA)	10	7	5	10	4
Air pressure (lb/in.^{2})	15	15	15	15	15
(N/m^2)	10^5	10^5	10^5	10^5	10^5
Acetylene flow instrument units	4	4	4	4	4

[a]To give the optimum absorption, the acetylene flow should be adjusted while a standard solution is being aspirated. It will be found that platinum, palladium, and gold require a lean flame whereas ruthenium requires a rich flame, and rhodium a normal flame, for maximum absorption. The height of the burner and the lateral adjustment should also be found experimentally. (Lateral adjustment need be carried out only once.)

Procedure

Transfer a suitable amount of the solution containing the noble metals to a 50-ml squat beaker. Add 1 ml of the lithium chloride solution (i.e., 10 mg of Li), and evaporate the solution just to dryness on a steam bath. Do not bake. This step is for the removal of excess nitric acid. (If a large volume of solution is to be evaporated transfer the solution first to a large beaker and evaporate it on a hotplate until the volume is reduced to approximately 10 ml, and then transfer quantitatively to a 50-ml beaker and evaporate just to dryness on the steam bath.) Add 2 ml of perchloric acid and 1 ml of the vanadyl chloride solution (i.e., 100 mg of V). Warm the mixture to dissolve the salts and then with water dilute it to 10 ml in an A-grade volumetric flask. Measure the absorption of the noble metals, and compare this measurement with suitable standards measured at the same time.

Calibration. Prepare a series of standard solutions for calibration by transferring the volumes of working solutions of noble metals given in Table 1.3 to six 100-ml volumetric flasks. Add 20 ml of perchloric acid and 10 ml of the vanadyl chloride solution, and dilute to the mark with water. Measure the absorbances of the noble metals, and construct a calibration curve of absorbance against concentration for each.

Calculations. From the calibration curves (see Table 1.3), read in parts per million the concentration of the metals present in the solutions measured.

$$\text{Total amount of metal } (\mu g) = \text{concn (ppm)} \times \text{dilution factor,}$$

where

$$\text{dilution factor} = \frac{\text{1st dilution vol (ml)} \times \text{2nd dilution vol (ml)}}{\text{aliquot portion taken (ml)}}$$

TABLE 1.3
Volumes of Standard Working Solutions Taken and Calibration Ranges[a] for the Noble Metals

Calibration std solution	Pt Vol (ml)	Pt Conc (ppm)	Pd Vol (ml)	Pd Conc (ppm)	Rh Vol (ml)	Rh Conc (ppm)	Ru[b] Vol (ml)	Ru[b] Conc (ppm)	Au Vol (ml)	Au Conc (ppm)
A	0	0	0	0	0	0	0	0	0	0
B	2	20	2	2	2	2	1	10	2	2
C	3	30	3	3	3	3	1.5	15	3	3
D	4	40	4	4	4	4	2	20	4	4
E	6	60	6	6	6	6	3	30	6	6
F	8	80	8	8	8	8	4	40	8	8
G[c]	10	100	10	10	10	10	5	50	10	10

[a] These calibration ranges are suitable for the analysis of noble metals in most solutions and also for calculation of results by computer techniques. If necessary, however, these ranges can be extended.

[b] The working solution for ruthenium is the stock solution.

[c] The absorbances of solutions G are: Pt, 0.3; Pd, 0.3; Rh, 03.; Ru, 04.; Au, 03.

Platinum, Palladium, Rhodium, Ruthenium, and Gold in Beads

Mallett *et al.* (5) developed a method for the determination of platinum, palladium, rhodium, ruthenium, and gold in fire assay prills. If silver is present in large amounts the determination should be done on solutions containing sufficient hydrochloric acid to keep it soluble.

The method is suitable for the determination of the noble metals in solution when base metals, other than small amounts of lead, are not present. The method can therefore be used for the analysis of prills obtained from the fire-assay procedure using lead as a collector. The method has been adapted for the analysis of prills cupelled at 1300°C, and for gold or silver prills where cupellation took place at lower temperatures and where small amounts of lead and large amounts of gold or silver are present. The method cannot tolerate large amounts of base metals. Mutual-interference effects are eliminated by the addition of uranium.

PROCEDURE 10 (5)

Reagents and Equipment

Uranium solution. Dissolve 59 g of pure U_3O_8 in 20 ml of 3:1 aqua regia. After the U_3O_8 has dissolved, dilute to 200 ml with water, 1 ml $\equiv$ 250 mg uranium.

Stock solutions. Stock solutions of each of the metals are prepared to contain 1 mg/ml except in the case of platinum which is 10 mg/ml.

Working solutions. Platinum. Dilute 10 ml of the stock solution to 100 ml with 4% aqua regia solution, 1 ml $\equiv$ 1 mg of platinum.
Palladium. Dilute 10 ml of the stock solution to 100 ml with 4% aqua regia solution, 1 ml $\equiv$ 100 μg of palladium.

TABLE 1.4

Instrumental Parameters for the Techtron AA4 Atomic Absorption Spectrophotometer

	Pt	Pd	Rh	Ru	Au	Ag	Pb
Wavelength (Å)	2659.5	2447.9	3434.9	3498.9	2428.0	3280.7	2170.0
Width of slit (μm)	50	50	100	50	100	50	100
Lamp current, (mA)	10	7	5	10	4	4	6
Air pressure (lb/in.^{2})	15	15	15	15	15	15	15
(N/m^2)	(10^5)	(10^5)	(10^5)	(10^5)	(10^5)	(10^5)	(10^5)
Acetylene flow instrument[a] units	4	4	4	4	4	4	4

[a]To give optimum absorption, the acetylene flow should be adjusted while a standard solution is being aspirated. It will be found that platinum, palladium, and gold require a lean flame while rathenium requires a rich flame for maximum absorption. Burner height and lateral adjustment should also be found experimentally. (Lateral adjustment need only be carried out once.)

Rhodium.   Dilute 10 ml of the stock solution to 100 ml with 4% aqua regia solution, 1 ml $\equiv$ 100 μg of rhodium.

Ruthenium.   Use the stock solution, 1 ml $\equiv$ 1 mg of ruthenium.

Gold.   Dilute 10 ml of the stock solution to 100 ml with 4% aqua regia solution, 1 ml $\equiv$ 100 μg of gold.

Stock solutions. Silver.   Dissolve 157.5 mg of $AgNO_3$ (A.R. grade) in 50 ml of 4% nitric acid. Dilute to 100 ml, store in an amber bottle, and keep in a dark place. 1 ml $\equiv$ 1 mg of silver.

Lead.   Clean pure lead shot in approximately 4% nitric acid. Rinse, and dry thoroughly. Dissolve 1.000 g of the cleaned lead shot in 100 ml of 25% nitric acid. Dilute to 1 liter with 4% nitric acid, 1 ml $\equiv$ 1 mg of lead.

Instrument parameters (See Table 1.4.)

Working solutions. Silver.   Dilute 10 ml of the stock solution to 100 ml with 4% nitric acid, 1 ml $\equiv$ 100 μg of silver.

Lead.   Dilute 10 ml of the stock solution to 100 ml with 4% nitric acid, 1 ml $\equiv$ 100 μg of lead.

Amount of Sample

The amount of sample taken through the fire assay procedure should be such that, after dissolution of the prill and subsequent dilution, the concentrations of noble metals present in the solution for measurement are in the ranges indicated in Table 1.5.

For example, if the estimate for platinum in the ore is 5 μg, and if the final solution for measurement is to be 10 ml with a suitable concentration of 80 mg/l, 800 μg of platinum is required. This means that 160 g of ore must be taken for the fire assay procedure. Solutions of less than 10 ml are not sufficient if all the noble metals are to be determined, because a solution volume of at least 5 ml is normally required for their measurement.

Procedure

Prills obtained after cupellation at 1300°C.   Transfer the prill(s) to a clean 50-ml beaker, and add 5 ml of 3:1 aqua regia. Cover with a watch glass, and warm gently on a hot plate or steam bath. Do not boil because boiling destroys the aqua regia without attacking the prill. The prill is usually dissolved readily by this technique. However it may be necessary to take the solution to near dryness and to repeat the addition of aqua regia

TABLE 1.5

Optimum Concentration Ranges of the Noble Metals and Lead in the Solutions for Measurement

Metal	Conc range (mg/l)	Metal	Conc range (mg/l)
Pt	20–100	Au	2–10
Pd	2–10	Ag	2–10
Rh	2–10	Pb	5–20
Ru	10–60		

several times. After dissolution is complete (the solution must be very carefully examined for any undissolved material), evaporate to near dryness. Add 0.4 ml of 3:1 aqua regia, 0.4 ml of the uranium solution, and 1 ml of water. Warm very gently to dissolve any salts or reduced gold. Cool the solution and transfer it to a 10-ml A-grade volumetric flask. The aqua regia content of the solution is now 4% by volume, and the uranium concentration is 10 g/l. Measure the absorption of platinum, palladium, rhodium, ruthenium, and gold, and compare it with that of suitable standards measured at the same time. Silver and lead, if present, are only there in trace amounts and are not usually measured unless such measurements are specifically required.

Gold prills. Follow the procedure given in the above section. The standard solutions for calibration must contain the same amount of gold as that in the sample solutions. For a gold prill, 0.5 mg of gold is usually added to the flux for the fire assay procedure. An amount of gold must then be added to each of the standard solutions for calibration, so that the final concentration matches that of the samples. In this instance, 50 mg/l of gold must be present in the standard solutions for measurement.

If the gold is to be determined, the sample solutions must be diluted to give a concentration of 2 to 8 mg/l of gold. When 0.5 g of gold has been added for the fire assay procedure, transfer 1 ml of the sample solution to a second 10 ml volumetric flask, add a further 0.4 ml of uranium solution, and dilute to the mark with 4% aqua regia. Compare the absorbance readings with those of a set of suitable standards.

Silver prills. Transfer the prill(s) to a 50-ml beaker, and add 5 ml of 3:1 aqua regia. Cover with a watch glass and warm gently on a hot plate or steam bath. Take to a low volume, and repeat the addition of aqua regia. The dissolution of silver prills can become very lengthy because of the formation of silver and lead chlorides, which retard the dissolution. Repeated additions of aqua regia will be necessary for complete dissolution of the prill. After the prill is completely dissolved, take just to dryness (do not bake) and add 0.4 ml of 3:1 aqua regia, 2 ml of 20% hydrochloric acid, and 0.4 ml of the uranium solution. Warm gently to dissolve the salts and the silver and lead chlorides. Transfer the solution to a 10-ml A-grade volumetric flask, and make up to the mark with 20% hydrochloric acid. The hydrochloric acid content of the solution is now about 20% by volume, and the uranium concentration is 10 g/l. Measure the absorbance of the noble metals and lead, and compare it with that of suitable standards (see the following). The standard solution must contain equivalent amounts of silver. For a silver prill, 0.5 mg of silver is usually added to the flux for the fire assay procedure. An amount of silver must then be added to each of the standard solutions for calibration, so that the final concentration matches that of the samples. In this instance, 50 mg/l of silver must be present in the standard solution for measurement.

If the analysis of silver is required, the sample solutions must be diluted to give 2 to 8 mg/l of silver. If 0.5 mg of silver was added, transfer 1 ml of the sample solution to a second 10-ml dry volumetric flask. Add 2 ml of concentrated hydrochloric acid and then 0.4 ml of the uranium solution. Finally dilute to the mark with water, and shake vigorously. The addition of the reagents in this order ensures that at all stages of the dilution, the solution is always in strong hydrochloric acid and silver chloride is not likely to precipitate.

Measure the absorbance of silver, and compare it with that of suitable standards.

Calibration

The preparation of suitable calibration standards for the various types of prills is described now. The volumes of working solutions to be taken for the standards and the calibration ranges are given in Table 1.6. Note that each calibration standard contains all the metals in the appropriate concentration.

For the measurement of noble metals in solutions of prills that have been cupelled at 1300°C. Transfer the appropriate amounts of the working solutions of the noble metals to 100-ml volumetric flasks. (See Table 1.6. Silver and lead are not usually present in these prills and these metals are therefore not added to the standards.) Add 4 ml of uranium solution and 4 ml of 3:1 aqua regia, and then dilute to the mark with water. Construct calibration curves for each of the noble metals of absorbance versus concentration.

For the measurement of noble metals and lead in solutions of gold prills that have been cupelled below 1300°C. With the exception of gold, transfer the appropriate amounts (Table 1.6) of the working solutions of the noble metals and lead to 100-ml volumetric flasks. Add 5 ml of the gold stock solution, i.e., 5 mg of gold, to each of the flasks containing standard solutions. Add 4 ml of the uranium solution and 4 ml of 3:1 aqua regia, and dilute to the mark with water. Construct calibration curves of absorbance versus concentration.

For the measurement of noble metals in solutions of gold prills that have been cupelled at 1300°C. Follow the calibration procedure given above, but do not add silver or lead to the standard solutions because these metals are unlikely to be present in the sample solutions. Construct calibration curves of absorbance versus concentration.

For the measurement of noble metals and lead in solutions of silver prills. With the exception of silver, transfer the appropriate amounts (Table 1.6) of noble metals and lead to 100-ml volumetric flasks. To each of the flasks, add 20 ml of concentrated hydrochloric acid and then 5 ml of the stock solutions for silver (i.e., 5 mg of silver). Shake to dissolve any silver chloride that may have formed. Add 4 ml of the uranium solution and 4 ml of 3:1 aqua regia. Dilute to the mark with water. Construct calibration curves of absorbance versus concentration.

For the measurement of gold in solutions of gold prills. Transfer the appropriate amounts (Table 1.6) of working solution for gold to a set of 100-ml volumetric flasks. Add 4 ml of the uranium solution and 4 ml of 3:1 aqua regia, and dilute to the mark with water. Construct a calibration curve of absorbance versus concentration.

For the measurement of silver in solutions of silver prills. Transfer the appropriate amounts (Table 1.6) of working solution for silver to a set of dry 100-ml volumetric flasks. Add 20 ml of concentrated hydrochloric acid and then 4 ml of 3:1 aqua regia. Dilute to the mark with water. Construct a calibration curve of absorbance versus concentration.

Calculations

From the calibration curves, read off the concentration in milligrams per liter of the metals present in the solutions measured.

TABLE 1.6
Volumes of Standard Solutions Taken and Calibration Ranges[a] for the Noble Metals, Silver and Lead

	Pt		Pd		Rh		Ru[b]		Au		Ag		Pb	
Calibration	Vol (ml)	Conc (mg/l)	Vol (ml)	Conc (mg/l)	Vol (ml)	Conc (mg/l)	Vol (ml)	Conc (mg/l)	Vol (ml)	Conc (mg/l)	Vol (ml)	Conc (mg/l)	Vol (ml)	Conc (mg/l)
A	0	0	0	0	0	0	0	0	0	0	0	0	0	0
B	1	10	1	1	1	1	1	10	1	1	1	1	2	2
C	2	20	2	2	2	2	2	20	2	2	2	2	5	5
D	5	50	5	5	5	5	3	30	5	5	5	5	10	10
E	8	80	8	8	8	8	4	40	8	8	8	8	15	15
F	10	100	10	10	10	10	6	60	10	10	10	10	20	20

[a]With the exception of those for silver and lead, the calibration ranges can be extended. However, the ranges shown here were found to be convenient for the analysis of the noble metals, silver, and lead in the prills used for the development of the method.
[b]The working solution for ruthenium is the stock solution.
[c]Optical densities are: Pt, 0.3; Pd, 0.3; Rh, 0.3; Ru, 0.4; Au, 0.3; Ag, 0.6; Pb, 0.5.

micrograms (μg) of metal in prill = conc (mg/1) $\times$ dilution factor

where

$$\text{dilution factor} = \frac{\text{1st dilution vol (ml)} \times \text{2nd dilution vol (ml)}}{\text{aliquot taken (ml)}}$$

Detection limits (solution) for the elements are as follows (in ppm): platinum, 0.5; palladium, 0.03; rhodium, 0.05; ruthenium, 0.5; and gold, 0.05.

Ruthenium and Osmium in Matte Residues

Kruger and Robért (33) described a barium peroxide sinter followed by distillation of osmium and ruthenium from the sinter solution for the determination of these two elements in matte residues. Ruthenium and osmium distillates were collected separately, with the former being determined by atomic absorption spectroscopy, Procedure 11, and the latter by the thiourea colorimetric method, Procedures 46 and 11.

PROCEDURE 11 (33)

Reagents and Equipment

Distillation apparatus.   The distillation apparatus is shown in Fig.1.2, using barium peroxide A.R. Grade. Use perchloric acid, 70%, and hydrochloric acid, 6 N.

Uranium solution.   Dissolve 59 g of pure U_3O_8 in 20 ml of 3-to-1 aqua regia. Filter, if necessary, and dilute to 200 ml with water:

$$1 \text{ ml} \equiv 250 \text{ mg of U}$$

Hydroxylamine hydrochloride (10% (w/v)).   Dissolve 10 g of hydroxylamine hydrochloride in 100 ml of distilled water.

Thiourea solution (10% (w/v)).   Dissolve 2.5 g of thiourea in water, and dilute to 25 ml with water.

Stannous chloride solution (10% (w/v)).   Dissolve 2.5 g of $SnCl_2 \cdot 2H_2O$ reagent grade in 8 ml of hydrochloric acid. Warm gently to dissolve. Cool, and dilute to 25 ml with water.

Procedure

Sample sinter procedure and distillation of osmium and ruthenium.   Transfer an appropriate amount of sample (see Table 1.7) to a 25 ml nickel crucible. Add 20 g of barium peroxide per gram of sample, and mix well by stirring with a thin plastic rod. Place the nickel crucible in a muffle furnace at 700°C for 16 hr. (At this temperature the mixture will not fuse but will form a sinter.) Cool the nickel crucible. To remove the sinter, invert the crucible onto glazed paper. Transfer the sinter to a clean, dry, 500-ml distillation flask A (see Fig. 1.2). Place 50 ml of 6 N hydrochloric acid in each of five 250-ml receivers and

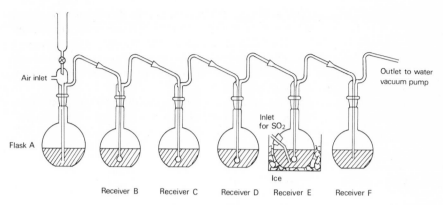

Air inlet

Outlet to water
vacuum pump

Inlet
for SO₂

Flask A

Ice

Receiver B   Receiver C   Receiver D   Receiver E   Receiver F

FIGURE 1.2.   Distillation apparatus.

connect the distillation apparatus as shown in Fig. 1.2. Cool receiver E in ice. Using a water vacuum pump, draw sulfur dioxide through receivers E and F and air through the complete distillation train for 10 to 15 min. (see Note 1) at an airflow between 0.2 and 0.5 ft³/hr. Increase the airflow to full pressure, and slowly add 75 ml of concentrated perchloric acid (see Note 2). When the reaction between the perchloric acid and the sinter has ceased, decrease the airflow to between 0.2 and 0.5 ft³/hr. Gently heat the contents of the flask with a burner until the sinter has disintegrated. Increase to the full heat of the burner, and boil the contents of the flask for 15 min. Remove the heat source from the distillation flask, and bring receiver B to the boil, using an alcohol or gas burner at a low heat. Disconnect the distillation flask from the distillation train and allow receiver B to boil for 5 min. Bring receiver C to the boil, remove heat source below receiver B, and allow receiver C to boil for 5 min. Bring receiver D to the boil, remove heat source below receiver C and allow receiver D to boil for 7 min (see Note 3). Remove the heat source and disconnect the distillation train by removing first receiver B and then the others in sequence. Combine the contents of receivers B, C, and D in a 400-ml beaker for the determination of ruthenium. Combine the contents of receivers E and F in a 400-ml beaker for the determination of osmium.

Determination of ruthenium.   Evaporate the ruthenium distillate until the volume is reduced to about 30 ml (see Note 4). Cool and dilute to 50 ml with water. Transfer an

TABLE 1.7
Mass of Sample and Dilution Required

Estimated Os (μg)	Mass of sample	Dilution of distillate (ml)	Aliquot portion (ml)
0–300	1.0	50	20
300–720	0.5	50	20

aliquot portion containing 100 to 500 μg of ruthenium to a 10-ml volumetric flask, and add 0.4 ml of the uranium solution and 0.8 ml of aqua regia (see Note 5). Dilute to the mark with 40% hydrochloric acid. If an aliquot portion of more than 3 ml is required, evaporate the respective aliquot portion to about 3 ml, add 0.4 ml of uranium solution and 0.8 ml of aqua regia, and after transferring the solution to a 10 ml volumetric flask, dilute to volume with 40% hydrochloric acid.

Using the following instrumental parameters on the atomic-absorption spectrophotometer, measure the absorption of the ruthenium and compare it with that of suitable standards measured at the same time:

Wavelength	3498.9 Å
Width of slit	50 μm
Lamp current	10 mA
Air pressure	15 lb/in.^{2}
(N/m^2)	(10^5)

Acetylene flow setting adjusted for optimum readings

In the same way, measure the absorption of a reagent blank and subtract the value obtained from the values obtained from the samples.

Calibration. Prepare a 1000 ppm standard ruthenium solution from metal or sponge. Dilute this solution so that it contains 100 μg of ruthenium per milliliter in 40% hydrochloric acid. Take aliquot portions of 1, 2, 3, 4, and 5 ml, add 0.4 ml of uranium solution and 0.8 ml of aqua regia, and dilute to 10 ml in a volumetric flask with 40% hydrochloric acid. This procedure gives standard ruthenium solutions of 10, 20, 30, 40, and 50 μg of ruthenium per milliliter. Measure the absorption of standards and samples relative to a blank solution containing uranium, aqua regia, and 40% (v/v) hydrochloric acid in a 10-ml volume.

Calculation. Calculate the amount of ruthenium in the sample, using the following formulas:

Ru in sample solution (μg) = Ru (μg/ml) × 10 × (dilution vol)/(aliquot portion taken)
Ru in reagent blank (μg) = Ru (μg/ml) × 10
Ru (%) = Ru in sample (μg) – Ru in reagent blank (μg)
× 100/mass of sample × 10^6

Determination of osmium. Add 2 ml of 10% hydroxylamine hydrochloride to the osmium distillate and allow the solution to stand for 16 hr. Evaporate until the volume is reduced to about 30 ml. Transfer the solution to a 50-ml standard volumetric flask, and dilute to volume with water. With a pipet add a suitable aliquot portion (see Table 1.7) into a 125-ml Phillips beaker. Add 10 ml of concentrated hydrochloric acid and 1 ml of 10% hydroxylamine hydrochloride, evaporate to 2.5 ml (visual comparison), and allow to cool. Add 2.5 ml of 10% thiourea solution and 0.25 ml of 10% stannous chloride solution. Transfer to a 25-ml volumetric flask, and heat in boiling water for 15 to 20 min. Cool slowly to room temperature, and dilute to volume with water. Centrifuge and on a Zeiss spectrophotometer measure the absorbance against that of water in a 2 cm cell at 480 nm.

Calibration. Prepare 1000 ppm standard osmium solution from the ammonium chloroosmate salt, $(NH_4)_2OsCl_6$. Dilute this solution so that it contains 20 μg of osmium per milliliter of 50% hydrochloric acid. Distill portions of the above standard osmium solution that contain 40, 100, 140, 200, and 240 μg of osmium in the presence of 2.25, 3.60, 4.50, 6.75, and 9.00 mg, respectively, of standard ruthenium solution as already described by adding the solutions to the distillation flask and distilling in the presence of 10 g of barium peroxide. Boil the contents of the distillation flask for 20 min at the full heat of the burner (instead of 15 min) and allow for removal of water. Draw a calibration curve of optical density against micrograms of osmium relative to a reagent blank solution carried through the distillation procedure.

Calculation. Calculate the amount of osmium in the sample, using the formula

$$\text{Os (\%)} = \frac{\text{Os}(\mu g)}{10^6} \times \frac{\text{dilution vol}}{\text{aliquot}} \times \frac{100}{\text{mass of sample}}$$

Notes. **1.** Before the distillation, the 6 N hydrochloric acid in receivers E and F must be saturated with sulfur dioxide so that the osmium tetroxide is reduced to osmium sulfite.

2. The perchloric acid oxidizes the ruthenium and the osmium to the volatile tetravalent state.

3. The successive boiling of receivers B, C, and D transfers the osmium into receivers E and F, where it is reduced to osmium sulfite by the sulfur dioxide. The ruthenium remains in receivers B, C, and D.

4. Because the perchloric acid fumes condense in the various receivers, it is not advisable to evaporate the ruthenium distillate to a volume less than 30 ml. The perchloric acid will oxidize the ruthenium to the volatile octovalent state.

5. The routine atomic-absorption analysis for ruthenium is usually done in the presence of other platinum-group metals and gold, and uranium is added for the elimination of interference effects. In the determination of ruthenium given here, uranium is added to the solutions for analysis so that standards and samples can be matched.

CHAPTER 2

SPECTROCHEMICAL ANALYSIS OF THE PLATINUM METALS

Clyde L. Lewis

INTRODUCTION

Optical emission spectrography is particularly applicable to the determination of concentrations lower than about one percent in materials where the major constituents (matrix) do not vary significantly from sample to sample. It has been used for many years, in combination with fire assaying, to determine trace or low concentrations of platinum metals in rocks, ores, and metallurgical materials (type 1).

Optical emission spectrography is equally useful in determining the concentrations of trace impurities in the refined platinum metals (type 2). In types 1 and 2, the matrix is relatively constant from sample to sample, or can be made so, and the concentrations determined are usually low.

X-ray emission spectrometry is generally more useful for the determination of higher concentrations (type 3), although the newer instruments allow trace determinations in favorable instances. This method is also subject to difficulties caused by variation in major constituents but is nevertheless of great use for the analysis of a wide variety of materials containing higher concentrations of platinum metals. These materials, which include alloys, concentrates, residues, refinery sweepings, scrap, etc., are usually analyzed by wet chemical methods, but often much time and work can be saved by applying X-ray spectrometry. Semiquantitative, or even qualitative, analyses of unknown samples or of precipitates, residues, and solutions during wet chemical analysis can be very useful to the chemist. In some instances it is possible to create an artificial matrix by dilution, thus providing a quantitative analysis without a great loss of accuracy.

In all these analyses, sample preparation and the preparation of the accurate standard samples required for comparison are extremely important. Sample preparation may in fact require much more time than the actual analysis if satisfactory accuracy is to be achieved.

The Determination of Low or Trace Concentrations of Platinum Metals in Other Materials

Spectrography has been applied to the determination of platinum metals in a wide variety of materials. Fothergill *et al.* (34) determined platinum and palladium in the atmosphere of a platinum refinery by a combined chemical and spectrographic method (35). Khrapaï determined platinum and palladium in silver (35) and in silver–gold alloys (36). Strasheim *et al.* (37) determined palladium and silver in high purity gold. Tomingas and Cooper (38) determined palladium in doré metal.

Naka and Naka (39) determined small amounts (5–500 ppm) of platinum in artificial fluorphlogopite fused in a platinum vessel. The sample was mixed with an equal weight of carbon powder as a buffer and ruthenium was added as internal standard. Direct current arc excitation was used. Yokoyama and Faris (40) found the vacuum cup technique satisfactory for determining ruthenium, rhodium, and palladium in complex U–Pu fission alloys. Talalaev (41, 42) applied a direct ac arc spectrographic method for determination of palladium and ruthenium in alumina catalysts. Khrapau (43) determined osmium, iridium, ruthenium, and rhodium in gold–silver alloys. The sample was melted with silver and dissolved in dilute nitric acid to remove base metals, silver, palladium, and most of the platinum. The residue was calcined in hydrogen and melted with nickel to provide a Ni–Au alloy which was arced against a copper counter electrode with an ac arc. Gut'ko *et al.* (44) applied an ac arc method to determine some 20 impurity elements, including platinum, palladium, rhodium, iridium, and ruthenium in refined selenium and tellurium. Egizbaeva *et al.* (45) determined small amounts of osmium in ores by a direct spectrographic method, using a 12-A dc arc with a chamber electrode and a high resolution spectrograph. Berenshtein *et al.* (46) and Livshits (47) reported improved methods for chemical–spectrographic determination of platinum, palladium, and gold in sulfide ores and for palladium in copper–nickel sulfide ores. Kawashima *et al.* (48) applied an iron powder-dc arc method for determination of 0.004–0.6% rhodium, 0.003–0.6% palladium, 0.004–0.6% iridium, and 0.001–0.4% platinum in iron and stainless steel. Terekhovich (49) determined 0.3 to 50 ppm of palladium in ores and technological products by a direct spectrographic method wherein the pulverized sample was mixed with a buffer (75% graphite and 25% $SrSO_4$ containing 0.05% of Ru or Zr as internal standard) and a 65 mg aliquot was arced with a 15 A dc arc.

An important and interesting aspect of spectrochemical methods is the determination of low or trace concentrations of precious metals in rocks, ores, minerals, or the intermediate products of extractive metallurgy. The amounts present are often so low that direct detection by spectrography is not possible. The classical procedure of *fire assaying* however, permits the precious metals to be collected from their original host and transferred to an artificial matrix where their concentrations are sufficiently high for spectrochemical analysis.

THE DEVELOPMENT OF COMBINED FIRE ASSAY AND SPECTROCHEMICAL METHODS

Although spectroscopy has been known for some hundreds of years, and fire assaying has been traced to biblical times (50), the combined use of these methods for the determination of precious metals is a comparatively recent development. de Lazlo (51) appears to have been one of the first, if indeed not the first, to apply a combination of the two methods to the determination of precious metals. In 1927, he concentrated gold, silver, and the platinum metals in a silver bead by fusion and cupellation, and sparked the bead to obtain the spark spectrum of platinum. Raies ultimes* were used to estimate concentration. In 1929, Schneiderhöhn (52) used a Zeiss quartz spectrograph directly on minerals of the South African Bushveld deposits. He used raies ultimes to place platinum metal concentrations in ranges, e.g., 1–0.1%, 0.1–0.01%, and translated the ranges, by relating sample amounts, to grams per ton. In 1930, Ida and Noddack (53) studied the abundance of the chemical elements, including the platinum metals, by both optical emission and X-ray spectroscopy. In the latter instance, the sample was placed at the target of a demountable X-ray tube within the tube. This

---

*Raies ultimes are the most sensitive lines, which are the last to disappear from the spectrum as the concentration of the element is diminished.

was a very slow procedure as the tube had to be evacuated each time the sample was changed. Schneiderhöhn and Moritz (54) used optical spectrography to study the distribution of platinum metals in minerals of the South African platinum deposits in 1931. Iwamura (55), working at Hyoto Imperial University, attempted to determine the concentration of gold in natural ores. In 1932 he described several ways of preparing electrodes so that reproducible spectra could be obtained. Noddack (56) described the application of X-ray spectra to the analysis of ores and metallurgical products. Goldschmidt and Peters (57) studied the geochemistry of noble metals, using a combined fire assay and spectrochemical procedure.

In 1933, Peters (58) described the fire assay and spectrographic technique used in studying the geochemistry of the noble metals. This was an important paper as it laid the foundation for most of the combination fire assay–spectrographic methods used today for the determination of precious metals in ores. Goldschmidt and Peters (59) also used spectrographic and X-ray fluorescence methods to study concentrations of rare elements, including gold and platinum metals, in coal. Nedler (60), in 1936, determined platinum, palladium, and rhodium in silver beads obtained from ore slimes and tailings. He used a spark discharge to excite the spectra. A log sector allowed the measurement of "fixation pairs" of lines (i.e., pairs of lines of the same type and same intensity under selected excitation conditions) and hence a correction for variation in discharge conditions. Concentration ranges were 1.0–0.02% for platinum and palladium. The accuracy obtained in determining these elements in the ore was estimated to be ±10%. When Seath and Beamish (61) were unable to obtain satisfactory results with de Lazlo's method of estimating concentrations from platinum spark lines, they prepared standards by evaporating platinum metal solutions in small lead trays and cupelling them to obtain silver or gold beads. They studied both spark and arc spectra, using a Hilger medium quartz spectrograph.

In 1940 Toisi (62) described two methods for the assay of minerals and applications of these to gold ores. In one method a powdered sample was streamed into the arc; in the other the electrodes were impregnated with gold solution. Pardo (63) described a method of concentrating precious metals by fire assay and a subsequent spectrographic determination of their concentrations in silver beads. In assaying Spanish ores (64) he used this method and an alternate one in which the gold was concentrated by electrolysis in the cavity of a carbon rod. The same rod was then used as cathode in a spectrographic analysis. Nedler and Efendiev (65) concentrated gold from 5 to 10 g of ore either by dissolving in aqua regia or by a fire assay fusion. In the latter instance, the lead bead was dissolved in nitric acid and the lead removed by filtration, the residue being dissolved in aqua regia. The solution was introduced into a spark at 2–3 ml in 5 min. Gold was determined by means of intensity ratios with platinum lines from added platinum. Nedler (66) simultaneously determined gold, platinum, palladium, and rhodium by the use of spark spectra. Gold in ores was dissolved to give a solution containing 0.001 to 0.0001% gold, and 1 mg of platinum per 100 ml of solution was added. A high voltage spark was used to produce spectra from the solution in a graphite electrode. Rare elements were determined in gold nuggets by introducing the gold, powdered to 200 mesh, into a carbon arc. In 1942, de Azcona and Pardo (67) claimed sensitivities of the order of 5×10^{-7} g for gold, platinum, palladium, and ruthenium, and 2×10^{-4} g for iridium, by a combination of either fire assay or electrolysis with spectroscopy.

In 1945, Scobie (68) developed a method for the determination of small amounts of gold and platinum metals in ores by fire assay followed by a spectrographic analysis of the silver beads. As this method is in use in some Canadian laboratories, it is described more fully in Procedure 12. Minguzzi (69) used lead for the collection of gold from pyrites. Silver beads were obtained and excited in the carbon arc. Chromium was chosen as internal standard in this determination. Hawley *et al.* (70), studying the platinum and palladium content of sulfide and arsenide minerals of the Sudbury District in Canada in 1951, developed a method whereby gold fire assay beads were mixed with gold amalgam. Standard samples were prepared by mixing platinum and palladium black with gold amalgam. Palladium appeared to dissolve in the amalgam while platinum formed a mixture. Hawley *et al.* (71) found that platinum and bismuth-rich fire assay beads did not always provide reliable results when mixed with gold amalgams. They returned to the lead collection method and developed a "lead bead method," wherein gold or silver fire assay beads were dissolved in molten lead to obtain lead beads. These beads were soldered onto the ends of copper rods and a high voltage spark discharge was used to excite the spectra. Standard samples were prepared by alloying platinum metals in lead and making successive dilutions with pure lead to obtain the desired concentrations. The mutual standard method of calculation (72) was used in developing analytical curves and in analyzing samples. Hawley and Rimsaite used the improved method to determine platinum metals in some Canadian uranium and sulfide ores (73).

Ginzburg *et al.* (74) in 1955, determined gold, platinum and palladium, without prior concentration, in ores, slags, etc., of the Russian copper smelting industry by introducing a regulus containing these metals into an ac carbon arc. A sensitivity of 30 ppm was found for gold and palladium and 10 ppm for platinum. The overall error, including preparation, was estimated to be ±20%. In 1956, Maslenitskii (75) studied the behavior of noble metals in the pyrometallurgical reduction of copper–nickel ores. He concluded that, if sulfides are in excess in the matte, all of the platinum group metals are collected there. Lewis (76) made a survey of 125 fire assays from three different laboratories on ore and metallurgical concentrates by using a modification of the "lead bead" method of Hawley, Rimsaite, and Lord. Livshits and Kashlinskaya (77) developed a method where by gold, platinum, palladium, and rhodium are coprecipitated with copper by sodium thiosulfate. All of the precious metals are then concentrated in a copper bead obtained from the precipitate. The copper bead is analyzed spectrographically in an ac arc. Van'kin *et al.* (78) obtained a Russian patent on this technique in 1957.

In 1958, Bufatin *et al.* (79) described a method for the chemical concentration of the platinum and palladium content of uranium and a subsequent spectrographic analysis. Losev (80) described a method for X-ray spectrographic determination of platinum in ore by a fire assay concentration and chemical treatment of the bead prior to analysis. P'yankov (81) described a method for collecting platinum, palladium, gold, and rhodium in copper after preliminary dissolution of the ores and chemical enrichment. The precious metals were determined spectrographically in the copper. Brooks and Ahrens (82) determined noble metals in silicate rocks by using ion exchange resins to remove the dissolved precious metals from solution. The solutions were evaporated to dryness and the residue collected in a sodium chloride matrix, which was analyzed qualitatively by spectrograph. They suggested that this procedure could be made quantitative. Miyamoto (83), in 1961, used a gold–silver alloy bead from fire assaying for the spectrographic determination of platinum and palladium in ores.

Ginzburg (84), in 1963, divided methods for spectrographic analyses of platinum metals into the following two groups: (1) direct determination and (2) the composite method in which use is made of preliminary enrichment of initial samples and separation of the elements to be determined as concentrate in a collecting metal. He rated the direct method as having no practical value for platinum metals in ores. The assay–spectrographic method is considered the most common of the composite methods with lead, silver, copper, and Cu–Ni alloys as collectors. He stated that the silver assay bead can be converted into a Cu–Ag alloy for determination of ruthenium and iridium. Ion exchange, extraction, and coprecipitation were mentioned as being promising chemical preconcentration methods.

Berenshtein *et al.* (46) and Livshits (47) reported improved methods for assay–spectrographic determination of platinum, palladium, rhodium, and gold in ores and rocks.

Barnett *et al.* (85) determined gold, platinum, and palladium in ores and slags by fire assay with silver bead collection, separation of the precious metals from silver by ion exchange, and spectrographic determination of the precious metals in silica after addition of silica sand to the ion exchange resin and ignition in a muffle furnace. Limits of detection were 0.03 ppm for gold and platinum and 0.01 ppm for palladium.

Haffty and Riley (86) used fire assay techniques to preconcentrate platinum, palladium, and rhodium from geologic samples into gold beads, dissolved the beads in aqua regia, evaporated 200 μl of solution on a pair of flat-end graphite electrodes and excited the spectra with an intermittent dc arc. They were able to determine palladium down to 4 ppb, platinum down to 10 ppb, and rhodium down to 5 ppb in 15 g of sample with satisfactory precision.

Whitehead and Heady (87) faced with the necessity to survey many samples for possible platinum and palladium content and finding neither the lead bead method nor the solution method of Haffty and Riley (86) sufficiently rapid for their purpose, developed a method whereby a 10 mg silver fire assay bead is arced at 8 A dc in a cupped graphite electrode along with 10 mg of either rhodium or platinum metal powder. They were able to determine about 1–30 ppb of gold, platinum, palladium, and rhodium in a one assay ton sample with precision varying from about ±5% coefficient of variation for platinum and rhodium, ±15% for palladium, and ±19% for gold.

Dorrzapf and Brown (88) determined platinum, palladium, and rhodium in geological samples by direct spectrographic analysis of a 4 mg gold bead obtained by fire assay and cupellation. The weighed bead as arced in a graphite electrode along with 2 mg of ammonium chloroosmate (to stabilize platinum volatilization) at 15 A in an argon–oxygen atmosphere. Detection limits were 5 ppb for platinum and less than 1 ppb for palladium and rhodium in a 1 assay ton sample.

A program sponsored jointly by the Canadian Department of Energy, Mines and Resources (Mines Branch) and the Spectroscopy Society of Canada since 1967 has resulted in the establishment of two standard samples for use in the determination of precious metals in ores and similar materials (89). One of these is a copper–nickel concentrate made from Sudbury ores and the other is an alluvial concentrate from the Tulameen River area of British Columbia. Work is also under way to establish the precious metal content of a copper–nickel matte for use as a standard sample.

Problems Involved in the Combined Methods

It may be seen from this history that the development of combined concentration–spectroscopic methods of determining precious metals in rocks and ores appears to have received its greatest impetus from the geochemists, who were primarily interested in determining the abundance and distribution of the elements. The platinum metals are rare in nature, and in general cannot be determined or often detected directly in naturally occurring ores. Some preliminary concentrating, such as that which may be achieved by fire assaying, must usually be done prior to the determination. Much work has been done to discover the magnitude of losses and where they may occur in the process (90). Recently, both direct chemical and fire assay determinations were made on samples of metallurgical concentrates with a view to comparing efficiencies (91).

An apparent deficiency of the classical fire assay concentration is that the lead collector is alien to the precious metals. Iron, nickel, or copper would seem to be better collectors as they often occur in nature with the precious metals. Lead, however, has the advantage of being easily eliminated during the cupellation stage by oxidation to litharge and absorption into the cupel. Other metals, while possibly better collectors, must be removed chemically during the second stage of concentration. This requires more time and involves a risk of losses which may be as great as that of the precious metals not collected by the lead during fusion.

THE DEVELOPMENT OF THE RICHVALE METHOD

The method used in the present authors' laboratory is called the Richvale method. During a study of the platinum and palladium content of sulfide and arsenide minerals of the Sudbury Nickel District (70), a method was developed whereby gold fire assay beads were dissolved in gold amalgam in preparation for spectrochemical analysis. Standard samples were prepared by mixing platinum and palladium black with gold amalgam. Palladium appeared to dissolve in the amalgam while platinum formed a mixture. This method did not give satisfactory results for platinum-rich and bismuth-rich ores in a later investigation by Hawley et al. (71). One of these researchers, Rimsaite, substituted lead for the gold–mercury amalgam and thus dissolved gold and silver assay beads in lead by using a blowpipe flame on a chemical block. Standard samples were prepared by dissolving gold, silver, and the platinum metals in lead in the same manner. The small lead beads were mounted on copper rods by soldering. As the spectrographic equipment available did not permit use of a low voltage alternating arc as described by Scobie (68), a high voltage spark was tried. An internal standard was not used because the precious metal concentrations varied considerably in fire assay beads from different ores and because some base metals were retained in the lead beads. The mutual standard method (72) was used for calculation of precious metal concentrations.

For the determination of precious metals in various ores and metallurgical products, the present author used the Rimsaite technique, with modification to suit the various tasks at hand or to save time when possible. Thus, standards were prepared in the same way by alloying the platinum metals with lead except that two series were made, one having a constant amount of gold and the other having a constant amount of silver. When gold or silver fire assay beads were prepared for analysis, these were alloyed with the proper amount of lead associated with one of the standard series. It was found that the somewhat cumbersome mutual standard method could be eliminated by establishing the standard series as ranges of weight of platinum metal percent and using the spark spectrum so that lead served satisfactorily as an internal standard. It was then easily possible to determine the concentration of gold, silver, or platinum metal in the sample by application of the original sample weight (before fire assay) and the weight of the lead bead used for analysis. The soldering technique for mounting the beads on copper rods was replaced by a technique more readily applicable to large numbers of samples; this is described now.

These modifications permitted determination of gold, platinum, palladium, and rhodium by means of the constant silver series of standards and of silver and the platinum metals by the constant gold series of standards, but they had the disadvantage of requiring at least two fire assay beads for the determination of both gold and silver. It has always appeared desirable to arrest cupellation in order to allow determination of both these elements simultaneously. Further, there is the possibility that ruthenium, iridium, and osmium, which were not successfully determined by this modified method, might be so determined. Finally, it would appear that time could be saved by arresting the cupellation, instead of cupelling to completion with added gold or silver as a collector and then adding lead again to allow the use of lead standards.

In 1958, Ott and Cornett (92) developed a method whereby the cupel is removed from the furnace and placed under a stream of nitrogen so that the lead bead is instantly frozen without oxidation. It is quite possible to hold the bead weight very close to a desired figure although some judgement is required as to the time when the cupel should be removed from the furnace. This technique requires only one series of standards for gold, silver, platinum, palladium, and rhodium in lead. It has not yet proved satisfactory for the quantitative determination of ruthenium, iridium, or osmium, but is used for most determinations of other precious metals in the present author's laboratory. Occasional use is still made of the constant gold and constant silver techniques, particularly for fire assay beads from outside sources. All three techniques are included in the method termed here the Richvale method.

The Richvale method provided the basis for an ASTM (Americal Society for Testing and Materials) standard method. Submitted in 1965 as a suggested spectrochemical method, it was accepted in 1968 after modification to include details of the fire assay method (93). After round robin testing, the method was accepted in 1970 as ASTM designation E400-70: Standard Method for Spectrochemical Analysis of Ores, Minerals, and Rocks by the Fire Assay Preconcentration–Spark Technique.

The Richvale Method for Platinum, Palladium, and Rhodium

Brief descriptions of the Richvale method have been given in the literature (94, 76). As it comprises three techniques whereby gold or silver fire assay beads or lead beads from arrested cupellation may be analyzed, three sets of standards are required. The *constant gold series* contains 5% gold and 95% lead. The *constant silver series* contains 10% silver and 90% lead. The *arrested cupellation series* contains only lead as a major constituent. All three series are prepared by diluting "master standards" with pure lead.

PROCEDURE 12 (28, 76)

Equipment

Graphite crucibles. The sizes should be as follows: large, 2.0 cm i.d. × 2.0 cm deep; medium, 1.2 cm i.d. × 0.5 cm deep; small, 0.5 cm i.d. × 0.5 cm deep.

Vycor crucibles. The 30-ml crucibles are equipped with Rose covers and inlet tubes.

Vycor tube. The tube should be 30 mm in diameter.

Meker burner. The burner should have wing tip.

Triangular files. The files should be 8 in. long.

Small bench vise.

Steel block. The block should be of a size suitable for clamping in the vise and having a ³/₁₆-in. hole drilled ¾ in. deep in one side.

Small ball peen hammer.

Sharp-pointed center punch.

Aluminum foil.

Forceps.

Excitation source. This source should provide a high voltage spark with the source parameters described. An Applied Research Laboratories (ARL) High Precision source unit is suitable.

Spectrograph. The spectrograph should have sufficient resolving power and linear dispersion to separate clearly the analytical lines in the wavelength region 2500–4000 Å. An instrument having a reciprocal linear dispersion of 5 Å or less per millimeter satisfies these conditions. An ARL 2-m spectrograph with a grating ruled at 36,600 lines per inch is suitable.

Photographic processing equipment. This equipment should provide developing, fixing, washing, and drying operations.

Microphotometer. The microphotometer should have a precision of ±1.0% for transmittance readings between 10 and 90%.

Calculating equipment. A calculating board with emulsion calibration scales is required for converting microphotometer readings to absorbance and concentration values.

The Preparation of Standards

Standards should be prepared from the purest materials that can be obtained. Use granular C.P. lead, free of silver, gold, and bismuth; spectrographically standardized palladium, platinum, and rhodium sponge; pure precipitated gold; filings from spectrographically standardized silver rods, in the standards described here. Machine the crucibles from high purity graphite.

Other equipment is standard.

(a) The preparation of master standards. For the constant gold and constant silver series, prepare as many master standards as are desired. They should contain only platinum, palladium, and rhodium in addition to the lead. The three platinum metals should be "scrambled" as to concentration. For example, three master standards could be as follows:

	I	II	III
Platinum	1.00%	0.75%	0.50%
Palladium	0.75%	0.50%	1.00%
Rhodium	0.50%	1.00%	0.75%

Master standards for the arrested cupellation series (lead beads) may be prepared similarly but should also include gold and silver in concentrations calculated to provide suitable analytical curves for these elements.

Procedure. Weigh the lead into one of the large graphite crucibles, along with the required amounts of platinum, palladium, and rhodium. Amounts taken should be large enough to minimize weighing errors. Place the graphite crucible inside a 30-ml Vycor crucible. Arrange the Rose crucible cover and inlet tube so that a gentle stream of hydrogen will play on the melt. Heat over a Meker burner until the melt appears clean and shiny; the platinum metals will appear as floating specks until they have dissolved completely. Allow the melt to cool to room temperature under hydrogen. Assign a small triangular file to each standard; keep each file in a properly labeled, clean, container. File each master standard to obtain a sufficient amount of filings for dilution.

(b) The preparation of working standards. Each of the constant gold standards contains 5% of gold, platinum, palladium, rhodium, and silver in concentrations suitable for establishing spectrographic curves and contains lead to make 100%. Use about 1 g of material for each standard. A silver master standard (10 mg silver in 990 mg lead) facilitates weighing the silver additions. Each master standard is usually diluted 10- and 100-fold. For example, if the master standard contains

Palladium	0.70%
Platinum	0.35%
Rhodium	0.15%

a 10-fold dilution with additions of gold and silver will yield the following standard:

Palladium	0.070%
Platinum	0.035%
Rhodium	0.015%

Silver	0.070%	
Gold	5.00%	(constant)

The mixture required to obtain this standard would be

> 100 mg master standard filings
> 50 mg gold
> 70 mg silver master standard
> 780 mg lead

A 100-fold dilution, with additions of gold and silver, would yield

Palladium	0.0070%	
Platinum	0.0035%	
Rhodium	0.0015%	
Silver	0.0070%	
Gold	5.000%	(constant)

Constant silver standards are prepared in the same manner except that silver is adjusted to 10% in each standard and a gold master standard (10 mg gold in 990 mg lead) is used to facilitate weighing the gold. The ranges covered are approximately 0.001–1.00% for all the elements being determined. The arrested cupellation master standards contain all five metals so that it is only necessary to dilute them with lead to obtain working standards for the arrested cupellation technique.

Melting is done in essentially the same manner for the working standards as for the master standards. Use the medium graphite crucibles; several standards may be melted simultaneously by the use of a large Vycor tube mounted horizontally over a Meker burner fitted with a wing tip. Place several graphite crucibles in the tube and flush the tube with hydrogen which is burned at the end of a Rose tube. Maintain a flow of hydrogen during the melting and cooling to room temperature. Continue the heating for 20 min after the lead becomes molten. When cool, file each standard, and remelt 50-mg portions of filings in small graphite crucibles by the same procedure. Mount these 50-mg lead beads for spectrography as described below. Prepare as many as are desired, and store in labeled capsules or vials. In this instance, the 50 mg may be approximate; for spectrographic purposes any convenient weight can be used.

The Preparation of Samples for Analysis

Judgement must be exercised in assaying and in the addition of gold or silver to the assays. One or two assay tons is usually a sufficient sample for normal low grade ores and similar materials. If the grade is suspected to be high, reduce the sample size accordingly, e.g., to 0.5 or 0.25 assay ton. (Results in grams per metric ton can be readily obtained by calculation but more easily by selecting a convenient metric sample weight.) It may be necessary to assay larger samples of rocks or to combine the lead fusion buttons from several assays to obtain an adequate concentration for detection.

For the conventional fire assay (see Chapter 6, Procedure 121), add 5 mg of gold or 10 mg of silver as a collector, depending on whether a gold or silver bead is desired. Weigh the assay bead, when ready, in a graphite crucible with sufficient lead to provide either a 5% gold–95% lead alloy or a 10% silver–90% lead alloy. The entire weight of the bead is assumed to be gold or silver for this purpose. Melt as described above for the standards. When cool, weigh the lead bead as accurately as possible. If the weight is greater than 85 mg, flatten the bead, divide, and remelt for duplicates. Care must be taken at this stage to avoid contamination.

Do not add gold or silver to the assay when the arrested cupellation technique is used. Move the cupel near the furnace door when the bead diameter appears to be 3–4 mm. Then allow cupellation to proceed until the diameter is 1–2 mm, at which time remove the cupel from the furnace, and place it in a gentle stream of nitrogen on the hearth.

Mounting the Lead Beads

The equipment required for mounting beads is illustrated in Figs. 2.1 and 2.2.

Mount the standards and sample beads in the same manner. Clamp the steel block securely in the vise. Prepare copper rods, $^3/_{16}$ in. in diameter, 1 ½ in. long, by cleaning in dilute nitric acid and filing or machining one end flat, normal to the length. Place a rod in the steel block, and then cover with a shallow tray made of folded aluminum foil to prevent loss of the bead if it should roll off the electrode during mounting. Prick 4 holes, roughly $^1/_{32}$ in. deep and evenly spaced, in the top of the electrode with the center punch. Place and center the lead bead on the electrode with forceps. Hold a folded aluminum foil strip over the bead while it is tapped with the hammer until it is flattened, taking care to

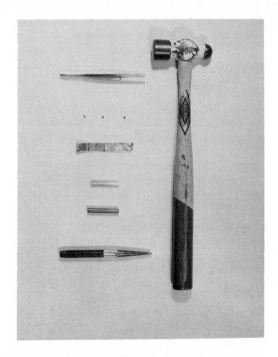

FIGURE 2.1.

FIGURE 2.2.

cover the end of the electrode uniformly. Force into the holes enough lead to secure the bead firmly to the electrode. With a little practice, this technique becomes easy and the bead can be formed into a circular sheet just covering the electrode, but not extending beyond the rim. It is desirable to maintain good technique but excellent reproducibility has been obtained even with poorly mounted beads.

Spectrochemical Procedure

(a) Electrode system. Place the copper rod carrying the lead bead in the lower electrode clamp. This is made the cathode. The anode is a graphite counter electrode similar to ASTM type C-5 (900), mounted across a 5-mm gap from the lead bead.

(b) Electrical parameters.

Capacitance	0.07 μF
Inductance	360 μH
Resistance in series with gap	none
Peak voltage output	20,000 V
Radio frequency current	5 A
Discharge trains	4 per cycle

(c) Exposure conditions.

Spectral region	2100–3900 Å
Slit width	0.050 mm

Slit length   2 mm
Prespark period   none
Exposure period   20 sec
Filters   Au 2424.9 Å, 50% transmittance

(d) Exposure index. Maintain the transmittance of the Pb 3118.92 Å line at approximately 30%.

(e) Calculation. The spectrum lines to be measured are listed in Table 2.1. The Pb 3118.92 Å line gives satisfactory results but lead lines at 3262.35, 3220.53, and 3240.19 Å have also been used in previous work. Most lead lines are diffuse, but reproducibility is found to be quite satisfactory despite this lack of sharpness. The data required for analysis include the following.

weight of sample assayed, in assay tons,
accurate weight of the lead (W),
percentage of each element as determined from the spectrochemical curves (%E).

Then,

$$\text{mg of element} = \frac{(\%E \times W)}{100}$$

If the original sample taken was one assay ton, then the element weights in mg may be stated directly as troy ounces per ton. If 0.5 assay ton was taken, the element weights must

TABLE 2.1
Line Pairs and Analytical Ranges for the Lead Bead Method

Element	Analytical line (Å)[a]	Concentration range (%)
Platinum	Pt 3064.71[b]	0.0045–0.140
	Pt 3042.63	
Palladium	Pd 3242.70	0.0045–0.140
	Pd 3373.00	
	Pd 3027.91	
Rhodium	Rh 3396.85	0.0045–0.070
	Rh 3323.09	
	Rh 3283.57	
	Rh 3372.25	
Gold	Au 2675.95	0.004–0.145
	Au 2427.95	
	Au 3122.78	
Silver	Ag 2437.79	030–1.40
	Ag 2357.92[c]	

[a] Internal standard line is Pb 3118.92 Å in all instances.
[b] May be used if Ni 3064.62 < Ni 3054.31.
[c] Beware of Pt 2357.57.

TABLE 2.2
Data on the Precision of the Lead Bead Method

Element	Average concentration (troy ounces per ton)	Coefficient of variation[a] (%)	No. of determinations
Platinum	0.180	6.1	356
Palladium	0.173	5.7	359
Rhodium	0.063	7.5	349
Gold	0.066	9.1	341
Silver	2.28	12.6	332

[a]The coefficient of variation, V, is calculated as follows:

$$V = \frac{100}{X} \left(\frac{\Sigma d^2}{n-1} \right)^{1/2}$$

where X is the average concentration in troy ounces per ton, d is the deviation of the determination from the mean, and n is the number of determinations made.

be doubled for ounces per ton; if 2 assay tons were taken, the element weights must be halved to obtain ounces per ton. See Table 2.2 for data on precision.

The Richvale Method for Iridium, Ruthenium, and Osmium

One of the problems yet to be solved is the determination of ruthenium, iridium, and osmium in fire assay beads. These metals are soluble in molten lead, but they appear to exsolve upon cooling.

The present author has rarely found lines of iridium or osmium in spectra from fire assay beads prepared from ore samples. It is known that some or all of the osmium may be lost as a volatile oxide during cupellation. In many ores, the content of iridium and osmium is probably so low that a much higher concentration factor than is practical would be required for their detection.

Ruthenium is occasionally detected in fire assay beads prepared from ores and often from metallurgical concentrates where high concentration has been achieved. Some work was done to determine this metal, but a satisfactory degree of reproducibility from sample to sample was not realized in practice. Also it was not found possible to prepare lead beads that were homogeneous with respect to ruthenium, even when homogeneity was achieved for gold, silver, platinum, palladium, and rhodium.

Hawley et al. (71) found that a spark discharge allowed the determination of ruthenium, iridium, and osmium in assay beads from high grade ores if a minimum of lead was added to the beads. Scobie (68) included iridium in the standard silver beads used in his method. Repeated and varied experiments in the present authors' laboratory with low concentrations of ruthenium and iridium in lead have shown that the former two metals are probably not soluble in lead at low temperatures and that it is very difficult to achieve sufficient reproducibility with a spark discharge for their determination with the other precious metals. It has been considered that a dc arc discharge might offer a solution to this problem if some element other than lead were used as an internal standard. For example, a counter electrode of copper and a dc arc applied to the residual bead after sparking might allow the determination of ruthenium plus any iridium and osmium that might be present.

THE DETERMINATION OF TRACE IMPURITIES
IN THE REFINED PLATINUM METALS

The difficulties encountered in dissolving platinum group metals and in analyzing them by wet chemical methods probably encouraged the early application of spectrochemistry to this problem. In 1922, Meggers *et al.* (95) published lists of rhodium lines in the spark spectra of platinum containing 0.001, 0.01, 0.5, and 1.0% of rhodium. Gerlach and Schweitzer (96) were working on the spectrographic analysis of the platinum metals in 1929 when they discovered the principle of homologous lines. This laid the foundation for the internal standard method which first allowed accurate quantitative spectrochemical analysis. In 1931 Gerlach and Ruthardt (97) published lists of the most sensitive lines of various impurity elements, including platinum metals, in platinum, iridium, and rhodium. Gerlach and Reidl (98) described spectrographic and electrical experiments on "pure" platinum in 1933. Two years later, Stauss (99) outlined the use of the spectrograph for the identification of alloys, the determination of minor impurities, and the control of the purity of platinum during production. In the following year Rollwagen and Ruthardt (100) determined arsenic and phosphorus in platinum to less than 0.005% with an intermittant arc and sulfur to less than 0.05% with a condensed discharge in a quartz vacuum tube.

In 1941, Hauser (101) determined beryllium in a platinum solution by an impregnated carbon electrode technique. Shortly afterwards, Babaeva *et al.* (102) quantitatively determined iridium and rhodium in intermediate products of platinum refining by applying a condensed spark discharge to dry salts. Ishitsuka (103) described the quantitative spectrographic analysis of platinum wire in 1943; Raper and Withers (104) described an arc method for the spectrographic examination of platinum in 1945. In the same year van der Voort (105) published a method for the determination of palladium in platinum. In 1947 Kheifits and Katchenkov (106) purified platinum chemically for use in standards and determined its purity spectrographically to be 99.999%. Babaeva *et al.* (107) described

an arc and a spark technique for determining iridium and rhodium in intermediate products of platinum refining. Babaeva and co-workers (108) also used a condensed spark to determine from 0.001 to 5% palladium in platinum, from less than 0.2 to 5% platinum in palladium, and from 0.001 to 1.0% rhodium in iridium. In 1948 Babaeva and Lapir (109) used an ac arc to determine iridium, platinum, and palladium in refined rhodium. The same authors (110) published a method for determining small quantities of iron in platinum and palladium in the following year. Pasveer (111) described a method for preparing precious metals for spectrographic analysis by pressing well-mixed powders into pellets and annealing them. For analyses, thin disks cut from these pellets were placed on mandrels, similar to the standard electrodes.

Kheifits and Katchenkov (112) described the spectrographic determination of small amounts of iridium, platinum, and rhodium in palladium in 1950. An excellent description of the status of spectrochemical analysis of the platinum metals in 1951 is contained in "Metal Spectroscopy" by Twyman (113) in a chapter entitled The Platinum Group of Metals, written by Withers. Hawley *et al.* (114) published a method for the quantitative spectrographic analysis of platinum and palladium in the same year. Oberländer (115) used homologous line pairs to determine traces (0.005–1.0%) and larger amounts (1–32%) of gold, palladium, rhodium, copper, iron, iridium, silver, and nickel in platinum. Qualitative and semiquantitative trace element analyses of platinum were described in Spectrographic Analysis of Pure Metals by D.M. Smith (116). In 1953 Lewis and Ott (117) described further work on analysis of palladium and Koehler (118) published methods for the determination of 24 elements in platinum and 22 elements in palladium. Lewis and co-workers (119) described a method for quantitative spectrographic analysis of refined rhodium in 1955.

Vorsatz (120), two years later, placed filter paper containing known amounts of impurities in contact with a globule of pure precious metal and arced to incorporate the impurities in the metal. Globules prepared in this manner were used as standards for spectrographic analysis. In 1958 Chentsova (121) published a method for the analysis of iridium in which iridium sponge was impregnated with solutions of platinum, palladium, rhodium, and iron. After drying, grinding, and calcining, these standards were analyzed with a low voltage spark discharge. Pankratova (122) prepared standards for the analysis of refined iridium and ruthenium by mixing solutions of iridium, platinum, ruthenium, rhodium, palladium, gold, and iron, evaporating to salts, and calcining in hydrogen. After mixing the powder with graphite powder, he vaporized it completely from a cratered electrode. Kuranov (123) described a method somewhat similar to these for determining the impurities in iridium and rhodium. In 1962, Lincoln and Kohler (124) published a comprehensive method for the spectrographic analysis of high purity platinum.

Gut'ko *et al.* (125), in 1963, analyzed refined osmium for some 15 impurity elements by evaporating most of the osmium as the tetroxide and mixing the residue with graphite powder. Standards were prepared by mixing oxides and metals with graphite powder. The standards covered the ranges: 0.005–0.1% for platinum, iridium, ruthenium, iron, and nickel; 0.0005–0.01% for palladium, rhodium, silver, gold, copper, silicon, barium, sodium, and aluminum; and 0.004% for manganese. Gut'ko and Pyatkova (126) described a direct method for determining 0.01% or more of platinum, rhodium, iridium, ruthenium, and iron and 0.002% or more of palladium, gold, silver, and copper in refined osmium in 1965.

Diehl (127), in 1965, applied dissolution methods to overcome the lack of suitable standard samples, for determining 1–100 ppm amounts of impurity elements in precious metals. The solution to be analyzed was concentrated on spectrographic carbon powder and the resulting mixture was pressed into pellets for ac arc analysis.

Tymchuk *et al.* (128) studied carrier distillation techniques for determination of trace impurities in copper by emission spectrochemistry. They found the halides of copper and silver, particularly the fluorides and chlorides, to be promising spectrographic carriers with sensitivities, in the case of copper fluoride, being equal to, or in many cases an order of magnitude better than, the sensitive globule arc method. They found indications that these carriers might be equally effective for impurities in silver, gold, and the platinum metals.

Vukanovic *et al.* (129) determined traces of rhodium, in solutions containing only rhodium, by depositing the rhodium electrochemically on a copper disk cathode. The disk was then used as a rotating disk electrode and the rhodium was excited with a 3.6 A interrupted arc. As little as 2.4×10^{-7} g of rhodium in 3 ml of solution could be determined.

Chwastowsk *et al.* (130) determined trace impurities in platinum and platinum–rhodium alloys by adsorbing the impurities on a cation exchange column while the platinum and rhodium passed through as anionic chloride complexes. The impurity elements were eluted with nitric acid, the solution was diluted, internal standards were added, and the solution was sparked on a carbon electrode.

In 1969, Kohler and Lincoln (131) published a comprehensive method for analysis of refined palladium.

RECOMMENDED METHODS

An investigation of the application of spectrochemistry to the analysis of the refined platinum metals was begun at Queen's University, Kingston, Canada, in

1950, under the direction of Professor J. E. Hawley. As this work proceeded, a collaboration between the present author, as spectrographer, and W. L. Ott, as chemist, began; it has continued until the present time. An understanding by each of the other's work has provided an arrangement whereby the best possible spectrochemical standards could be developed. Grateful acknowledgment is made to Mr. Ott, without whom it would have been difficult indeed to write this chapter. The investigation was carried out at Queen's until 1953, when it was continued in the metallurgical laboratories of Falconbridge Nickel Mines Limited, Thornhill, Canada.

Methods for the spectrochemical analysis of platinum (114), palladium (114, 117), and rhodium (119) were developed and published. Work was also done on the analysis of ruthenium and iridium, but this was discontinued before completion and hence was not published. These methods are described in detail here.

The excellent work of Lincoln and Kohler on the analysis of platinum (124), which has recently been published, provides for the determination of many more impurity elements at much lower concentrations than the author's method for platinum and has therefore been included. Similarly, Kohler and Lincoln's method for analysis of palladium (131) is considered the best for this element.

The analysis of refined platinum metals may involve several problems. These metals are usually highly pure, but may contain many impurity elements at quite low concentrations. It is difficult to obtain information as to the physical form in which the impurity elements are present. Refined platinum metals are usually produced as powders called *sponge*. The distribution of impurities in these powders may be heterogeneous so that samples must be treated before analysis. Great care must be taken to avoid loss during such sample preparation. Melting, for example, would drive off the more volatile elements so that the results would be falsely low. Finally, there is almost a complete lack of commercially available standard samples of known impurity content, and there is the necessity to develop satisfactory working standards. There are, however, two platinum reference standards available. Produced by the National Bureau of Standards, Washington, D.C., and offered for sale in 1967, these are designated SRM#680, High Purity Platinum, and SRM#681, Doped Platinum (132). Both are in the form of wire 0.020 in. (0.51 mm) in diameter and are available in two lengths, 4 in. (10.2 cm) designated L-1 and 39.4 in. (1.0 m) designated L-2. Both have been extensively tested and found satisfactory with respect to homogeneity, using the following methods: optical emission spectrography, spark source mass spectrography, and electrical measurements, including EMF, Thermal Coefficient of Resistivity (T.C.R.), and Residual Resistivity Ratio (R.R.R.). The recommended values of the concentrations of the various elements given on the provisional certificate were obtained by one or more of the following methods: optical emission spectrography, spark source mass spectrography (isotopic dilution), polarography, spectrophotometry, nuclear activation analysis, and vacuum fusion.

The necessity of preparing accurate working standard samples for low concentrations of impurity elements is probably the most serious problem involved in platinum metal spectrochemistry. When the impurity concentration ranges desired are high—hundredths of one percent, for example—standard samples may be produced by dissolving the platinum metal, adding aliquots of solutions of the impurity element solutions, coprecipitating the platinum metal and impurities, and reducing the precipitate to metal powder. Grinding in a mortar will then usually produce a homogeneous mixture. Because many of the impurity elements may be present in the platinum metal in trace amounts, the additions are usually made in such a way that these residual traces can be determined by the extrapolation of working curves.

For the analysis of purer platinum metals wherein impurity concentrations are in parts per million or less, the preparation of standard samples by the solution reduction method is less satisfactory because such factors as nonhomogeneous precipitation and crystallization become highly critical and prevent satisfactory homogeneous mixtures being attained, even with considerable grinding after reduction. It is then found more satisfactory to work entirely with powders. Both methods are described here.

The general method of spectrochemical analysis may be outlined as follows. Pellets of the platinum metal are subjected to dc arc and/or controlled spark excitation along with pellets of standard samples. The intensity ratios of selected pairs of analytical lines and platinum metal *internal standard lines* are determined photometrically. Photographic emulsions are calibrated by means of iron spectra obtained with stepped filters having known transmission ratios. The concentrations of the impurity elements are read from analytical curves relating intensity ratio to concentration. The analytical curves are checked by means of intensity ratios from the standard samples. The equipment used is described with the appropriate procedure. The spectrochemical procedure and other conditions peculiar to the individual methods are listed or described in each section.

In all instances, photographic processing, photometry, emulsion calibration, the preparation of analytical curves, and the calculation of intensity ratios are carried out according to standard procedures (133).

The Method of Lincoln and Kohler for High-Purity Platinum

The most comprehensive spectrochemical method known to the present author for the analysis of high purity platinum is that described by Lincoln and Kohler (124). It provides for the determination of 27 elements in the concentration ranges presented in the accompanying tabulation.

Element	Concentration range (ppm)	Element	Concentration range (ppm)	Element	Concentration range (ppm)
Aluminum	1–160	Copper	0.1–20	Osmium	80–800
Antimony	20–800	Gold	0.5–160	Palladium	1–400
Arsenic	20–400	Iridium	10–800	Rhodium	0.6–80
Bismuth	5–40	Iron	2–160	Ruthenium	5–800
Boron	4–100	Lead	5–400	Silicon	5–160
Cadmium	10–160	Magnesium	0.1–160	Silver	0.1–400
Calcium	0.1–80	Manganese	0.5–40	Tellurium	40–800
Chromium	0.5–80	Molybdenum	10–160	Tin	2–100
Cobalt	0.5–800	Nickel	1–80	Zinc	10–800

SUMMARY OF THE METHOD

Platinum standards are prepared by adding stock solutions of impurity elements to high-purity ammonium platinum chloride, drying, blending, and reducing to platinum sponge. The sponge is then ground to powder and pressed into pellets. Samples of sponge for analysis are ground to powder and pressed into pellets in a similar manner. The spectra are obtained by using a dc arc excitation in an argon–oxygen atmosphere.

PROCEDURES 13 AND 14 (124)

Equipment

Polypropylene bottle.

Mixer mill. (Spex Industries Model No. 5000).

Rotary mill. The mill should have platinum baffle (Engelhard Industries).

Quartz boat.

Tube furnace. The furnace should be equipped for special atmospheres.

Polyethylene vials and balls.

Pellett press and die assembly. (Engelhard Industries).

Molds. See Figs. 2.3 and 2.4.

Excitation source. This source should provide a dc arc with an open circuit voltage of 300 V. (An ARL Multisource, Model No. 5700, was used by Lincoln and Kohler.)

Stallwood jet. This jet should provide an annular curtain of 70% argon–30% oxygen gas blown upward at a rate of 4 l per min around the arc column. (The Spex Industries Model No. 9014 was used by Lincoln and Kohler.)

FIGURE 2.3.

Procedure

The preparation of standard samples.   Make ammonium platinum chloride by dissolving platinum (99.999+ % purity) in aqua regia, fuming with hydrochloric acid to remove the excess of nitric acid and precipitating with ammonium chloride. Add appropriate amounts of stock solutions of impurity elements to the ammonium platinum chloride in a polypropylene bottle at 100°C, and when dry blend the mixture mechanically by shaking the bottle in a mixer mill. Then transfer to a rotary mill and grind for a minimum of 72 hr.

Reduce the standards thus prepared by heating them in a quartz boat at 200°C for 2 hr, at 300°C for 1 hr, and at 600°C for 30 min in an atmosphere 93% in nitrogen and 7% in hydrogen, flowing at 14 l per hour. Grind the platinum sponge to a powder in a mixer mill.

Impurity elements are divided into two groups as shown in the accompanying tabulation to minimize wavelength interference in the standards.

FIGURE 2.4.

Group A		Group B	
Cadmium	Magnesium	Aluminum	Manganese
Calcium	Molybdenum	Antimony	Nickel
Copper	Palladium	Arsenic	Osmium
Gold	Ruthenium	Bismuth	Rhodium
Iridium	Silicon	Boron	Silver
Iron	Tellurium	Chromium	Zinc
Lead	Tin	Cobalt	

At least six standards spanning the range 0.01–10 ppm each contain all of the impurities. A separate standard is made for each group at concentrations above 10 ppm.

The preparation of samples for analysis. Grind samples of platinum sponge for analysis to a −325 mesh powder in a polyethylene vial, containing a polyethylene ball, in a mixer mill. Mix a 200 mg portion of the sieved platinum powder with 50 mg of graphite briquetting powder in the mixer mill. Make a 50-mg portion of the mixture into a cylindrical pellet 0.157 in. in diameter by 0.166 in. long by pressing in a special mold on 100 mg of graphite briquetting powder at a pressure of 40 tons/in^2.

Spectrochemical Procedure

(a) Electrode system. Place the pellet of the sample or standard in the crater of a high purity graphite electrode (National Carbon Co. Special Graphite Spectroscopic Electrodes 0.242 in. in diameter) (133). Place the electrode in the Stallwood jet attachment, and make this the anode. As a counter electrode use a ⅛-in.-diam rod of high purity graphite 1½ in. in length and with a flat end.

(b) Electrical parameters.

Open circuit dc voltage	300 V
Current	12 A

(c) Exposure conditions.

Spectral region	2180–4600 Å
Slit width	0.020 mm
Slit length	2 mm
Exposure period	30 sec
Atmosphere	70% argon–30% oxygen at 4 l/min
Filters	as noted in Table 2.3
Gap	3 mm

(d) Exposure index. Maintain transmittance of Pt 2814.0 at approximately 40%.

(e) Replicate exposures. Make triplicate exposure of each sample. Run two or more control standards in duplicate.

TABLE 2.3
Line Pairs and Analytical Ranges for the Lincoln and Kohler Method

Element	Analytical line (Å)	Filter (% transmission)	Internal standard line	Filter (% transmission)	Concentration range (ppm) Low	High
Aluminum	3961.5	10	a	10	1	20
	2373.1	20	b	20	20	160
Antimony	2311.5	20	b	20	20	800
Arsenic	2349.8	20	b	20	20	400
Bismuth	3067.7	10	a	10	5	40
Boron	2497.7	10	b	20	4	100
Cadmium	3466.2	10	a	10	10	160
Calcium	4226.7	10	a	10	0.1	5
	4226.7	2	a	10	1	80
Chromium	4254.3	20	a	10	0.5	10
	4254.3	10	a	10	5	20
	2762.6	10	c	10	20	80
Cobalt	3453.5	10	a	10	0.5	10
	3395.4	2	a	10	10	800
Copper	3247.5	10	a	10	0.1	1
	3247.5	1	a	10	1	20
Gold	2676.0	10	b	20	0.5	20
	3122.8	10	a	10	10	160
Iridium	3220.8	10	a	10	10	160
	2924.8	10	c	10	80	800
Iron	3020.6	10	a	10	2	20
	2788.1	10	c	10	20	160
Lead	2802.0	20	c	10	5	80
	2802.0	10	c	10	40	400
Magnesium	2802.7	10	c	10	0.1	10
	2779.8	10	c	10	10	160
Manganese	2801.1	20	c	10	0.5	10
	2801.1	10	c	10	1	40
Molybdenum	3132.6	10	a	10	10	160
Nickel	3493.0	10	a	10	1	10
	3437.2	10	a	10	10	80
Osmium	2909.1	10	c	10	80	800
Palladium	3404.6	10	a	10	1	10
	2447.9	20	b	20	10	400
Rhodium	3434.9	10	a	10	0.6	10
	3396.9	2	a	10	10	80
Ruthenium	3498.9	10	a	10	5	40
	2678.8	10	b	20	40	800
Silicon	2881.6	10	c	10	5	25
	2528.5	10	b	20	10	160
Silver	3382.9	10	a	10	0.1	10
	3382.9	1	a	10	10	400

TABLE 2.3 *cont.*

Element	Analytical line (Å)	Filter (% transmission)	Internal standard line	Filter (% transmission)	Concentration range (ppm) Low	High
Tellurium	2385.8	20	b	10	40	800
Tin	2840.0	20	c	10	2	25
	2840.0	10	c	10	10	100
Zinc	3345.0	10	a	10	10	80
	3345.6	10	a	10	80	800

^{a}Pt 3110.1 Å. ^{b}Pt 2578.4 Å. ^{c}Pt 2814.0 Å.

Precision and Accuracy

The overall precision for total impurities is estimated at a coefficient of variation of 10.1%. Accuracy is unknown as few data are available for comparing analyses obtained from other chemical methods with spectrochemical analyses.

Powder Technique for Refined Platinum

The powder technique described here does not provide for the determination of as many impurity elements nor do the analytical ranges extend as low as in the Lincoln and Kohler method, but the method has nevertheless been found very useful for the routine control analysis of refined platinum in the present author's laboratory. It provides for the determination of eleven elements in the concentration ranges shown in the accompanying tabulation.

Element	Concentration range (%)	Element	Concentration range (%)
Palladium	0.005–0.25	Iron	0.02–0.12
Rhodium	0.003–0.040	Magnesium	0.0015–0.010
Gold	0.002–0.015	Copper	0.0005–0.012
Iridium	0.002–0.015	Silicon	0.0005–0.005
Nickel	0.002–0.015	Silver	0.0002–0.005
Calcium	0.002–0.12		

SUMMARY OF THE METHOD

Platinum standards are prepared by screening a selected sample and blending impurity elements with the fine powder to provide a master standard which is diluted with platinum powder for working standards. Standards and samples are mixed with graphite briquetting powder and pressed onto short lengths of graphite rod to form pellets. Both dc arc and high-voltage spark modes of excitation are employed to obtain spectra. Experimental conditions are given in Table 2.4, and data on precision in Table 2.5. Further details are given in Beamish (133a).

The Method of Kohler and Lincoln for High-Purity Palladium

Kohler and Lincoln (131) have published a spectrographic method for determination of 28 elements in palladium sponge and metal over the concentration ranges shown in the accompanying tabulation.

Element	Concentration range (ppm)	Element	Concentration range (ppm)	Element	Concentration range (ppm)
Aluminum	6–400	Gold	3–400	Platinum	40–800
Antimony	50–1000	Iridium	10–160	Rhodium	4–800
Arsenic	100–1000	Iron	20–800	Ruthenium	2–100
Bismuth	1–160	Lead	3–800	Silicon	40–400
Boron	4–160	Magnesium	4–800	Silver	6–100
Cadmium	6–800	Manganese	1–160	Tellurium	20–800
Calcium	15–800	Molybdenum	1–160	Tin	10–160
Chromium	3–160	Nickel	3–100	Titanium	3–100
Cobalt	3–100	Osmium	20–160	Zinc	20–160
Copper	2–800				

SUMMARY OF THE METHOD

Palladium standards are prepared by adding stock solutions of impurity elements to high purity palladium nitrate solution, drying, blending, and reducing to palladium sponge. The sponge is then ground to metal powder and pressed into pellets. Samples of sponge for analysis are ground to powder and pressed into pellets in a similar manner. Metal samples are sawed or filed using tungsten carbide cutting tools. The fine particles are cleaned in hydrochloric acid, washed, dried, and pressed into pellets. Spectra are obtained by using dc arc excitation.

PROCEDURE 15 (131)

Equipment

Teflon beaker.

Teflon stirring rod.

Quartz boat.

Agate pestle.

Tube furnace.   The furnace should be equipped for special atmospheres.

TABLE 2.4

Line Pairs and Analytical Ranges for the Platinum Powder Technique

Element	Analytical line (Å)	Internal standard Line (Å)	Concentration range (%) Low	High
		Spark spectra		
Palladium	Pd 3242.70	Pt 3256.43	0.005	0.025
Iridium	Ir 2924.79	Pt 2904.0[b]	0.002	0.015
		Arc spectra		
Rhodium	Rh 3396.85	Pt 3335.83	0.003	0.020
	Rh 3434.89	Pt 3455.6[b]	0.008	0.040
Gold	Au 2675.95	Pt 3004.15	0.002	0.015
Nickel	Ni 3492.95	Pt 3455.6[b]	0.002	0.015
Calcium	Ca 3933.66[a]	Pt 4034.17	0.002	0.012
Iron	Fe 2611.87	Pt 3004.15	0.002	0.012
Magnesium	Mg 2852.12	Pt 2818.83	0.0015	0.010
Copper	Cu 3273.96	Pt 3335.83	0.0005	0.006
	Cu 3247.54	Pt 3335.83	0.002	0.012
Silicon	Si 2881.57	Pt 3004.15	0.0005	0.005
Silver	Ag 3382.89	Pt 3335.83	0.0002	0.005

[a]Filter 25% transmittance.
[b]Unlisted line.

TABLE 2.5

Data on the Precision of the Platinum Powder Technique

Element	Average concentration (%)	Coefficient of variation (%)	No. of determinations
Palladium	0.007	11.9	5
Iridium	0.0038	9.6	6
Rhodium	0.011	7.2	5
Gold	0.006	10.8	5
Nickel	0.006	8.8	5
Calcium	0.002	18.5	5
Iron	0.003	5.8	5
Magnesium	0.0015	4.3	5
Copper	0.0015	11.3	5
Silicon	0.0007	12.2	4
Silver	0.0017	6.3	5

Drying oven.

Polyethylene vials and balls.

Spectrograph.   The spectrograph should be ARL 2-meter with 24,400 lines/in.; grating dispersion 5.2 Å/mm.

Excitation source.   (ARL Multisource Model No. 5700)

Densitometer.   (ARL Model No. 3400)

Mixer mill.   (Spex Industries Model No. 5000)

Roll mill.   (Engelhard Industries)

Pellett press and die assembly.   (Engelhard Industries)

The Preparation of Standard Samples

Prepare palladium nitrate solution from comercially available palladium sponge having a purity of 99.99%. Prepare master solutions by adding calculated volumes of standardized stock solutions of impurity elements to a volume of palladium nitrate solution equivalent to 25 g of palladium. Use aliquots of the master solutions to prepare each set of standards. The stock solutions contain the equivalent of 1 g of metal per liter calculated from a reagent grade salt dissolved in water or dilute acid. Divide the impurity elements into three desired groups and use the salts shown in the accompanying list.

Group A	Group B	Group C
Iridium chloride	Rhodium chloride	Gold chloride
Tin chloride	Platinum chloride	Platinum chloride
Gold chloride	Antimony chloride	Rhodium chloride
Ruthenium chloride	Arsenic chloride	Lead nitrate
Titanium chloride	Osmium chloride	Copper nitrate
Calcium nitrate	Tellurium chloride	Magnesium nitrate
Iron nitrate	Manganese nitrate	Calcium nitrate
Copper nitrate	Chromium nitrate	Silver nitrate
Cadmium nitrate	Aluminum nitrate	Sodium metasilicate
Lead nitrate	Silver nitrate	
Magnesium nitrate	Bismuth nitrate	
Sodium metasilicate	Cobalt nitrate	
($NaSiO_3 \cdot 9H_2O$)	Zinc nitrate	
Ammonium molybdate	Nickel nitrate	
	Boric acid	

Use Groups A and B to prepare standards of concentration range 0.6–6 ppm. For concentration levels from 12 to 160 ppm, add only those impurity elements in Group A or B to a single standard. Use the impurity elements in Group C for concentrations from 200 to 800 ppm. Mix each solution standard well in a Teflon beaker with a Teflon stirring rod and transfer it to a quartz boat. Dry it initially in a tube furnace at 70°C for 12 hr under nitrogen flowing at 8 l/min. Complete the drying in an oven at 90°C. Grind the standard manually to a fine powder in the quartz boat with an agate pestle. (A thoroughly dried standard will powder readily.)

Reduce the thoroughly blended standard in a tube furnace at 50°C, 100°C, 150°C, and 200°C for 0.5 hr; and at 400°C for 1 hr in an atmosphere of 93% nitrogen and 7% hydrogen. Maintain a flow rate of 8 l/min throughout the reduction cycle. Introduce carbon monoxide* for an additional 0.5 hr at 400°C also with an 8 l/min flow. Turn off the heat and allow the standard to cool in carbon monoxide. Transfer the final sponge to a plastic vial containing a ⅜ in. plastic bead and reduce to a metal powder with a Spex mixer mill.

The Preparation of Samples for Analysis

Convert samples of palladium sponge, as received, to powder in a polyethylene vial containing a polyethylene ball using a mixer mill for 60 sec. Mix a 200 mg portion of the palladium thoroughly with 50 mg of National Spectrographic graphite powder (SO-1) in the mixer mill for 60 sec.

Make a 50 mg portion of the mixture into a cylindrical pellet 0.157 in. in diameter by 0.166 in. long by pressing in a special mold on 100 mg of graphite briquetting powder at a pressure of 40 tons per sq in. For metal samples such as bar, sheet or wire, use a No. 0 Jeweler saw blade or a No. 5 file (once only for each sample) to obtain fine metal particles. Boil samples taken in this manner in a 1:1 solution of hydrochloric acid to remove iron contamination, decant the solution, and wash repeatedly with distilled water, and dry the particles. Use only -100 mesh particles for analysis.

Spectrochemical Procedure

(a) Electrode system. Place the pellet of the sample or standard in a cylindrical high-purity graphite electrode (National Carbon Co.), 0.242 in. diam, undercut for 0.093 in. to leave a neck of 0.125 in. and a platform 0.125 in. thick recessed 0.032 in. to receive the pellet. Maintain a friction fit of the pellet in the electrode. Use a counterelectrode 1.5 in. long by ⅛ in. diam with a flat end. Make the analytical gap 3 mm.

(b) Electrical parameters.

Open circuit dc voltage	300 V
Current	11 A

(c) Exposure conditions.

Spectral region	2180–4600 Å
Slit width	0.030 mm
Exposure period	30 sec
Filters	primary filtering at split-split field 50T/100T at camera position with selected lines independently filtered as shown in Table 2.6

---

*When palladium in finely divided form is exposed to hydrogen at elevated temperatures, it should be cooled in nitrogen, carbon monoxide, or some other inert gas to remove the hydrogen adsorbed on the metal. Failure to do so may result in ignition or explosion of the material on exposure to air. Although the possibility of catalytic ignition is greatly reduced by using annealing gas for reduction of the palladium standards, carbon monoxide is preferred during the cooling cycle as an extra precaution. Use a cycle of 30 sec on–30 sec off to avoid overheating.

TABLE 2.6
Line Pairs and Analytical Ranges for the Kohler and Lincoln Method

Element	Analytical line (Å)	Filter (% transmission)	Internal standard line (Å)	Filter (% transmission)	Concentration range (ppm) Low	High
Ag	3382.9	0.8	Pd 3346.1	4	6–30	
	3382.9	0.4	Pd 3346.1	4	8–80	
	3382.9	0.3	Pd 3320.1	3	30–100	
Al	3082.2	4	Pd 3346.1	4	6–30	
	2373.1	8	Pd 3346.1	4	20–160	
	2373.1	4	Pd 3346.1	4	30–400	
As	2349.8	8	Pd 3346.1	4	100–1000	
Au	2676.0	8	Pd 3346.1	4	3–250	
	2676.0	4	Pd 3346.1	4	10–100	
	3122.8	4	Pd 3346.1	4	100–400	
B	2497.7	8	Pd 3346.1	4	4–50	
	2497.7	4	Pd 3346.1	4	10–160	
	2496.7	4	Pd 3346.1	4	50–160	
Bi	3067.7	8	Pd 3346.1	4	1–10	
	3067.7	4	Pd 3346.1	4	4–40	
	2938.3	8	Pd 3346.1	4	50–160	
Ca	4226.7	0.8	Pd 3346.1	4	15–30	
	3158.9	8	Pd 3346.1	4	40–400	
	3158.9	4	Pd 3346.1	4	160–800	
Cd	2288.0	8	Pd 3346.1	4	6–100	
	3466.2	4	Pd 3346.1	4	100–800	
Co	3453.5	4	Pd 3346.1	4	3–10	
	3044.0	4	Pd 3346.1	4	10–100	
Cr	4254.2	8	Pd 3346.1	4	3–6	
	4254.2	4	Pd 3346.1	4	3–20	
	2835.6	8	Pd 3346.1	4	20–100	
	2835.6	4	Pd 3346.1	4	20–160	
Cu	3273.9	0.8	Pd 3346.1	4	2–10	
	3273.9	0.4	Pd 3346.1	4	3–100	
	3273.9	0.3	Pd 3320.9	3	30–160	
	2618.4	8	Pd 3346.1	4	160–800	
Fe	3020.6	4	Pd 3346.1	4	20–40	
	2994.4	4	Pd 3346.1	4	30–100	
	2788.1	8	Pd 3346.1	4	30–400	
	2788.1	4	Pd 3346.1	4	400–800	
Ir	3220.8	8	Pd 3346.1	4	10–100	
	2924.8	4	Pd 3346.1	4	50–160	
Mg	2795.5	4	Pd 3346.1	4	4–10	
	2779.8	4	Pd 3346.1	4	10–60	
	2776.7	4	Pd 3346.1	4	160–800	
Mn	2794.3	4	Pd 3346.1	4	1–10	
	2801.1	4	Pd 3346.1	4	3–30	

TABLE 2.6 *cont.*

Element	Analytical line (Å)	Filter (% transmission)	Internal standard line (Å)	Filter (% transmission)	Concentration range (ppm) Low High
	2949.2	8	Pd 3346.1	4	10–100
	2933.1	4	Pd 3346.1	4	50–160
Mo	3132.6	8	Pd 3346.1	4	1–6
	3132.6	4	Pd 3346.1	4	3–25
	3085.6	4	Pd 3346.1	4	50–160
Ni	3493.0	4	Pd 3346.1	4	3–10
	3050.8	8	Pd 3346.1	4	3–25
	3050.8	4	Pd 3346.1	4	15–100
	3054.3	4	Pd 3346.1	4	15–160
	3031.9	4	Pd 3346.1	4	100–1000
Os	2909.1	8	Pd 3346.1	4	20–80
	2909.1	4	Pd 3346.1	4	20–160
Pb	2833.0	8	Pd 3346.1	4	3–10
	2833.0	4	Pd 3346.1	4	6–160
	2873.3	4	Pd 3346.1	4	160–800
Pt	2659.4	8	Pd 3346.1	4	40–100
	2659.4	4	Pd 3346.1	4	50–200
	2830.3	4	Pd 3346.1	4	100–800
Rh	3323.1	4	Pd 3346.1	4	4–100
	3191.2	8	Pd 3346.1	4	30–160
	3191.2	4	Pd 3346.1	4	100–800
	2729.0	8	Pd 3346.1	4	400–800
Ru	3498.9	8	Pd 3346.1	4	2–80
	3498.9	0.8	Pd 3346.1	4	5–100
	3498.9	0.4	Pd 3346.1	4	25–100
Sb	2598.1	8	Pd 3346.1	4	50–400
	2311.5	8	Pd 3346.1	4	100–500
	2311.5	4	Pd 3346.1	4	250–1000
Si	2881.0	4	Pd 3346.1	4	40–70
	2519.0	4	Pd 3346.1	4	40–400
Sn	3175.0	4	Pd 3346.1	4	10–30
	2863.3	4	Pd 3346.1	4	30–160
Te	2385.8	8	Pd 3346.1	4	20–800
Ti	3361.2	4	Pd 3346.1	4	3–20
	3349.0	4	Pd 3346.1	4	10–100
Zn	3345.0	8	Pd 3346.1	4	20–160

Precision and Accuracy

The precision of this method was estimated from a sample of palladium containing 21 impurity elements. For single elements, the coefficient of variation ranged from 2.4% for silicon to 7.2% for platinum. The average coefficient was 4.9%. Accuracy is unknown but is assumed to be similar to the precision.

The Analaysis of Refined Rhodium

The general procedure described here is that of Lewis *et al.* (119). Twelve elements in the concentration ranges in the accompanying tabulation may be determined in rhodium by the method described.

Element	Concentration range (%)	Element	Concentration range (%)	Element	Concentration range (%)
Iridium	0.50–1.50	Ruthenium	0.01–0.05	Silver	0.005–0.02
Iron	0.02–0.10	Copper	0.005–0.03	Platinum	0.002–0.015
Nickel	0.02–0.10	Lead	0.005–0.02	Cobalt	0.005–0.01
Palladium	0.02–0.10	Silicon	0.005–0.02	Gold	0.005–0.01

SUMMARY OF THE METHOD

Rhodium standards and samples are prepared as powders from the sponge and the finely divided black forms of the metal. The powder is mixed with graphite and pressed onto a graphite rod to form a pellet.

Both dc arc and high voltage spark modes of excitation are employed. The method described is for ordinary refined rhodium; the powder technique used for high purity platinum and palladium, although not tested for rhodium by the present authors, is recommended for high purity rhodium. Line pairs and precision data for the analysis of rhodium are given in Tables 2.7 and 2.8.

PROCEDURE 16 (119)

Equipment

Nylon screen. (325 mesh.)

Alundum mortars and pestles.

Mixing box. The box should provide a dust-free atmosphere.

Glass vials.

Stainless steel scoop. The scoop should have a volume equivalent to 40 mg of sponge–graphite mixture.

Gelatin capsules. (No. 00)

Hardwood storage boards. These boards should be drilled to fit the gelatin capsules.

Pellet molds. These molds should have a 0.250 in. bore.

Laboratory hydraulic press. The press should have a 10,000 psi capacity.

Brass electrode holders. These should be as described by Wark (134).

Spectrographic equipment. This should include an exciting source providing a dc arc with an open circuit voltage of 300 V.

TABLE 2.7
Line Pairs for Rhodium Analysis

Element	Analytical line (Å)	Filter (% transmission)	Internal standard line (Å)	Filter (% transmission)
		Spark spectra		
Cobalt	Co 3453.50	100	Rh 3485.0[a]	100
Copper	Cu 3273.96	25	Rh 3381.44	64
Iridium	Ir 2924.79	80	Rh 2892.22	80
Iron	Fe 2599.57	64	Rh 2688.10	64
Lead	Pb 2833.07	100	Rh 2903.31	100
Nickel	Ni 3414.76	64	Rh 3381.44	64
Palladium	Pd 3404.58	64	Rh 3381.44	64
Silver	Ag 3382.89	25	Rh 3381.44	100
		Arc spectra		
Gold	Au 2748.26	100	Rh 2747.63	100
Platinum	Pt 2830.30	100	Rh 2892.22	100
Ruthenium	Ru 3728.03	50	Rh 3721.2[a]	50
Silicon	Si 2528.52	50	Rh 2629.91	50

[a]Unlisted line.

TABLE 2.8
Data on the Precision of Rhodium Analyses

Element	Average concentration (%)	Coefficient of variation (%)	No. of determinations
Cobalt	0.004	7.1	5
Copper	0.007	5.7	10
Gold	0.006	2.9	5
Iridium	0.94	2.6	10
Iron	0.025	6.2	9
Lead	0.009	11.4	5
Nickel	0.025	4.0	10
Palladium	0.031	3.2	10
Platinum	0.006	7.0	5
Ruthenium	0.014	13.8	5
Silicon	0.004	7.1	9
Silver	0.010	6.3	5

The excitation source should be capable of providing dc arc, high voltage spark, or multi-source discharges.

The Preparation of Standards and Samples

Dissolve rhodium and iridium by the pressure technique described in Procedure 115, see Figs. 2.5–2.8. Treat 1.000 g of rhodium or iridium sponge in the reaction tube with 0.9 ml of nitric acid and 15.4 ml of hydrochloric acid at 300°C for 36 hr. Dissolve ruthenium sponge in the same manner, but with 1.5 ml of 60% perchloric acid in place of the nitric acid. Obtain the other impurity elements as high purity metals, which are easily dissolved in open vessels, except for silica, which is added as a colloidal suspension. Prepare the standards as solutions by adding stock solutions of the impurity elements to aliquots of the rhodium solution. Then reduce these solutions to sponge or black as required. Determine iridium, palladium, copper, iron, nickel, and silica in sponge, and cobalt, silver, gold, lead, ruthenium and platinum in black as given below.

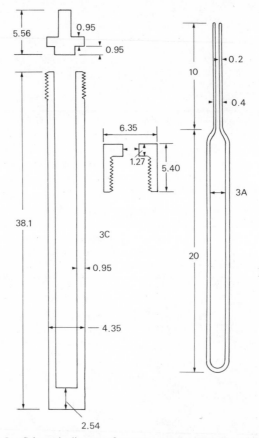

FIGURE 2.5.   Schematic diagram of pressure vessel: glass tube and metal sheath.

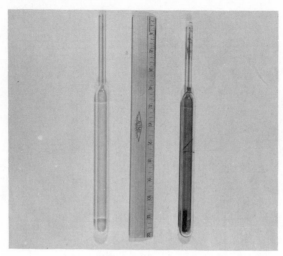

FIGURE 2.6.   Pressure vessels for samples of 0.2 to greater than 1 g. All dimensions are in centimeters.

Preparation of Standards for the Sponge Technique

Rinse the pipets with dilute hydrochloric acid and then with tap water. Soak in a hot detergent solution for about 10 min. Rinse thoroughly with tap water and then again with distilled water. Allow the pipets to dry in a drying oven at 105°C. When ready to use, rinse the pipet with the solution to be measured, discard, and refill before transferring the correct volume into the beaker. Check the amounts on the note sheet as they are added. When all the impurity elements solutions have been measured out, add the platinum solution, using the 20-ml pipet, in the same manner.

Wash the wall of the beaker with dilute hydrochloric acid, and evaporate the solution on a hot plate under an infrared lamp until the volume is 6 ml or the solution level is less than 0.25 in. from the bottom. Transfer the solutions to clean silica crucibles (reserved for platinum standards) using a glass rod. Wash the beaker several times with dilute hydrochloric acid. Evaporate the solution completely in the crucibles on a hot plate under infrared lamps. The salt will be reddish. Place the crucibles on a hot plate at a high heat, and bake until the salts become a dark chocolate color. Place the crucibles in a desiccator until ready for reduction.

Pulverize the salts with a clean aluminum spatula by scraping down the sides of the crucible and breaking any large pieces; keep a separate spatula for use only with platinum. Reduce the salts to metal by allowing a hydrogen flame (burning at the end of a Rose inlet tube) to play on the sides and bottom of the crucible, stirring with the spatula, until the salts become dark gray. White flashes in the flame will be noted while the reduction is proceeding; these will cease when reduction is complete. The platinum sponge should be dark gray. Remove the hydrogen flame, and rub the sponge on the sides of the crucible with the spatula to remove the material deposited on the sides. Place the crucible in a crucible furnace, and cover it with a Rose cover with the Rose inlet tube in place and

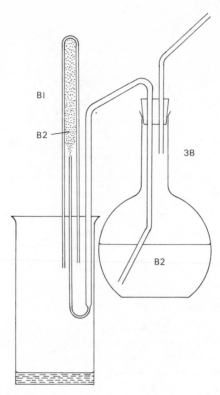

FIGURE 2.7.   Method of rinsing solutions from frozen tubes after reaction. B1 is frozen solution containing liquid chlorine. B2 is warm wash solution. 3B is air inlet.

connect hydrogen. Use a hydrogen regulator pressure of 10 psi and a flow rate of about 0.2 l/min. Heat the sponge to red heat for 20 min. Remove the crucible from the furnace, with the flame in contact with the sponge until the redness disappears. Place the crucible in a desiccator in a labeled position until cool. Check the weight of the sponge, and add briquetting graphite in the ratio of 2:1 by weight of sponge to graphite. Mix thoroughly in a mortar and make into pellets.

(a) Sponge technique.   Prepare rhodium sponge as described in Procedure 15 for platinum at a temperature of 800–900°C for the reduction. Evaporate the solutions to dryness, crush the salts to powder, and then reduce the powdered salts to sponge by heating in hydrogen. Add colloidal silica (Ludox) to the sponge standards in amounts determined by assaying the suspension. Mix the sponge thoroughly and add graphite powder in the ratio 2 parts by weight of sponge to 1 part of graphite. Grind the mixture in an Al_2O_3 mortar until it is homogeneous.

(b) Black technique.   Neutralize the chlororhodic(III) acid solution, then adjust the pH to 13.5 or 14.0 with potassium hydroxide, the concentration of the solution being con-

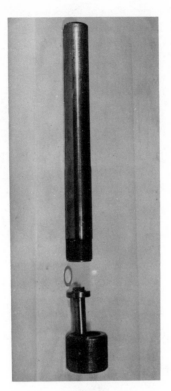

FIGURE 2.8.

trolled at 2.5 mg of rhodium per milliliter. Add solid hydrazine dihydrochloride until a reaction begins. Heat the solution gently during the process. Separate the finely divided rhodium black by centrifuging or decanting. The completeness of reduction may be estimated by the absence of the colored chlororhodate ion. (This has been checked gravimetrically after reduction in hydrogen.) Dry the rhodium black and add it to an equal weight of briquetting graphite powder in a small mullite mortar. Add methanol as a lubricant, and treat the mixture in the same manner as sponge except to use a shorter mixing time.

The Preparation of Samples for Analysis

Quarter samples to be analyzed to obtain representative portions, and dissolve these by the pressure technique in Procedure 115. Prepare the aliquot of each sample solution as sponge and one as black. Make mixtures with graphite powder; add the black to its own weight of graphite powder, and grind in an alundum mortar for 1 hr, using C.P. methanol as a lubricant. Regulate the alcohol addition so that the mixture finally has a pasty consistency. Evaporate to dryness under infrared lamps, and mix manually for a few minutes. Measure out approximate 40-mg portions for pelleting; and prepare pellets by pressing 40-mg portions of the sponge–graphite mixture and black–graphite mixture onto the ends of 1-cm lengths of 0.242-in.-diam graphite rods in pelleting molds.

PROCEDURE 17 (133)

(a) *Electrode system.*   The electrode system is the same as that described for platinum (133). Adjust the analytical gap to 2 mm for the high voltage spark and 3 mm for the dc arc. Make the sample the cathode for the high voltage spark and the anode for the dc arc.

(b) *Electrical parameters.*
(1) *High voltage spark.*

Capacitance	0.007 μF
Inductance	360 μH
Resistance in series with gap	none
Peak voltage output	20,000 V
Radio frequency current	5 A
Discharge trains	4 per cycle

(2) *Direct current arc.*

Open circuit voltage	300 V
Current	5 A

(c) *Exposure conditions (for both spark and dc arc).*

Spectral region	2300–4050 Å
Slit width	0.050 mm
Slit length	2 mm
Pre-burn period	none
Exposure period	20 sec
Filters	10% transmission

(d) *Exposure index.*   Maintain the transmittance of Rh 3381.44 Å (spark) and of Rh 2892.22 Å (arc) at approximately 30%.

(e) *Replicate exposures.*   Make triplicate exposures with the spark and the arc for all samples, along with two or more control standards in duplicate.

The Analysis of Refined Iridium

This method provides for determination of fourteen elements in the ranges indicated.

Element	Concentration range (%)	Element	Concentration range (%)	Element	Concentration range (%)
Iron	0.015–0.03	Lead	0.005–0.015	Palladium	0.005–0.010
Silica	0.010–0.025	Silver	0.005–0.015	Platinum	0.005–0.010
Rhodium	0.005–0.020	Bismuth	0.005–0.010	Ruthenium	0.005–0.010
Copper	0.005–0.015	Cobalt	0.005–0.010	Zinc	0.005–0.010
Gold	0.005–0.015	Nickel	0.005–0.010		

SUMMARY OF THE METHOD

Standards are prepared by adding measured aliquots of stock solutions of impurity elements to iridium sponge, evaporating, and reducing by heating in hydrogen at carefully controlled temperatures. Samples of iridium are dissolved and reduced to sponge under the same conditions. The sponge is ground, mixed with graphite powder, and pressed into pellets for spectrochemical analysis in the manner described for the other metals. Line pairs and precision data for analysis of iridium are given in Tables 2.9 and 2.10.

PROCEDURE 18 (133)

The equipment is as described for Procedure 16.

The Preparation of Standard Samples

Dissolve the iridium by the pressure technique (Procedure 115). Prepare iridium sponge from solution in the manner described for platinum (Procedure 16) except that a temperature of 1000°C must be attained for complete reduction to metal. (The method of preparing standards and samples as solutions, evaporating to dryness, and reducing to sponge is not satisfactory for some of the impurity elements determined in iridium; the chlorides of such

TABLE 2.9
Line Pairs for Iridium Analysis

Element	Analytical line (Å)	Internal standard line (Å)
	Spark spectra	
Bismuth	Bi 3067.71	Ir 3074.76
Cobalt	Co 3453.50	Ir 3421.76
Copper	Cu 3273.96[a]	Ir 3200.03
Gold	Au 2675.95	Ir 2682.45
Iron	Fe 3719.93	Ir 3688.16
Lead	Pb 3683.47	Ir 3676.65
Nickel	Ni 3414.76	Ir 3421.76
Palladium	Pd 3421.24	Ir 3421.76
Rhodium	Rh 3396.85	Ir 3421.76
Ruthenium	Ru 2678.75	Ir 2736.41
Silver	Ag 3382.89[a]	Ir 3371.44
Zinc	Zn 3345.02	Ir 3326.10
	Arc spectra	
Platinum	Pt 2650.85	Ir 2593.13
Silica	Si 5032.24[b]	Ir 5030.72[b]

[a]Filter, 25% transmission.

[b]Second order of Si 2516.12 and Ir 2515.36.

TABLE 2.10
Data on the Precision of Iridium Analyses

Element	Concentration (%)	Coefficient of variation (%)	No. of determinations
Bismuth	0.005	10.7	11
Cobalt	0.005	3.6	7
Copper	0.012	4.9	6
Gold	0.012	4.7	11
Iron	0.027	4.2	6
Lead	0.009	4.6	7
Nickel	0.014	5.1	7
Palladium	0.0075	4.5	7
Platinum	0.006	2.3	7
Rhodium	0.018	3.7	7
Ruthenium	0.0075	3.4	5
Silica	0.013	2.2	10
Silver	0.007	9.1	7
Zinc	0.005	3.6	7

elements as bismuth, lead, and zinc may be volatilized and partially lost at 1000°C.) Add stock solutions of the impurity elements by pipet to weighed portions of sponge in Al_2O_3 mortars. After drying at about 100°C under infrared lamps, grind the standards thoroughly, and transfer to crucibles for careful reduction first at a low temperature and then gradually heat to 600°C under hydrogen.

The Preparation of Samples for Analysis

Grind the samples, as sponge, for analysis to obtain a representative portion with a particle size equivalent to that of the standards. Then mix with half its weight of graphite powder and make into pellets.

Spectrochemical Procedure

(a) Electrode system. The electrode system is the same as described in Procedure 33. Use a gap of 2 mm for the spark discharges and 3 mm for the dc arc. Make the sample the cathode for the high voltage spark and the anode for the dc arc.

(b) Electrical parameters.
   (1) High voltage spark.

Capacitance	0.007 μF
Inductance	360 μH
Resistance in series with gap	none
Peak voltage output	20,000 V
Radio frequency current	5 A
Discharge trains	4 per cycle

(2) Direct current arc.
Open circuit voltage	300 V
Current	5 A

(c) Exposure conditions.
Spectral region	2400–4150 Å
for silica	4600–6300 Å
Slit width	0.050 mm
Slit length	2 mm
Preburn period	none
Exposure period	
Spark	30 sec
Direct current arc	20 sec
Filters	25% transmission

(d) Exposure index. The Ir 2593.13 and Ir 3421.76 lines are maintained at approximately 30% transmittance.

(e) Replicate exposures. Make triplicate exposures with the spark and the arc for all samples, along with two or more control standards in duplicate.

The Analysis of Refined Ruthenium

The spectrochemical analysis of ruthenium is hampered by the complexity of the spectra. One method worthy of further examination involves the distillation of ruthenium from sulfuric or perchloric acids, leaving the nonvolatile impurities in solution. These can then be determined by one of the standard dissolution methods such as the porous cup technique or impregnation of carbon electrodes, but substituting another element in solution as an internal standard. This method indicated promise when tested in the present authors' laboratory.

The unpublished method described below was used by the senior author as early as 1953. It provides for the determination of eight elements in the concentration ranges shown in the accompanying tabulation.

Element	Concentration range (%)	Element	Concentration range (%)
Platinum	0.02–0.25	Nickel	0.02–0.10
Iridium	0.05–0.20	Iron	0.02–0.10
Silica	0.03–0.15	Palladium	0.005–0.05
Rhodium	0.02–0.15	Copper	0.005–0.03

SUMMARY OF THE METHOD

Standards are prepared by adding aliquots of impurity elements to aliquots of ruthenium solution, evaporating to dryness, and reducing by heating the salts in a stream of hydrogen at controlled temperatures. Standards for silica are prepared by adding weighed amounts of a silicic acid–graphite powder mixture to portions of ruthenium sponge. Samples are prepared for analysis by dissolving and reducing to sponge. The untreated ground sponge is taken for a silica determination. In all instances, the ruthenium sponge is mixed with graphite powder and pressed into pellets for spectrochemical analysis. Line pairs and precision data for analysis of iridium are given in Tables 2.11 and 2.12.

PROCEDURE 19 (133)

Equipment

Glass reaction tubes and steel protective shells.   These are described for the pressure dissolution technique in Procedure 16.

Dewar flask.

Oxygen–gas burner.   The burner should provide a fine, pointed flame.

Volumetric chemical glassware.

Silica crucibles.

Electric hotplate.

TABLE 2.11
Line Pairs for Ruthenium Analysis

Element	Analytical line (Å)	Internal standard line (Å)
	Spark spectra (20-sec exposure)	
Copper	Cu 3273.96[a]	Ru 3364.82
Nickel	Ni 3012.00	Ru 3056.05
Palladium	Pd 3421.24	Ru 3364.82
Rhodium	Rh 3323.09	Ru 3364.82
	Spark spectra (45-sec exposure)	
Iron	Fe 3475.45	Ru 3403.77
Platinum	Pt 2830.29	Ru 2934.18
	Arc spectra	
Iridium	Ir 3220.78	Ru 3143.65
Silica[b]	Si 5048.23	Ru 5053.65

[a]Intensity reduced by a 10% transmittance filter.
[b]Second order of Si 2524.11/Ru 2526.82.

TABLE 2.12

Data on Precision of Ruthenium Analysis

Element	Concentration (%)	Coefficient of variation (%)	No. of determinations
Copper	0.031	4.7	5
Iridium	0.042	12.5	5
Iron	0.035	9.6	5
Nickel	0.044	8.2	5
Palladium	0.006	4.5	5
Platinum	0.059	10.0	5
Rhodium	0.068	5.4	5
Silica	0.06	8.8	11

Desiccator.

Infrared lamps.

Other equipment. This is as described in Procedure 16.

The Preparation of Standard Samples

Conditions for dissolving ruthenium by the pressure technique are given in Procedure 115. Prepare ruthenium sponge from solution in the manner described for platinum except to place the ruthenium salts, in a porcelain boat, in a quartz tube, and heat externally while a stream of hydrogen is passed through the tube. Maintain a temperature of 680°C during the reduction. Prepare standards by adding aliquots of impurity solutions, and then reducing to sponge as described. Prepare standards for the determination of silica by making controlled additions of a standard silicic acid–graphite powder mixture to portions of ruthenium sponge. Take one portion of sponge as the *x*, or residual silica, standard, while the others have additions of one half, the same, and twice the estimated residual concentration. Alternatively, make additions of colloidal silica. Mix the ruthenium sponge with half its weight of graphite powder by grinding in a mortar.

The Preparation of Samples for Analysis

Dissolve the samples for ruthenium sponge for analysis, then evaporate to dryness, and reduce to sponge under controlled conditions as described above. Carefully grind a portion of the untreated sponge in an alundum mortar for the determination of silica.

Spectrochemical Procedure

(a) Electrode system. Place the pellets in the brass holders for excitation. Use as the counter electrode a high purity rod, 0.242 in. in diam sharpened to a cone of 30° included angle with a hemispherical tip of 0.062 in radius (ASTM designation C-S). Maintain a gap of 2 mm for spark discharges and 3 mm for the dc arc discharge. Make the sample the cathode when using the high voltage spark and the anode for the dc arc method.

(b) Electrical parameters.
  (1) High voltage spark.
Capacitance	0.007 μF
Inductance	360 μH
Resistance in series with gap	none
Peak voltage output	20,000 V
Radio frequency current	5 A
for Ir, Ni	8 A
Discharge trains	4 per cycle

  (2) Direct current arc.
Open circuit voltage	300 V
Current	5 A

(c) Exposure conditions.
Spectral region	2200–3950 Å
for silica	4600–6300 Å
Slit width	0.050 mm
Slit length	2 mm
Preburn period	none
Exposure period	20 sec
for Pt, Fe	45 sec
Filters	10% transmission

(d) Exposure index.   The Ru 3364.82, Ru 3056.05, Ru 2934.18, and Ru 5054.65 lines are maintained at approximately 30% transmittance.

(e) Replicate exposures.   Make triplicate exposure with the arc and spark exposures for all samples, along with two or more control standards in duplicate.

THE DETERMINATION OF HIGHER CONCENTRATIONS
OF PLATINUM METALS

Higher concentrations of platinum metals, i.e., over about one percent, are most often encountered in materials such as alloys, refinery concentrates, residues, or scrap. Most of these materials may be expected to differ radically in composition from sample to sample and hence are most often analyzed by wet chemical methods. Such methods, however, may be lengthy and tedious, and should, therefore, be complemented by spectrochemical analysis whenever possible.

Alloys with constituents varying only slightly from sample to sample are usually amenable to spectrochemical analysis, especially by direct reading optical or X-ray methods. It is also possible to obtain qualitative, semiquantitative or even quantitative information from some of the nonuniform materials as a substitute or an aid to wet chemical analysis.

Koehler (118) published a technique for the spectrochemical analysis of platinum–rhodium alloys in 1950. Ayres and Berg (134) used the porous cup technique to determine palladium, platinum, iridium, and rhodium in solutions. Bardocz and Varsanyi (135) determined 1–25 atomic percent of rhodium in platinum–rhodium alloys in 1954. Five years later, Kuranov (136) published a method which he had used since 1954 for the spectrochemical analysis of alloys of platinum, palladium, and rhodium, containing from 2 to 5% of platinum and palladium. Spark spectra were used in this analysis. Kuranov *et al.* (137) used a spectrochemical method for the routine analysis of high purity gold and silver and their alloys. Kuranov (138) also published data on the preparation of standards for the spectrochemical analysis of alloys.

In 1956, Gunn (139) determined platinum in the range 0.05–1.0% in reforming catalyst by X-ray fluorescence. MacNevin and Hakkila (140) described a method for the estimation of palladium, platinum, rhodium, and iridium in solutions in 1957. In 1959, Lincoln and Davis (141) determined platinum quantitatively in alumina-base reforming catalyst, and Rabillon and Griffoul (142) outlined briefly the determination of rhodium in platinum–rhodium alloys by X-ray spectrography. Neeb (143) in 1961 dropped osmium solutions onto paper containing a reagent and analyzed the resulting spot by X ray.

QUALITATIVE SPECTROCHEMICAL ANALYSIS

Qualitative spectrochemical analysis is desirable for every unknown sample coming into the laboratory. It is often the basis for deciding what elements should be determined quantitatively, and it also yields information on possible interfering elements.

Optical emission spectrography, with a dc arc, may provide information as to the presence or absence of about seventy elements within an hour, depending on the complexity of the sample. X-ray spectroscopy with a scanning spectrometer requires about the same time, and will detect all elements above titanium in the periodic table, using an air atmosphere, for all elements above sodium when the spectrometer is evacuated. Each method allows the detection of some elements not ordinarily, or easily, detected by the other. The optical method provides better sensitivity for trace concentrations whereas the X-ray method permits better estimations of major constituent percentages. An experienced spectrographer, maintaining systematic control of sample and analytical conditions, can often provide surprisingly accurate estimates of element concentrations by both methods.

Both methods are subject to line or wavelength interferences, and great care must be exercised to avoid reporting elements as present on the basis of an interfering spectrum line. Several major wavelengths should be checked for every element.

Qualitative analysis may provide all the information required in some instances. For example, precipitates, solutions, filtrates, or residues from wet chemical analysis may be checked for the presence or absence of specific elements or for the presence of possible interfering elements.

CHAPTER 3

SPECTROPHOTOMETRY

INTRODUCTION

The growing interest in analytical trace techniques for the noble metals is a constituent of an increased general appreciation of the functions of trace elements. These functions are associated with such properties as physiological reactions and with physical properties of alloys such as hardness, resistance to corrosion, and electrical conductivity. Although some of the noble metals are involved in studies of these functions, perhaps the greater interest in this field is associated with their occurrence in natural deposits. These contain the noble metals in very small proportions, very often of the order of $10^{-3}\%$, and of this total only a small fraction may consist of the more insoluble metals, viz., rhodium, iridium, and ruthenium. The proportions of osmium and perhaps of ruthenium in most of these deposits remain unknown notwithstanding the advances in trace analytical methods. Indeed, it is a fact that, despite more than a century of analytical effort, no procedure has been recorded that can be used with confidence for the direct determination of each of the noble metals in primary deposits.

These small amounts of platinum metals require some type of concentration prior to any determination. For this purpose the classical extraction with lead as the collector remains the most generally effective procedure, and the resulting concentrate, either in the form of the lead button or the silver or gold bead, is readily treated to allow for the application of spectrophotometric methods of determination.

During the past decade two new approaches to the problem of wet separation of the platinum metals from each other have been proposed. Each of these, useful in its own right, will facilitate the examination of the efficiency of variations in the older methods and indeed they may ultimately replace the classical procedures. One process involves the use of selective extraction and the other, specific "adsorption." In many instances these separations are especially suited to subsequent spectrophotometric determinations, the number of which has been increasing continuously.

It may be helpful here to make reference to a difficulty that, to some degree, characterizes many of the existing spectrophotometric methods. Because so little is known about the identities of the dissolved platinum metals constituents, most of these methods are empirical. The study of ion exchange and chromatographic techniques has emphasized our very great ignorance of the identities of dissolved metal constituents, not only of the noble metals, but also of the associated base metals. Particularly little is known about how their stability depends upon acidity, salt content, aging, temperature, etc. It is because of this deficiency that a proper interpretation of the kind and degree of interference in platinum determinations is so difficult. Many of the reports that deal with the problem of interference involve determinations made by merely adding the interfering elements to a previously prepared platinum metal solution with no attention to the effect of the prior methods of dissolution, etc. In some instances this procedure has invalidated the data.

A specific problem peculiar to the metals of Periodic Group VIII arises from their ease of hydrolysis to form insoluble hydrated oxides. Those spectrophotometric methods that involve heating at pH 4–8 are particularly susceptible to this source of error. This tendency, together with the resulting sensitivity to the presence of electrolytes, accounts for much of the trouble encountered by chemists who seek to apply the various spectrophotometric procedures for platinum metals. An adequate recognition of these variables requires some familiarity with the processes for dissolving the platinum metals. Of these metals, only palladium and rhodium may be quantitatively attacked by single mineral acids. Rhodium may sometimes be dissolved in concentrated sulfuric acid, an acid that introduces an interference into the determination of platinum metals, the character and intensity of which is seldom fully realized.

The more general method of dissolving platinum and palladium, ineffective for the other platinum metals, is treatment with aqua regia. Here the presence of nitric acid will interfere with most of the classical as well as with many of the instrumental methods of analysis. The use of hydrochloric acid to remove the oxides of nitrogen encourages the formation of various chloro complexes, in which form osmium and ruthenium may resist distillation. The difficulties associated with all of these methods of dissolution are often very much magnified by the presence of the associated base metals.

In so far as dry fusions are concerned, there is the additional and almost unpredictable source of trouble arising out of the amphoteric character of the oxides of the platinum metals, e.g., in the sodium peroxide fusion of iridium in a silver crucible.

Consideration of all these factors should convince analytical researchers, particularly in the field of platinum metals, that general statements regarding the degree and kind of interferences should arise only from investigations far more intensive than merely measuring in simple solutions the proportions of contaminant producing a percentage departure from the known value.

Concerning the references to interfering cations, the authors have been guided by the fact that, with the exception of iridosmines, the most important natural deposits contain platinum and palladium as major constituents with rhodium, iridium, ruthenium, and osmium as minor constituents. Of the nonplatinum metals, the associated cations are generally the base metals of Periodic Group VIII, gold, copper, and chromium. For artificial products, the major constituents are platinum, palladium, and less frequently rhodium. In addition there are the hardening metals, iridium and ruthenium, which are sometimes added in appreciable proportions. Other interferences arise from the various anions peculiar to the preliminary methods of separation. It should be noted that the interference of one platinum metal may vitiate the advantages gained by the noninterference of another. For example, if palladium and osmium interfere with the determination of ruthenium, there is little advantage in freedom from interference by rhodium and platinum.

GOLD

5-(p-Diethylaminobenzylidene)rhodanine Method*

This reagent is one of the more acceptable reagents for the spectrophotometric determination of gold. Its sensitivity is 0.01 μg cm^{-2}. Merejkovsky (144) used

*Extraction spectrophotometric determination of gold(III) with 5-(p-dimethylaminobenzylidene)-rhodanine (143a).

the reagent for the microdetermination of gold in organic tissues. Poluektov (145) added the reagent, in a solution containing ethanol, chloroform, and benzene, to a solution of gold containing a few drops of nitric acid. The pink-violet organic layer permitted the visual detection of 0.1–0.2 μg in 5 ml of solution. To isolate gold, the author coprecipitated it with mercury(II) and tin(II) chloride. Sandell (146) used this rhodanine method in a colorimetric procedure for gold following its isolation with tellurium precipitated by tin(II) chloride. In contrast to such collectors as mercury salts, tellurium did not interfere in subsequent operations, and as little as 0.2 mg in 50–100 ml of solution served to isolate 0.2–0.3 μg of gold in the presence of 0.5 g of iron, copper, and lead.

The reductant tin(II) chloride was found to be superior to sulfur dioxide when iron and other reductants were present. In weakly acidic solutions the rhodanine reagent produced an insoluble red-violet product which, by analogy with known salts, contained the rhodanine complex of gold (I). Although strongly oxidizing substances reacted with the rhodanine reagent to produce a violet-red product, the latter, in contrast to the gold complex, was soluble in carbon tetrachloride.

The colloidal red-violet gold(I) complex is formed in 0.12 M and 0.075 M hydrochloric acid; at the former acidity, there is greater precision, but at the lower acidity as little as 0.5 μg of gold can be determined in a final volume of 4 or 5 ml, using a 1-cm cell. The full color intensity of the suspensions is attained in 1 or 2 min at the lower acidity and more rapidly at the higher acidity. With the latter the average error is 3% in the range 1–8 μg of gold. The interference of small amounts of palladium can be avoided by the addition of dimethylglyoxime. Platinum(IV) does not interfere, but platinum(II) in amounts of the order of one or more parts per million may interfere, particularly on standing.

For amounts of gold in excess of about 0.4 μg, visual comparisons can be made; for lesser amounts a photometric measurement with a green filter (500 nm) is recommended. The times of color development must be measured exactly, but the preparation of standards need not include a tellurium collection. Beer's law is obeyed up to at least 2 ppm. The method is designed for small amounts of gold only and is acceptably free of interferences provided the initial isolation with tellurium or other suitable reductants and collectors is used. As would be expected with a colloidal dispersion, the presence of electrolytes affects the color distribution; the method is particularly sensitive to the acidity of the reaction medium. For the low concentrations of gold for which the method is suited, however, the dispersions are stable, and normally a protective colloid is not required.

Natelson and Zuckerman (147) applied the method to the estimation of gold in biological materials. These authors introduced a modification of Sandell's ashing technique and applied the centrifuge to avoid filtration. The claim for greater simplicity and an increase in sensitivity by a factor of six is not substantiated. The authors found Beer's law to apply only to gold concentrations below 1 μg/ml rather than the 2 ppm reported by Sandell (146). Differences of opinion with methods involving colloidal suspensions are not surprising.

Hara (148) used 5-(p-dimethylaminobenzylidene)rhodanine for the determination of 0.03 – 0.3 ppm of gold. The red color reached a maximum intensity after 5 min and was stable for 30 min. A 562-nm filter was used. The interference of iron was removed by sodium metaphosphate. Cotton and Woolf (149) recorded a modified rhodanine method for the determination of gold in thin films. These authors rejected the o-tolidine method because of poor sensitivity, the dithizone method as lacking in precision and the recorded rhodanine methods because of inaccuracy. The proposed modification involves a solvent extraction by isoamyl acetate. To attain the desired accuracy, a minimum excess of reagent is required, and to this end the authors use a dilute solution of isoamyl alcohol. Because the colors fail to develop at acidities greater than 1 M and fade with time, the pH, the periods of extraction, and the time of color development require careful control. Experimental data indicate a gold – rhodanine ratio of 1:1 and the explanation alternative to that invoking the presence of gold(I) in the complex is a reaction with gold(III) to form a complex anion.

Compared to the aqueous method (146), the procedure provides greater precision, and in the present authors' opinion it is one of the most reliable methods for the determination of small amounts of gold.

Two procedures are included here. Procedure 20 is that described by Sandell (150), and includes an acceptable method for isolating gold. Procedure 21 is that described by Cotton and Woolf (149) which has been applied to gold films. One may hope that the latter will receive further attention with a view to the including of a prior isolation.

PROCEDURE 20 (150)

Reagents

5-(p-Diethylaminobenzylidene)rhodanine solution. Dissolve 0.05 g in 100 ml of absolute ethanol. The solid dissolves slowly, and the solution is stable.

Tin(II) chloride solution. Dissolve 20 g of the dihydrate in 100 ml of 2 M hydrochloric acid. Prepare fresh at reasonable intervals. Remove insoluble material by filtration.

Tellurium tetrachloride solution. This solution is to contain 1 mg of tellurium per ml. Treat 100 mg of precipitated tellurium with 1 or 2 ml of nitric acid, and evaporate to dryness. Add 1 ml of hydrochloric acid, and again evaporate to dryness. Dissolve the residue in 10 ml of concentrated hydrochloric acid, and dilute to 100 ml with water.

Sodium fluoride solution. Use 1 g in 100 ml of water.

Standard gold solution. Use a 0.001% solution of chloroauric acid in 0.10 M hydrochloric acid. This solution is conveniently obtained by diluting a 0.010% gold stock solution in 1.0 *M* hydrochloric acid with water. Prepare the stock solution by dissolving pure gold in a few milliliters of aqua regia, evaporating just to dryness on a water bath, adding 1 ml of concentrated hydrochloric acid, evaporating almost to dryness, and making up to volume.

Procedure

The isolation of the gold.   The sample solution, containing 0.1–10 μg of gold, may conveniently have a volume of 50 ml. It should not contain strong oxidants such as nitric acid. Add enough hydrochloric acid to make its concentration 2.5 *M* and 0.2 ml of the tellurium solution. Mix and add 5 ml of the tin(II) chloride solution, or more as required to reduce iron and copper and to produce a brown, colloidal precipitate of tellurium. Then add an excess of 3–5 ml. Heat to boiling, and keep near the boiling point for 30 min, or until the precipitate is well coagulated. Collect the precipitate in a 7-ml porous, porcelain filter crucible. Wash the precipitation beaker and the crucible with five 5-ml portions of 1:4 hydrochloric acid. Wash carefully to remove all iron and other foreign substances.

Add 1 ml of aqua regia (1 vol of nitric acid to 3 vol of hydrochloric acid) to the precipitation beaker, moisten the wall with the aid of a stirring rod, heat almost to boiling, and pour the solution into the crucible. By the use of a stirring rod, bring the acid into contact with the tellurium on the wall of the crucible, and when all or most of the precipitate has been dissolved, draw the solution through the crucible, receiving the liquid directly in a 2-ml Pyrex beaker. Add 1 ml more of aqua regia to the beaker, and repeat the operation as described to complete the dissolution of any remaining tellurium. Then wash the beaker and crucible with two portions of water of a few milliliters each. Evaporate the solution to dryness on a steam bath, avoiding prolonged heating of the dry residue. Allow the beaker to cool, and add 0.01 ml of aqua regia. By means of a stirring rod, moisten the bottom and lower sides of the beaker with the drop of acid, so that all the residue is wetted. Set the beaker aside to permit the acid to evaporate at room temperature. (Conveniently, the beaker is allowed to stand overnight under a large inverted beaker to protect from dust.) Prepare standards containing 0.0, 0.2, and 0.4 μg gold by transferring the proper amount of standard gold solution to the glass-stoppered tubes, adding 0.30 ml of 2.0 *M* hydrochloric acid and water (redistilled) to make the volume almost 4 ml, followed by 0.25 ml of the sodium fluoride solution, and mixing. (In principle, it would be more correct to prepare the standards by evaporating to dryness after the addition of aqua regia and tellurium, and further treating in the same manner as the sample solution, but this is believed to be an unnecessary refinement.)

The determination of the gold.   When the tellurium residue is entirely dry, add exactly 0.30 ml of 2.0 *M* hydrochloric acid to the beaker and bring into contact with all of the solid by the use of a stirring rod. Add 1 ml of water, and stir. If the solution is clear, transfer it to a tube and rinse the beaker carefully with small portions of redistilled water to make the total volume in the tube 3.5 ml. If the solution is turbid (silver chloride), filter through a small, porous, porcelain filter crucible, receiving the filtrate directly in a tube, and wash with small portions of water to give a total volume of 3.5 ml (or 4 ml if the final volume is to be 5 ml).

Add 0.25 ml of the sodium fluoride solution to the sample, and mix both sample and standards by inversion. Now add 0.30 ml of the rhodanine solution to each tube, and mix at once by inverting three times. Bring the volume of each solution to the mark (4.5 or 5 ml) with water, and again mix by inverting three times. In a few minutes the colors will reach their full intensity. If the sample is less strongly colored than the 0.4-μg standard, find the gold content by visual comparison, examining the tubes axially against a white background. If the gold content of the sample falls between two standards, a more exact comparison can be made by mixing the two standards in a small beaker and dividing the mixture between the two tubes, and again comparing. If more than 0.4 μg of gold is present in the sample solution, transfer the latter to a dry absorbance cell, and obtain the transmittance with a green filter (500 nm) exactly 10 min (or other fixed times) after the addition of the rhodanine reagent. To establish the standard curve take 0, 0.5, 1.0, 2.5, 5, 7.5, and 10 μg of gold, add enough 2.0 M hydrochloric acid to make the total volume 0.30 ml (bear in mind that the standard gold solution is 0.10 M in hydrochloric acid), treat with the sodium fluoride solution and rhodanine reagent as described, and dilute to volume. A plot of absorbance versus concentration should give a straight line. The absorption cell should be cleaned with dilute hydrochloric acid after each reading to dissolve any precipitate that may have been deposited on the walls.

The Determination of Gold in Films

PROCEDURE 21 (149)

Isoamyl acetate is the best extractant among chloroform, carbon tetrachloride, isoamyl alcohol, diethyl ether, benzene-chloroform (1:3), and isoamyl acetate. It gives the quickest extraction and has the advantages of a low volatility and a clean separation from the aqueous phase. Most of the color of the organic phase is caused by the yellow reagent (absorbance maximum: 455 nm). The red-orange gold complex can be observed with more dilute solutions of the reagent, and has an absorption peak at 510–515 nm. In order to estimate small amounts of gold accurately, the minimum excess of reagent compatible with the complete formation of the complex is required. The dilute solution of reagent can be made by extracting a portion of the ethanolic solution, diluted with water, into isoamyl acetate, or directly by diluting a stronger solution of the rhodanine derivative in isoamyl acetate. On the basis of a 1:1 gold-rhodanine complex, 1 ml of a 0.004% (w/v) solution can complex 30 μg of gold.

Reagents

Standard gold solutions.   Prepare a stock solution (500 μg/ml) by dissolving gold wire in a minimum of aqua regia, twice evaporating *almost* to dryness with hydrochloric acid, and making up to a standard volume. Prepare the working solution, containing 5 μg/ml, by diluting immediately before use. Use deionized water for all solutions.

Constant-boiling hydrochloric acid.   Use the standard laboratory grade without purification.

Isoamyl acetate.   Use the standard laboratory grade without purification.

5-(p-Diethylaminobenzylidene)rhodanine solution.   Prepare a 0.0044% (w/v) solution in isoamyl acetate, as described above, from a stronger stock solution. The diluted solutions are stable. The reagent may be purified by dissolving it in acetone and partially precipitating with water. Trial extractions of 5 μg of gold from a 0.12 M hydrochloric acid solution may be made using 0.04% (w/v) ethanolic solution of the pure reagent.

Procedure

Dissolve the gold film from its Terylene support in a minimum of aqua regia, and evaporate almost to dryness on a water bath with some additional hydrochloric acid. Remove the last traces of volatile acid by blowing air over the surface. Suitably dilute the residual chloroauric acid to a standard volume, and take an aliquot containing between 4 and 10 μg. Add 0.10 $\pm$ 0.005 ml of constant-boiling hydrochloric acid, and adjust the total volume to 5.00 ml; add 0.30 ml of the rhodanine solution and 10.0 ml of isoamyl acetate. Shake the container (a test tube stoppered by a B14 cone covered with a Teflon cone to prevent solvent leakage is suitable) mechanically for 15 min. Filter the organic layer through cotton wool directly into a 4-cm absorption cell (volume, 10 ml), and compare the absorbance of the gold solution with a blank (5.00 ml of water taken through the procedure) at 515 nm after 5 min.

Tin(II) Chloride Method

Tin(II) chloride has a long history of application for the detection of gold. From time to time, techniques were recorded which allowed the *purple of Cassius* test to be applied semiquantitatively. Bettel (151) made visual comparisons with standards after the isolation of gold from cyanide solutions by precipitation with a zinc–copper couple. Excess of cyanide was added to ensure the dissolution of ferrocyanides and zinc cyanide. To compensate for the impurities usually present in the zinc, the latter was measured accurately. Ten minutes were required for a determination, and for more accurate results the author used a modified procedure which required 17–20 min.

Brodigan (152) also used the tin(II) method for barren cyanide solutions from which the gold was removed by reduction with zinc in the presence of excess of cyanide and lead nitrate. Depending upon the quantity of gold, the color varied from yellow to purple. Seven minutes were required for a determination, and considering that no standards were used the accuracy was surprisingly high. With suitable reductants and more precise techniques, this method for cyanide solutions would provide increased accuracy and applications. Stanbury (153) applied the *purple of Cassius* method to certain excretions from patients subjected to gold salt treatments.

The various factors affecting the accuracy and precision of the tin(II) method were first recorded by Fink and Putnam (154). It was found that the tint of the color was related to the acid concentration, and solutions below 0.05 N in acid produced only yellow to light brown colors irrespective of the gold content. In 0.64 N acid, the purple color appeared. The time of color development was also dependent upon the acid strength, proceeding more rapidly with the weaker acidic solutions; with the latter the color intensity was not a function of the reagent strength. The yellow colloidal form was acceptably stable, whereas the purple form precipitated rapidly in strongly acidic media. As would be expected, the salt content of the solution must be kept to a minimum.

The authors provided a procedure for the colorimetric determination of gold in cyanide solutions. It was stated that "when less than 0.04 mg of gold is to be determined, the low acid stannous chloride test is believed to be superior to the gravimetric assay procedure with respect to accuracy and speed." It is disappointing that no data were provided to substantiate this dubious claim. One cannot doubt, however, that colorimetric methods competitive with the classical assay are within the realm of probability. It is nevertheless a surprising fact that no one has yet recorded data to prove that any colorimetric, volumetric, or spectrographic method for gold or any noble metal offers the speed, accuracy, and precision of the classical assay when applied to ores. Sandell (150) recorded the sensitivity of the tin(II) method as 0.05 μg cm^{-2}. A procedure was provided for solutions containing 10–100 μg of gold per 20 ml or less. An acidity of 0.04 N was preferable although acceptable results could be obtained in 1 N acid. Although a filter was not necessary, slightly lower absorbances could be obtained with a green light. Platinum, palladium, ruthenium, tellurium, selenium, silver, mercury, etc., interfered.

An interesting application of the tin(II)–gold reaction was described by Cole (155), who used textile fibers previously immersed in a solution containing pyrogallol and tin(II) chloride to produce a range of colors when in contact with gold solutions. The method offers potential applications for chromatographic separations as well as for semiquantitative determinations.

PROCEDURE 22 (150)

Reagents

Tin(II) chloride solution. Make up a 10% solution of the dihydrate in 1 M hydrochloric acid. This solution should be prepared fresh weekly.

Standard gold solution. Use a 0.001% solution of gold as chloroauric acid in 0.1 M hydrochloric acid. This solution is best prepared fresh at reasonable intervals by diluting a 0.01 or 0.1% solution.

Procedure

The sample solution may contain 10–100 μg of gold(III) in a volume of 20 ml or less, and should preferably be only slightly acidic (0.05 M in hydrochloric acid). If necessary, however, one may work with a 1 M solution in hydrochloric acid. Dilute with water to 20 ml in a 25-ml volumetric flask, and add all at once 2 ml of the tin(II) chloride solution. Make up to the mark with water, mix, and allow to stand for 20 min. Then determine the transmittance against a reagent blank. It is hardly necessary to use a filter, although slightly lower transmittances will be obtained with green light. The standard solutions for the reference curve must have the same acidity and the same concentration of extraneous salts as the sample solution.

Bromoaurate Method*

Perhaps the most useful separation and spectrophotometric method for gold is the bromoaurate extraction into isopropyl ether. Because of interferences due to impurities in the isopropyl ether the gold must be returned to the aqueous phase prior to the spectrophotometric determination (156).

There is very little interference from coexisting platinum metals except for osmium. The presence of this element is very unlikely because of its volatility as the tetroxide during the usual dissolution steps. Iron is the most serious potential interference because of the high coloration which results from its reaction with bromide. This problem can be eliminated through the addition of phosphoric acid.

The following method is similar to that proposed by McBryde and Yoe (157) with the exception that gold is brought into the aqueous phase prior to the determinative step.

PROCEDURE 23

Reagents

48% Hydrobromic acid.   It must be reagent grade. It must be redistilled before use to remove bromine traces.

Isopropyl ether.   It must be reagent grade. If it contains peroxide, a yellow color may occur in both phases obscuring the phase boundary. In this case the ether should be shaken with ferrous sulfite and distilled over this salt.

Procedure

Add 5 ml of concentrated hydrobromic acid to 10–12 ml of the gold sample solution containing a minimum amount of hydrochloric acid. Extract twice with 15-ml portions of isopropyl ether. Combine the extracts, and shake with 5 ml of 4 M hydrobromic acid. Discard this acid. Evaporate the isopropyl ether to dryness on a steam bath. Add a few milliliters of aqua regia, evaporate, then add 2 ml of a 2% sodium chloride solution and a few drops of concentrated hydrochloric acid, and evaporate. Repeat the addition of hydrochloric acid and the evaporation in order to remove nitrate. Add one drop of hydrochloric acid and 10 ml of water. Transfer with suitable washing to a volumetric flask of an appropriate volume, add 5 ml of hydrobromic acid, and dilute to volume. For samples

---

*Photometric determination of gold as bromoaurate in gold plating electrolytes and in gold–silver alloys (Yaskevich *et al.* (155a)).

known to contain iron the solution should contain 5–10 drops of concentrated phosphoric acid. Measure the transmittance at 380 nm.

Dithizone Method

Although sufficiently accurate quantitative data concerning the optimum range for dithizone (diphenylthiocarbazone) are not available, the sensitivity appears to be about 0.01 μg cm^{-2}. It is one of the few spectrophotometrically applicable, colored complexes of gold.

Fischer and Wey (158) extracted these complexes with carbon tetrachloride. Bleyer et al. (159) and Beaumont (160) used dithizone for the quantitative estimation of gold and many other metal cations. It was used by Shima (161) for the determination of gold in ores and base metals.

The first explanation for the reaction between dithizone and gold was recorded by Erdey and Rady (162). These authors ascribed the yellow-brown color of the carbon tetrachloride extract to the compound Au(HDz)$_3$ (H$_2$Dz, dithizone) which is formed in a 0.5–0.1 N acid solution.

Interference from palladium is eliminated by thiocyanate; silver is removed as the bromide and iron is complexed with phosphoric acid. Platinum(IV) does not interfere. In the presence of copper the proposed method requires two optical filters, one of which is used for the sum of the copper and gold content, and a second for the copper content alone.

PROCEDURE 24 (162)

Reagents

Standard gold solution. Dissolve C.P. gold in aqua regia, and subsequently evaporate in the presence of hydrochloric acid to a few milliliters. Blow air over the solution to remove nitrogen compounds. Standardize the solution with hydroquinone (Procedure 50).

Carbon tetrachloride. Redistill several times.

Ammonia solution. Distill ammonia from a warm sodium hydroxide solution, and collect the gas in water which has been redistilled several times.

Dithizone solution. Dissolve 15–25 mg of dithizone in 100–200 ml of carbon tetrachloride, and then add ammonia solution (1:200). The dithizone passes into the ammoniacal solution, and the oxidized impurities are extracted by the carbon tetrachloride. Acidify the aqueous solution, and extract with carbon tetrachloride. Wash the green solution repeatedly with water, and dilute with carbon tetrachloride to the required volume. Because air oxidizes dithizone in solution, keep this solution in a brown Winkler buret and covered with a layer of 1% sulfuric acid. This ensures stability for 3 weeks. The stability can be increased by a protective carbon dioxide atmosphere.

Procedure

Transfer the gold solution to a 100-ml separatory funnel, and add water to a total volume of 50 ml. Acidify the solution with 5 ml of 1 *N* sulfuric acid, and add 5 ml of dithizone solution and 5 ml of carbon tetrachloride. Shake vigorously for 1 min, and after 2 min place the organic layer into a second separatory funnel which contains a 1:1000 ammonia solution. Add dithizone to the first separatory funnel until the green color of the last portion of the added solution shows no change. Transfer all of the organic phase to the second separatory funnel containing the ammonia solution. Extract the excess of dithizone into the aqueous phase. To ensure the completeness of this extraction, repeat the ammonia extraction in another separatory funnel, and then wash with water. The carbon tetrachloride solution should have the brown-yellow color of the gold dithizonates. Transfer this solution to a 10–20 ml volumetric flask, and rinse with carbon tetrachloride. Measure the absorbance in a 10- or 20-ml cell, using an S47 filter, against a blank of pure carbon tetrachloride. Pour the solution into the cell through a dry filter paper.

Trioctamine Diphenylcarbazide Method

Adam and Pribil (163) extracted gold(III) from sulfuric or hydrochloric acid into a chloroform solution of trioctamine or trioctylmethylammonia chloride and measured the yellow extract at 325–330 nm. Extracts from either acid have the same sharp maximum. The absorbance is constant when the solutions are at least 1.2 *M* in sulfuric acid or 3 *M* in hydrochloric acid. The color is stable for at least 40 min. Only a single extraction is needed for quantitative extraction of gold. Under optimal conditions the Beer–Lambert law is obeyed up to 0.22 mg of gold per 5 ml of extractant for sulfuric acid media, and up to 0.5 mg per 5 ml for hydrochloric acid solutions. If measurement is made against water as reference, the calibration curve is linear but does not pass through the origin of the graph. There is no interference from up to 3% (w/v) of potassium nitrate. There is interference from chromium(VI), molybdenum(VI), uranium(V), platinum(IV), and palladium(II). Large amounts of silver must be removed as silver chloride (otherwise silver chloride is dispersed in the chloroform phase and makes filtration impracticably slow). In the presence of platinum the latter is stripped into 0.1 *M* sodium hydroxide. In this case, the chloroform phase containing the gold becomes cloudy after a while, but this can be avoided by transferring the chloroform to another separatory funnel, and shaking it with 10 ml of 5 *M* sulfuric acid, 1 ml of concentrated hydrochloric acid, and 15 ml of water, then filtering after 5 min and measuring the absorbance of the chloroform phase.

A second more sensitive method involved the reaction of gold with diphenylcarbazide after the separation of the gold from a trioctylamine. The violet color is stable and can be extracted into a chloroform solution of trioctylamine. The absorbance at 560 nm becomes constant (as a function of acidity) if extraction is made from $4-5\ M$ sulfuric acid medium. This has the advantage that a simpler spectrophotometer, suitable for visible light and with glass optics, can be used and that the reaction is at least ten times as sensitive as the trioctylamine reaction. Diphenylcarbazide reacts with some other metals and therefore cannot be used directly.

PROCEDURE 25 (163)

Reagents

Gold(III) chloride solution $(0.05\ M)$.   Standardize with hydroquinone (Procedure 50) and dilute 10- and 100-fold with water to give solutions containing 570 and 57 μg of gold per milliliter.

Trioctylamine or trioctylmethylammonium chloride solutions (5% (w/v) in chloroform). Both compounds are products of General Mills Inc., Kankakee, Illinois, marketed as Alamine S-336 and Aliquot S-336 respectively.

Diphenylcarbazide soludion (1% (w/v)).   Dissolve 0.5 g of reagent grade chemical in 5 ml of concentrated acetic acid (with warming) and dilute to 50 ml with water. Filter after 30 min. Prepare fresh daily.

Procedure

Extract the gold(III) chloride solution with 5 ml of 5% trioctylamine–chloroform solution. Transfer the chloroform layer into a silica dish. Extract the remaining aqueous phase with 5 ml of pure chloroform and add this to the first extract. Evaporate chloroform (on a heated sandbath), then add 0.3–0.5 g of hydroctylamine hydrochloride, and ammonia until free ammonia can be smelled. Evaporate to dryness and heat over a burner to remove organic matter. Cool, then add 0.3–1.0 ml of aqua regia to the residue and evaporate again. Dissolve this residue in water and transfer to a 150-ml separatory funnel. Add 10–20 ml of 5 M sulfuric acid and 1 ml of 0.05% of diphenylcarbazide solution. Warm the contents of the funnel to 40–45° (hot tap water), then after cooling add 5 ml of trioctylamine solution and shake the mixture for 2 min. Filter the chloroform layer and measure its absorbance at 560 nm. After 15 min the absorbance becomes constant and remains so for at least 6 hr. Platinum gives the same color and must be separated as described above.

Procedure for Gold in Ores and Silicates

Weigh 10 g and transfer to a Teflon beaker. Add 50 ml of hydrofluoric acid and 2 ml of concentrated nitric acid. Evaporate and then add a further 2 ml of concentrated sulfuric acid and evaporate to white fumes. To the residue add 5 ml of aqua regia and evaporate to dryness. Add 50 ml of 2.5 M sulfuric acid and warm on a sand bath to dissolve the residue. Transfer the solution to a 150-ml separatory funnel and extract with 5 ml of 5% chloroform solution of trioctylamine. Measure the absorbance in a 10-mm silica cell at 330 nm.

PALLADIUM

Tin(II) Methods

For the range 0.5–2.5 ppm of palladium, tin(II) salts provide useful spectrophotometric procedures. It is a surprising fact that, until recently, little quantitative application was made of this reagent, which is one of the oldest and most generally useful reagents for the detection of palladium, platinum, rhodium, and gold. Knyazheva (164) used this reagent in a method for the determination of palladium and platinum in silver products. The procedure involved the simultaneous extraction by either of the reaction products of tin(II) chloride with palladium and platinum, and the destruction of the palladium color by sodium hypophosphite. Platinum was then determined by visual colorimetry. Solutions for comparison, containing the amount of platinum thus ascertained, were prepared by adding palladium to match the color of the original platinum–palladium solution. The pink color of the palladium complex was stabilized by copper(II) chloride, and although silver chloride was formed it did not interfere. Whereas this method will be of restricted application, it may find some use in the analysis of silver assay beads. Color comparisons cannot be made by artificial light, however, and the method is subject to the difficulties inherent in methods of determination by difference.

Data pertinent to this procedure are provided by Ayres and Meyer (165), who used amyl acetate to extract the colored constituents formed by tin(II) chloride in solutions containing platinum and palladium. These authors found that the amount of palladium extracted by the amyl acetate along with the platinum depended upon the concentration of hydrochloric acid and tin(II) chloride. Interference by the palladium in the extract increased up to concentrations of 5 ppm and then remained constant up to 50 ppm. Consequently, a deliberate adjustment of the palladium content was arranged to fall within this range.

In a later publication, Ayres and Alsop (166) recorded two procedures for the application of tin(II) salts. For the range 0.5–2.5 ppm of palladium, tin (II) phosphate was used in a medium of phosphoric and perchloric acids. For the range 8–32 ppm, using 1-cm cells, tin(II) chloride was used in a hydrochloric–perchloric acid medium. For the latter reagent, the absorbance peak occurs at 635 nm. The green color reaches maximum intensity in about 20 min at room temperature, and remains stable for about 30 min. The applicable range of acidity for color development is 1.5–2.1 M, the optimum perchloric acid concentration is 0.5 M, and the maximum concentration of phosphoric acid is 2 M. Beyond these proportions an unstable color is developed. Tin(II) chloride concentrations in the final solution should be 0.02–0.04 M. The presence of chloride is essential, and should not fall below 0.92 M. Lower concentrations result in low absorbance values. The order of addition of reagents should be as follows: palladium solution, dilute mixed acids, tin(II) chloride. The total volume of solution is somewhat critical. For a final volume of 25 ml, the addition of tin(II) chloride to 17 ml or less of the palladium and acid mixture gives reproducible absorbances.

The rate of addition of reagent and temperature variations are not critical. There are objectionable interferences from all the noble metals, but relatively large amounts of iron and cobalt can be tolerated. Nitrate and sulfate do not interfere. No method of color extraction has yet been recorded. The palladium–tin(II) phosphate system produces a red-violet color in the absence of chloride and in the presence of perchlorate. The narrow absorbance band has its maximum at 487 nm, and the color system obeys Beer's law. Maximum color develops within 10 min at room temperature and is stable for 15 hr. The absorbance for a fixed amount of palladium is dependent upon the concentrations of perchloric and phosphoric acids (see the preceding). The identity of the colored tin(II) complex is not known.

PROCEDURE 26 (166)

Reagents

Chloride-free palladium solution.   Heat the required volume of the standard palladium solution, containing 1 g of palladium per liter, with 10 ml of 70–73% perchloric acid, and add 10 ml of concentrated nitric acid and sufficient 85% phosphoric acid to make the final solution 1.6 M in the latter. After the period of vigorous boiling, cool, and dilute to known volume with distilled water.

Tin(II) chloride solution.   Prepare a solution 2 M in tin(II) chloride and 3.6 M in hydrochloric acid. Dilute as required. Air oxidation can be avoided by storage in a desiccator containing carbon dioxide.

Hydrochloric–perchloric acids.   Make up a solution 2.4 M in hydrochloric acid and 2.3 M in perchloric acid.

Procedure

To the aliquot of the palladium solution in a 25-ml volumetric flask, add 10 ml of the mixed hydrochloric–perchloric acid. Add 2 ml of a 0.5 M tin(II) chloride solution, 1 M in hydrochloric acid. Dilute to volume, and allow to stand for 30 min to develop the green color. Measure the absorbance at 635 nm against a reagent blank.

The most recent tin(II) method was described by Pantani and Piccardi (167), who used tin(II) bromide for the determination of platinum, rhodium, iridium, gold, and palladium. For palladium, the yellow-brown color shows an absorbance maximum at 385 nm and a plateau between 440 and 460 nm. The latter is used to avoid the absorbance of tin(II) bromide complexes below 400 nm. The color obeys Beer's law over the range 1–10 ppm of palladium and is particularly sensitive to variations in acidity and tin(II) concentration. The optimum acidity is 3 M and tin(II) bromide concentration should be greater than 0.1 M. The color can be extracted by isoamyl alcohol producing a spectral curve on which the peak at 385 nm is absent. The stability of the extracted color is improved when perchloric acid is present in the aqueous phase. The separation of palladium from platinum and rhodium involves the use of standard methods. Palladium can be determined in the presence of iridium. Ayres's phosphate method (Procedure 27 (166)) is preferred to this bromide method.

PROCEDURE 27 (166)   TIN(II) PHOSPHATE METHOD

Reagents

Chloride-free palladium solution.   Prepare as described below. Heat the required volume of the standard palladium solution containing 1 g of palladium per liter with 10 ml of 70–73% perchloric acid and add 10 ml of concentrated nitric acid and sufficient 85% phosphoric acid to make the final solution 1.6 M in the latter. After the period of vigorous boiling, cool, and dilute to known volume with distilled water.

Tin(II) phosphate solution.   Dissolve 8 g of pure tin metal by boiling with 400 ml of 85% phosphoric acid. Cool, and dilute to 1 liter. The supernatant liquid over the white solid should be 0.03 M in tin(II) salt. Protect from oxidation by storage in a carbon dioxide atmosphere.

Procedure

Transfer a suitable aliquot of the chloride-free palladium solution to a 25-ml volumetric flask, and dilute perchloric and phosphoric acids, and then 2.0 ml of the tin(II) phosphate solution which has been adjusted to 0.0065 M in tin(II) salt and 1.6 M in phosphoric acid, to make the final volume 1.16 M in perchloric acid and 0.77 M in phosphoric acid. Allow to stand for 10 min to develop the red color, and measure absorbance at 487 nm against a reagent blank.

Nitrosoamines

Overholser and Yoe (168) stated that organic compounds containing the p-nitrosophenylamino group produced colored reaction products suitable for the spectrophotometric determination of palladium(II) salts. From among the potential reagents the authors recommend p-nitrosodiphenylamine and p-nitrosodimethylaniline or diethylaniline. Although the diphenyl derivative is the most sensitive of the reagents, the dimethyl compound has some advantages, among which is the solubility of the resulting complex in the color developing medium and the more favorable application to spectrophotometric techniques.

The diphenyl derivative has the disadvantage of producing a colloidal, colored constituent with its accompanying sensitivity to salts; thus although the p-nitrosodiphenylaniline complex has a maximum color intensity at pH 3.0, it cannot be used at this acidity because of the formation of turbid solutions. The very high sensitivity available with this complex, however, accounts for its greater acceptance as a colorimetric species in spite of a greater instability of color, a slower reaction rate, and a larger temperature effect. In so far as interference from associated metals is concerned, there is little to choose between the two derivatives.

The effectiveness of both methods is reduced by the deleterious effect of neutral salts and the interference of oxidizing agents and of gold. The latter may be isolated by solvent extraction with ethyl acetate or diethyl ether. Silver and large proportions of platinum also interfere and must be removed. The effects of platinum and rhodium on determinations made with this reagent have been discussed by Ryan (169), who used the reagent to detect palladium, platinum, and rhodium in the presence of large amounts of one another and of other metals. The reaction with palladium is specific in the sense that there exists a large difference in the rate of reaction of the salts of the three metals. Neutral rhodium solutions react with p-nitrosodiphenylamine when heated to give an orange to orange-red color. Platinum(IV) and other platinum metals in large amounts interfere because of their colored salts. Platinum(II) interferes through a sensitive color reaction similar to that for palladium.

Large proportions of the associated base metals interfere. Oxidizing agents interfere by giving colored solutions; cyanide and iodide prevent the formation of the colored compound. No other species are known to interfere with the palladium-p-nitrosodiphenylaniline reaction.

The colored constituent is a complex of composition $[Pd(C_6H_5NHC_6H_4NO)_2Cl_2]$ yielding a red solution in dilute hydrochloric acid. The reagent can only be used with a sodium acetate–hydrochloric acid buffer solution, and little variation from the optimum pH of 2.1 is permissible. Beer's law applies over the range 0.05 – 0.3 ppm, and these concentrations are considered to be the optimum range.

Measurements may be made by a photoelectric colorimeter, but the Duboscq cannot be used. Yoe and Overholser (170) also provided a useful procedure for the determination of palladium in the presence of silver by color development with p-nitrosodiphenylamine in a nitrate-containing medium, and this may find useful applications for the analysis of silver assay beads. The authors' procedures (170) are included below. The diphenyl reagent has been used for the determination of palladium after extraction by butanol. Przheval'skit *et al.* (171) developed the color from the reaction with 0.5–5 μg of palladium in 10 ml of an aqueous ethanol solution at 55–60°C at pH 1.8. The colored constituent was isolated by three extractions with butanol, and the absorbance was measured at 510–530 nm. The ethanolic medium provided improve stability.

p-NITROSODIPHENYLAMINE METHOD

The applicable concentration range for this reagent is 0.05–0.3 ppm, and little variation from the optimum pH of 2.1 is permissible. Like the dimethyl derivative, the color intensity is very sensitive to the presence of salts. A solution as little as $10^{-4}\,M$ in sodium chloride may decrease the color intensity, and in a 1 M solution the color may fail to develop. In the recommended procedure the maximum salt concentration is 0.03 M. The acidity of the solution affects the intensity of the color of the reagent and of the palladium complex. The reagent color is particularly sensitive at the recommended pH for color development. For as little as 20 μg of palladium turbidity appears at pH's greater than 2.1. Whereas the latter value is not the most favorable for color intensity, it affords the maximum sensitivity without producing a turbidity. With the conditions recommended below, precipitates will appear on standing longer than a few hours. Within this time the solution should remain clear and the color stable. Figure 3.1a shows the absorption curve for the reagent and its palladium(II) complex.

PROCEDURE 28 (150)

Reagents

p-Nitrosodiphenylamine solution.   Dissolve 50 mg of *p*-nitrosodiphenylamine in 500 ml of 95% ethanol, dilute, filter, and make up to 1 liter with water.

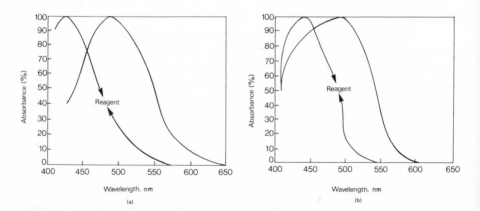

FIGURE 3.1.   Absorption curves of (a) *p*-nitrosodiphenylamine and its complex with PdCl$_2$, and (b) *p*-nitrosodimethylaniline and its complex with PdCl$_2$ (151).

Buffer solution.   Add 240 ml of 1 M hydrochloric acid to 200 ml of a 1 M sodium acetate solution, and dilute to 1 liter with water.

Standard palladium solution.   The final solution should contain 1 mg of palladium per milliliter. Dissolve pure palladium metal in a few milliliters of 1:3 aqua regia, evaporate on a steam bath to a moist residue; repeat three times with hydrochloric acid, add a few milliliters of water and filter. Burn the filter paper, and repeat the above treatment on the residue with aqua regia and hydrochloric acid. Filter, add to the main palladium solution, and dilute to the required volume with 1 M hydrochloric acid.

Procedure

Transfer to a 100-ml flask a slightly acidic solution of palladium, containing not more than 10 μg of metal, add 25.0 ml of the sodium acetate–hydrochloric acid buffer solution and 2.0 ml of the reagent solution, and dilute to 100 ml with water. Allow to stand for 30–45 min, and measure the absorbance at 510–525 nm against a reagent blank.

p-NITROSODIMETHYLAMINE METHOD

The sensitivity of the color reaction for palladium is 0.0015 μg cm^{-2}, the practical sensitivity being 0.0067 ppm, measured in a 1-cm cell. The optimum range of concentration is 0.25–1.0 ppm. With a large excess of reagent there are two absorbance maxima, at 505 and 525 nm. With a slight excess there is a single peak about 495 μm. In a sodium acetate–hydrochloric acid buffer solution the color remains stable on heating only within the narrow pH range 2.2 $\pm$ 0.2. Figure 3.1b shows the absorption curve for the reagent and its palladium(II) complex.

PROCEDURE 29 (150)

Reagents

p-Nitrosodimethylaniline solution.   Dissolve 25 mg of a *p*-nitrosodimethylaniline in 50 ml of 95% ethanol, dilute to 75 ml with water, filter, and make up to 100 ml. Standardize gravimetrically.

Buffer solution.   Add 80 ml of 1 M hydrochloric acid to 200 ml of a 1 M sodium acetate solution, and dilute to 1 liter with water. The pH of the buffer is about 4.8.

Procedure

Add to a 50-ml volumetric flask an aliquot of weakly acidic palladium solution, containing about 10 μg of metal, and 10 ml of the sodium acetate–hydrochloric acid buffer solution. Add 0.5 ml of the *p*-nitrosodimethylaniline solution diluted with water to 50 ml, and shake. Allow to stand for 5 min, and determine the absorbance at 525 nm against a reagent blank.

Yoe and Overholser (170) discuss the separation of platinum from gold, silver, and palladium. For milligram samples of the metals, these separations are accomplished with ease, and suitable methods are suggested in Chapter 7.

The separation from palladium of microgram amounts of gold, platinum and silver involves difficulties, and most of the older methods require separation by solvent extraction. Some of these procedures still provide the most efficient separations. To remove amounts of platinum exceeding 2 μg, Yoe and Overholser (170) rejected the diethyl ether and ethyl acetate extractions of the tin(II)–platinum-containing complex, and were able to apply satisfactorily the hydrolytic precipitation (172).

A solvent extraction method for the separation of less than 1 mg of palladium from platinum was described by Yoe and Kirkland (173). The colored palladium–p-nitrosodimethylaniline complex is extracted by purified C.P. chloroform. The extraction is made at room temperature and the aqueous phase can be used for the determination of platinum. The chloroform extract of palladium is evaporated, and the residue treated with nitric acid and hydrogen peroxide. Subsequently the palladium is determined by p-nitrosodimethylaniline. The authors also provide a method for the simultaneous spectrophotometric determination of platinum and palladium when these are present in microgram amounts (see Procedure 37).

To remove gold in concentrations as great as 10,000 to 1 of palladium, Yoe and Overholser (170) extracted with ethyl acetate or diethyl ether by the following method:

PROCEDURE 30 (170)

To 15 ml of the gold and palladium solution in a separatory funnel, add hydrochloric acid to produce an approximately 3 M solution. Add 10 ml of ethyl acetate, shake and draw off the aqueous layer. Shake the organic layer with a few milliliters of 3 M hydrochloric acid, and combine this wash solution with the aqueous extract. Repeat if necessary. Evaporate the ethyl acetate solution to dryness on a steam bath, dissolve the residue in a few drops of hydrochloric acid, then evaporate to a moist residue. Add water, and determine the gold gravimetrically with hydroquinone (Procedure 50) or spectrophotometrically.

PLATINUM

Tin(II) Methods

Probably the earliest application of tin(II) chloride to the colorimetric determination of platinum was made by Davis (174). Visual comparisons were made over the concentration range 2.5–100 ppm. The full color intensity developed within 10 or 15 min and remained unchanged for several hours. Preliminary to the determination, platinum was precipitated, isolated, converted to a salt, and finally dissolved in hydrochloric acid. For amounts of platinum less than 0.2 mg the colorimetric method yielded more accurate results than the gravimetric method, but for larger amounts the reverse was true.

Wölbling (175) added tin(II) chloride to ammonium chloride-containing solutions of platinum, palladium, rhodium, and iridium, and subsequently extracted the platinum with ethyl acetate. The procedures were qualitative in nature, however, and no data were provided.

Sandell (150) used the tin(II) chloride method for samples containing 0.5–2 ppm of platinum. The color intensity depended upon the hydrochloric acid concentration, and a final acidity of about 0.25 M was recommended. The color development was very rapid, and the color remained stable. Of the platinum metals, palladium provided the greatest interference. Ruthenium interfered to a lesser degree, and rhodium, iridium, and gold in low concentrations did not interfere seriously. Small quantities of iron and copper did no harm. Reference was made to the fact that "chloroplatinous acid" could be isolated by extractions with ethers and with esters from 1 M hydrochloric acid.

The most generally useful information concerning this method has been recorded by Ayres and Meyer (165, 176, 177), and with slight modifications their data and procedures have been used successfully by the present authors for a wide variety of isolated and partially isolated platinum salts.

The optimum concentration range is 3–25 ppm, over which the relative error is within 1%. The minimum error occurs at about 10 ppm. By measuring solutions containing 50–60 ppm of platinum against a 50 ppm standard, the error may be reduced to 0.1%. The sensitivity of 0.025 μg cm^{-2} of platinum at 405 nm, the wavelength of minimum transmittance. The color system conforms to Beer's law up to a concentration of 30 ppm of platinum. With a proposed differential method, the system follows Beer's law up to about 70 ppm (165).

The interferences from associated noble metals increase in the following order: palladium, gold, rhodium, osmium, ruthenium, iridium. Some variation is found in the reported interference from rhodium. Milner and Shipman (178), however, found the interference greater in 0.3 N than in 2 N acid. At the lower acidity, the absorbance increases rapidly with time whereas in the 2 N acid the opposite effect is observed. On standing for 3 hr the rhodium interference is minimized. With lower acidities, however, when small amounts of rhodium are present and a separation is not advisable one can avoid the interference by measuring the transmittance of the platinum complex immediately after mixing.

In so far as the remaining platinum metals are concerned their isolation is readily accomplished by an oxidizing hydrolysis. Gold may be separated by any one of the organic reductants and other reagents. Whereas some of the frequently associated base metals can be tolerated, there is interference from nickel, and it is not always easily removed. Chromium and tellurium interfere, but these are readily eliminated. There is little interference from sulfate and nitrate. The analytical chemist would do well to note that these data and, in general, similar data, were obtained by the direct addition of the interfering substance to determine the transmittance deviations. Thus one must not expect that the degree of interference will be comparable to those instances where the mixture of metal salts have been subjected to the complete dissolution processes. One need only recall that a solution of platinum fumed with sulfuric acid will not respond quantitatively to a hydrolytic separation. In the presence of the above interfering cations, solvent extractions of platinum can be applied; this method is discussed later.

Although the identity of the yellow-to-orange constituent remains unknown, there is agreement that it is not a colloidal constituent and thus the stability of the complex is not critically affected by the presence of salts, acidity, etc. The color development is almost instantaneous, and remains stable to within 0.1% for days; on continued standing the color fades, but can be restored by addition of a tin(II) salt. The influence of the proportions of tin(II) chloride and of hydrochloric acid has been the subject of some investigation. Milner and Shipman (178) have shown clearly that the composition of the tin(II) reagent is not without effect. Arising from the discovery that transmittances in 0.3 N acid were appreciably different for reagent grade and for technical grade reagent it was found that some tin(IV) was necessary for good precision. Because tin(II) chloride produced some color development and tin(IV) no discernible reaction, and because mixtures of these salts produced satisfactory results it was concluded that the complex contained tin in both oxidation states. The difficulties concerning the origin of tin(II) chloride are avoided by maintaining an acid strength of 0.7 M in hydrochloric acid before the addition of the tin(II) chloride. One must avoid the addition of reagent to weaker acid solutions because the subsequent addition of acid may fail to produce the maximum color.

With the acidities recommended below, the tin(II) concentration has little effect on the color intensity.

Some considerable data have been recorded on the character of tin(II) chloride reaction. Ayres (176) studied the chemistry of the reaction with platinum, and found evidence for a variety of reaction products. Precipitation tests with phenylarsonic acid, which may selectively precipitate quadripositive ions, indicated the existence of $[PtSn_4Cl_4]^{4+}$. They concluded that "chemical evidence at hand is insufficient to warrant an attempt to explain the manner in which the platinum, tin and chloride are bound in the complex."

Milner and Shipman (178) stated, without proof, that at acidities of 1.5–2.5 N, "a true stannous–platinum chloride complex is formed." At the low activity originally recommended by Sandell (150), the reaction was affected by the presence of tin(IV) to the degree that reagent grade tin(II) chloride produced colors that were not reproducible and were of lesser intensity than colors from the technical grade reagent.

One may conclude from the existing literature that the platinum–tin(II) chloride color reaction proceeds most satisfactorily at acidities of the order of 2 N, and that the identities of the colored dissolved constituents are unknown. The optimum wavelength remains at 403 nm.

Berman and Goodhue (179) introduced a variation in the tin(II) method to achieve a sensitivity five times greater than the method proposed by Ayres and Meyer (165). This was done primarily by a larger reduction in the amount of tin(II) chloride and the addition of sulfuric and perchloric acid to stabilize the colored constituent and effect a hypsochromic shift in the absorption band.

In the presence of perchloric acid there is a marked fading caused, at least partially, by light. In the presence of 5 ml of concentrated perchloric acid the yellow color develops within a few minutes and remains stable for at least 4 hr. The amounts of sulfuric acid are not critical, but fuming should be avoided. If this inadvertently takes place, the addition of several ml of hydrochloric acid and boiling to remove the volatile acid restores the color system. Too little sulfuric acid gives low results. The absorbance is measured at 310 nm, and the recommended range is 0.4 – 6 ppm with 1-cm cells, and 0.08 – 1.2 ppm with 5-cm cells. Quartz cells are used because 310 nm is beyond the 320 nm limit recommended for Corex glass cells. To avoid the necessity of ultraviolet accessories, however, the authors ascertained that a tungsten lamp source and Corex glass cells could be used with a threefold increase in slit width. The method, Procedure 32, is described below. It is particularly suitable for solutions containing nitric acid because the procedure accomplishes its removal. The interferences from associated metals have been properly examined by carrying the constituents through the complete treatment prior to measurements. In general, the platinum metals interfere, with iridium having the least effect. Of the associated base metals, chromium, iron, and copper interfere least. It is recommended here that, as with all other spectrophotometric methods for platinum, a prior separation of the latter be effected.

The stability of the unknown tin(II) – platinum complex is not critically affected by the proportions of the reagent, although minimum amounts are recommended. For the applicable range of platinum, 2 ml of the freshly prepared reagent is the optimum amount.

Faye and Inman (180) introduced a useful variation in the tin(II) method. The procedure followed a separation of platinum from palladium by using an anion-exchange column, then stripping with perchloric acid, and subsequently fuming with sulfuric acid.

Following treatment in the presence of hydrochloric acid, the platinum was extracted by a solution of tin(II) chloride and a tributylphosphate–hexane mixture. The organic phase was centrifuged, and the absorbance measured in a 1-cm Corex cell.

It should be noted that the efficiency of the tin(II) method for platinum after stripping from an exchanger may be vitiated by the presence of organic matter. Coburn *et al*. (181) have recorded that nitric–sulfuric acid mixtures failed to destroy the organic matter but mixtures of perchloric and nitric acids were effective.

The tin(II) method was applied to binary alloys containing uranium and less than 5% of platinum, Large proportions were determined gravimetrically. Wagner (182) and Shmulyakovskii (183) used the tin(II) method for platinum in catalysts, and a procedure for determining platinum in cathode slimes with and without extraction was recorded by Struszynski and Chwastowska (184).

PROCEDURE 31 (150)   METHOD A

Reagents

Standard platinum solution.   To 0.500 g of pure platinum sponge, add 5 ml of 1:3 aqua regia. Warm slightly if necessary, and evaporate the solution over steam. Moisten with hydrochloric acid, and evaporate as before. Repeat once, and add 10 ml of twice-distilled water. Filter into a 500-ml calibrated flask. Burn the filter paper to a white ash, add 1 ml of aqua regia, and repeat the previous evaporations and filtrations. Make up to the mark with twice-distilled water and hydrochloric acid to 0.1 M, and standardize with thiophenol as described.

To the sample containing 10–25 mg of platinum add 4 drops of concentrated hydrochloric acid, and dilute to 200 ml with water. Add to the cold solution 1 ml of freshly prepared 10% (v/v) solution of thiophenol in 95% ethanol, and boil for 2 hr. Filter through a medium porosity A2 filter crucible, and wash with 50 ml of 0.012 M hydrochloric acid. Transfer the crucible to a muffle, and heat at 235–275°C for 1.5–2 hr, and weigh as $(C_6H_5S)_2Pt$. The gravimetric factor is 0.4719.

Tin(II) chloride solution.   Prepare a solution 1.0 M in reagent grade tin(II) chloride dihydrate and 3.5 M in hydrochloric acid. Filter the final solution through an A2 porous-bottomed filtering crucible. If the solution is to stand for some days, oxidation may be avoided by covering the surface with a layer of xylene.

Procedure

To 25–100 μg of platinum in a 50-ml volumetric flask add 5 ml of concentrated hydrochloric acid and 10 ml of the tin(II) chloride solution (see Note 1), and dilute to the mark (see Note 2). Measure the transmittance at 403 nm. The color is stable for some days.

If extraction is applied, shake the whole volume with 5.0 ml of isoamyl acetate, and measure the transmittance of the clear organic phase at 398 nm against a blank carried through the procedure. The color is stable for an hour.

Notes 1.   If extraction is to follow, add 10 ml of a 20% ammonium chloride solution to speed up clarification of the phases.

2.   In the presence of appreciable amounts of rhodium, add only 1 ml of hydrochloric acid and 1 ml of the tin(II) chloride solution, and measure the transmittance at once.

PROCEDURE 32 (179)   METHOD B

Reagents

Standard platinum solution.   Prepare as described in Procedure 31.

Tin(II) chloride solution.   Prepare a 2% (w/v) tin(II) chloride solution as described in Procedure 31, from tin(II) chloride dihydrate and 1.5 M hydrochloric acid.

Procedure

Transfer the platinum solution to a small Erlenmeyer flask, adding 6 ml of concentrated sulfuric acid and 5 ml of concentrated hydrochloric acid. Heat gently to avoid fuming, cool, and transfer to a 25-ml volumetric flask, rinsing carefully with several small volumes of water. Add 5 ml of concentrated perchloric acid and then 2.0 ml of the tin(II) chloride solution. Dilute with water, and mix by shaking thoroughly for several min. Measure the absorbance at 310 nm against a reagent blank.

Note.   Although a hydrogen lamp source and quartz cells are recommended, one may use a tungsten lamp and Corex cells provided the slit width of the Beckman Model DU quartz spectrophotometer is increased sufficiently.

A third tin(II) method involves a degree of isolation by solvent extraction. Various liquid extraction methods have been proposed. Ayres and Meyer (165) used amyl acetate, which retained the spectral characteristics of the tin(II) complex in the aqueous solution, but the color faded rapidly. Stability for 1 hr could be achieved by the addition of resorcinol, but the extracted colors were reproducible only to about ± 0.5% absolute transmittance. This and other extraction procedures can be used to secure a suitable concentration of platinum, but a reduction in interference from other constituents is not always achieved. Indeed the reverse effect may be experienced. With an amyl acetate extraction there is some slight gain in tolerance toward associated constituents, but interference from rhodium is thus increased, and despite a proposed variation, the interference of palladium remains a problem.

Ethyl acetate has been used successfully as an extractant of platinum by Figurovskii (185) and Poluektov and Spivak (186). The latter extracted platinum following a coprecipitation of platinum with mercury produced by a tin(II) chloride reduction. The sensitivity was reported as 0.03 ppm with a 10-g sample. Dithizone extractions were used by Young (187) to isolate platinum, palladium, and gold followed by volumetric determination. Pollard (188) produced a volumetric method that involved an extraction of the diethyldithiocarbamates of platinum(II) from a hydrochloric acid solution containing mercury(II) chloride and traces of the tin(II)–platinum complex. A modification of this method was applied by Yoe and Kirkland (173), who used chloroform to extract simultaneously the diethyldithiocarbamate complexes of platinum and palladium.

Faye and Inman (180) introduced an extraction method applied to the method proposed by Berman and Goodhue (179). The method, Procedure 33, was adapted to follow a separation from palladium by anion exchange and a subsequent stripping by perchloric acid.

PROCEDURE 33 (180)

Add to the perchloric acid eluate 2 ml of concentrated sulfuric acid, and evaporate to the appearance of sulfur trioxide. To the cooled solution, add 5 ml of 12 M hydrochloric acid, and again evaporate to fumes. Add 5 ml of 12 M hydrochloric acid, wash into a 60-ml separatory funnel, and dilute to 50 ml with water. Add 5 ml of a tin(II) chloride solution (1 M in tin(II) chloride and 2 M in hydrochloric acid) then 10 ml of a tributyl phosphate solution (10 ml of tributyl phosphate in 200 ml of hexane), and shake for 1 min. Allow the phases to separate, and discard the aqueous phase. Centrifuge the organic phase, and measure the transmittance at 350 nm in a 1-cm quartz cell against a blank carried through the complete procedure. Similarly prepare a calibration curve.

The following tin(II) chloride procedure is particularly applicable to platinum solutions obtained from chromatographic separations (189) and hydrolytic separations (190) and to solutions containing only traces of platinum metals. The method has been applied to the platinum section of chromatographic strips (171).

PROCEDURE 34 (191)

Add the platinum section of the strip to 10 ml of 3 M hydrochloric acid in a 30-ml beaker, and let stand for 10 min with intermittent stirring. Decant the mixture through a Whatman No. 40 filter paper into a 50-ml volumetric flask containing 5 ml of a 1 M tin(II) chloride solution in 3.5 M hydrochloric acid. Use a further 10 ml of 3 M hydrochloric acid to extract the platinum from the pulp. Continue the extractions with successive 5-ml portions of water to reach 50.00 ml. Measure the absorbance at 402 nm against a reagent blank.

p-Nitrosodimethylaniline Method*

p-Nitrosodimethylaniline (173, 192) forms with platinum an orange-red color with a sensitivity of 0.0029 μg cm^{-2} and a working sensitivity of 0.015 ppm when measured in a 1-cm cell. The optimum range is 0.7–2.4 ppm, over which the average and maximum precision are 1.4 and 3.0%, respectively. The colored solutions show absorbance maxima at 550 and 525 nm, the latter being used for the absorbance measurements. The color is believed to be due to two different

---

*Conditions of photometric determination of platinum with p-nitrosodimethylaniline (Rabinovich *et al.* (191a)).

complexes, one of which is $Pt[(CH_3)_2N \cdot C_6H_4NO]_4Cl_2$. The dark green crystals of this complex gave a single absorbance maximum at 518 nm.

There is evidence to indicate that the color reaction is confined to platinum(II), that platinum(IV) is first reduced by the reagent, and that this reduction is one of the rate determining factors for the color development. The color remains constant for at least 24 hr. In order to obtain uniform color development it is necessary to use a large molar excess of reagent over platinum and to standardize the volume of the solution, the time of heating and the concentration of the buffer reagent. The color development is slow, and heating is required. Two techniques may be used; one at 85°C in the presence of ethanol, the second at 100°C in its absence. Heating times in excess of 90 min at 85°C and of 45 min at 100°C result in a discoloration, owing perhaps to oxidation of the excess of reagent. The color development is about 93% complete during the recommended heating time. With a 60-min heating time at 85°C, a deviation of ±5 min is permissible. The 20-min heating time at 100°C may deviate by ±2 min.

A specific difficulty with the method is the tendency for excess of reagent to precipitate during the final cooling period. The prescribed addition of ethanol avoids this difficulty. The volume of the reacting medium is kept as small as is convenient to encourage a more rapid color development.

The method is sensitive to pH changes and the proportion of buffer solution. Maximum color development occurs with a sodium acetate–acetic acid buffer at pH 3.8 or with a sodium acetate–hydrochloric acid buffer at pH 3.0. The latter is used at pH 2.2 ± 0.2 in the recommended procedure as a useful compromise to allow the separation of palladium. The amount of buffer solution must not vary from the optimum by more than 0.5 ml. This method is subject to interference from all the platinum metals and from many of the naturally associated base metals. An alternative procedure applicable to solutions of platinum and palladium is described below. Although there is some tolerance to small amounts of sulfate, the degree of interference from sulfate present after fuming is not known.

PROCEDURE 35 (192)

Reagents

Standard platinum solution. With gentle heating prepare a standard chloroplatinate solution by dissolving exactly 1 g of "pure" platinum in 10 ml of aqua regia. Evaporate the solution almost to dryness, and treat with three successive 2-ml portions of (1 + 1) hydrochloric acid, each time evaporating almost to dryness in order to free the solution from nitrogen oxides. Transfer the final material to a 100-ml volumetric flask, add 4 ml of concentrated hydrochloric acid, and dilute the solution to volume. This gives a stock solution that is about 0.5 M in hydrochloric acid and contains 10 mg of platinum per milliliter. Standardize gravimetrically by precipitating the platinum from 5-ml aliquots.

To 150 ml of the platinum solution, from which oxides of nitrogen have been removed by prior evaporation with hydrochloric acid, add 5 g of sodium acetate and 1 ml of formic acid. Digest on a steam bath for 5–6 hr, cool, and filter through a good grade of ashless paper. Use a small sector of paper to remove any platinum adhering to the beaker wall. Wash with hot water to the absence of chlorides. Fold the paper to enclose the precipitate, ignite slowly, then finally at 800°C. Cool over a saturated solution of calcium nitrate, and weigh as the metal.

Note. This procedure is not recommended for very small amounts of platinum.

Prepare less concentrated platinum solutions by appropriate dilutions. Make up standard solutions of less than 100 ppm fresh daily because such dilute platinum solutions slowly decrease in effective concentration.

p-Nitrosodimethylaniline solution. Dissolve 50 mg of *p*-nitrosodimethylaniline (Distillation Products Industries No. 180), purified by crystallization from 25% ethanol, in 50 ml of freshly distilled absolute ethanol. Ethanol previously exposed for extended periods to air produces discolorations when the mixed solutions are subsequently heated.

Buffer solution. Add 50 ml of a 4 *M* sodium acetate solution to 53 ml of 4 *M* hydrochloric acid. The pH of this solution should be 2.2 ± 0.2.

Procedure

Adjust the solution containing 35–120 μg of platinum as hexachloroplatinate in a 50-ml borosilicate flask to between pH 2 and 3 by adding a 0.1 *N* sodium hydroxide solution 0.1 *M* hydrochloric acid. Add 1 ml of the buffer solution and 1 ml of the *p*-nitrosodimethylaniline solution, and wash the wall of the container with 5 ml of freshly distilled ethanol. The solutions should now have a pH of 2.2 ± 0.2. Heat the solution on a steam bath for 60 min at 85 ± 2°C. Cool immediately by any precise technique, and make up to 50 ml with 95% ethanol. Prepare a blank solution in precisely the same manner, omitting only the platinum. Measure the absorbance at 525 nm versus this blank.

Note. The 100°C modification requires the above technique except that the flask is rinsed with distilled water instead of ethanol. Any turbidity will disappear during the heating or the subsequent dilution with ethanol. The heating period at 100°C is 20 min. The 100°C modification was used successfully by Conrad and Evans (193) after a chloroform extraction of iron by cupferron.

A second modification of the *p*-nitrosodimethylaniline method was proposed by Westland and Beamish (194), who found that an aqueous solution of the reagent afforded increased precision at the expense of greater instability of reagent. The reagent is not readily soluble in water, and the preparation in hot water requires subsequent filtering to remove a precipitate which slowly continues to form after standing for a few days. Thus the concentration could not be closely reproduced and standards were required for each lot of reagent.

PROCEDURE 36 (194)

Prepare a solution containing 165 mg of *p*-nitrosodimethylaniline in 100 ml of water by stirring the hot mixture for 30 min and filtering. Add 3 ml of this solution to the sample of the chloroplatinate at pH 2.2. Heat the solution for 25 min at 85°C, and dilute to the 50-ml mark. Measure the transmittance in a 5-cm cell at 515 nm versus a reagent blank carried through the whole procedure.

PLATINUM AND PALLADIUM

Additive Absorbance Method

The application of the p-nitrosodimethylaniline method was extended by Yoe and Kirkland (173) to include an additive method of determining both platinum 'and palladium in a single sample. A solvent extraction method for the separation of the two metals was also described, with a subsequent method of determining platinum. The simultaneous determination depended upon the fact that no reaction occurs between platinum(II) and p-nitrosodimethylaniline for several hours at room temperature, while palladium gives an immediate color under the same conditions. Furthermore the absorption spectra of the red dissolved constituents from both metals are similar (see Fig. 3.2).

The technique involves the determination of the palladium content in one aliquot by color development at room temperature, and in a second portion a development of color from both metal constituents by heating. The directions are given in Procedure 37. The method can be applied to samples in which the palladium to platinum ratio ranges from 10:1 to 1:60. The sodium acetate–hydrochloric acid buffer described in Procedure 37 is used to maintain a pH of 2.2 ± 0.2, at which acidity the colored constituent with palladium is sufficiently stable, although a greater sensitivity for platinum could be achieved with the sodium acetate–acetic acid buffer at pH 3. The working sensitivity for palladium by the simultaneous method is 0.0067 μg cm^{-2}.

PROCEDURE 37 (173)

(a) The Determination of Palladium

Adjust an aliquot of the solution of platinum and palladium, containing 12–40 μg of the latter to pH 2–3, and make up to a total volume of 10 ml. Add 1 ml of the sodium acetate–hydrochloric acid buffer and to the solution at pH 2.2 ± 0.2 add 1 ml of a 5 mg/ml ethanolic p-nitrosodimethylaniline solution and 5 ml of distilled water, if a temperature of 100°C is to be used subsequently for platinum, or 5 ml of absolute ethanol, if the 85°C technique is to be used for platinum. Allow the solution to stand for 5 min at room temperature, then dilute to 50.0 ml with 95% ethanol. Measure the intensity of the orange-red color at 525 nm against a reagent blank and determine the palladium content from a standard curve similarly prepared.

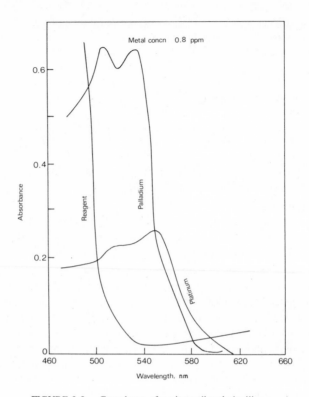

FIGURE 3.2.   Complexes of p-nitrosodimethylaniline.

(b) The Determination of Total Platinum and Palladium

Use an aliquot of the sample containing palladium and platinum which will give a total color intensity falling in the range of most accurate spectrophotometric measurement (about 20–60% transmittance or 0.2–0.7 absorbance units). Adjust the solution to pH 2–3, and make up to a total volume of 10 ml. Then treat with 1.0 ml of the sodium acetate–hydrochloric acid buffer (pH 2.2 ± 0.2) and 1.00 ml of 5 mg/ml p-nitrosodimethylaniline reagent solution. Add 5 ml of distilled water, and heat the solution for 20 min at 100°C, or add 4 ml of freshly distilled ethanol and heat the solution for 60 min at 85°C. Cool immediately to room temperature, dilute to volume with 95% ethanol, and measure the absorbance at 525 nm against a reagent blank which has been heated in exactly the same manner.

The color intensity of the platinum may be calculated from the relationship

$$A_{Pt} = A_{Pt + Pd} - A_{Pd},$$

where $A_{Pd + Pt}$ is the absorbance attributable to platinum and palladium (determined in part (b)), A_{Pd} the absorbance attributable to palladium (obtained in part (a)), note that a concentration factor must be used if aliquots used in parts (a) and (b) are not the same), and A_{Pt} the absorbance attributable to platinum.

PLATINUM

3,4-Diaminobenzoic Acid Method

3,4-Diaminobenzoic acid reacts with platinum when the mixture is heated at pH 9.8–12.2. The reaction was used as a spectrophotometric method for platinum, first by Johnson and Ayres (195) and in 1971 by Keil (196). The resulting blue to blue-green solution has maximum absorption at 715 nm. The system conforms to Beer's law over the optimum range of 0.35–1.20 ppm of platinum. The absorption is not highly sensitive to reagent concentration, pH, or normal room temperature fluctuations. Developed samples are stable for at least 4 hr. The only ions which interfere enough to require separation are palladium, osmium, and ruthenium. The reaction product with 3,4-diaminobenzoic acid is 2 to 1 of platinum. It is interesting to note that the complex in alkaline solution is anionic and cationic in acid solution. Separation procedures are easily applied. Palladium is easily removed by two extractions in its complex with 1-(2-pyridylazo)-2-naphthol into chloroform, and the layer of chloroform is suitable for spectrophotometric determination of palladium. Osmium and ruthenium can be removed as the octavalent oxides from a dilute hydrochloric acid solution by oxidation, preferably by ammonium paraperoxydisulfate.

Keil (196) discussed the selectivity by using masking agents containing EDTA, urea, potassium iodide, and triethanolamine. Interferences were observed with SCN^-, $S_2O_3^{2-}$, BrO_3^-, CO_3^{2-}, $C_2O_4^{2-}$ at > 5 mg/l; fluorine at > 25 mg/l and at > 30 mg, strongly oxidizing and reducing agents. Tolerable amounts of 54 cations were recorded.

PROCEDURE 38 (195, 196)

Reagents

3,4-Diaminobenzoic acid. The acid should be obtained from the Aldrich Chemical Co. and used as received. Prepare solutions by dissolving the desired amount (usually 0.3 g) in 10 ml of dimethylformamide and diluting to 50 ml with distilled water.

N,N-Dimethylformamide (DMF). This should be Eastman No. 5870, used as received.

Potassium tetrachloroplatinate(II). This can be obtained from the Fisher Scientific Co.

Standard platinum(IV) solution.   Prepare by dissolving 1.000 g of Grade 1 platinum thermocouple wire, 99.99% pure, in hot aqua regia. Take the resultant solution almost to dryness, add a small amount of hydrochloric acid, and again take the solution to dryness. Repeat this treatment three times in order to destroy any nitroso complexes. After the final evaporation, add 10 ml of hydrochloric acid and make up the solution to 1 liter with distilled water. Prepare more dilute platinum solution as needed by volumetric dilution of this stock. Standardize with any suitable procedure.

Procedure

Place 35 to 120 μg of platinum in a volume of 25 ml or less in a 100-ml borosilicate glass volumetric flask. Add 2 ml of the 3,4-diaminobenzoic reagent and adjust the pH of the solution to below three. Float the flask for 15 min in a large beaker of boiling water and then cool to room temperature. At this point the sample is light amber in color and sometimes contains a precipitate. Make the sample alkaline by adding an excess of either 0.1 or 1 *M* sodium hydroxide depending on the amount of acid present. The sodium hydroxide should be added from a buret until a green color appears in the solution. Then add an amount of sodium hydroxide equal to one third the amount already added. At this point the sample is blue-green and the precipitate is completely dissolved. Make up the sample to 100 ml with distilled water and measure the absorbance of 715 nm within 1 hr. The reagent blank is negligible at 715 nm, but should be determined as a precaution.

RHODIUM

*"sym-*Diphenylcarbazone" Method

A 1-to-1 mixture of 1,4-diphenylcarbazide and 1,4-diphenylcarbazone reacts with rhodium(III) salts to form a purple complex of unknown structure (197). The optimum range is 0.3–1.5 ppm with a 1-cm optical path, and the optimum wavelength for absorbance measurements is 565 nm. The color is developed in a perchloric acid medium of pH 3.0. Both the acidity and the volume of water are critical. The excess of reagent, amount of buffer, and the period of heating beyond 30 min may be varied within reasonable limits.

The preparation of the sample involves fuming with perchloric acid, which process removes interfering chloride and nitrate, volatilizes osmium and ruthenium, and precipitates platinum and gold. Iridium, iron, cobalt, nickel, lead, and copper interfere. Chromium(VI) and mercury(II) may be removed by volatilizing their chlorides.

PROCEDURE 39 (197)

Reagents

Color-forming reagent.   Use the *"sym-*diphenylcarbazone" as a 0.025 *M* solution in acetone. Slight variations in absorbance resulting from the use of different preparations of the reagent are eliminated by the addition of a small amount of acetic acid (1 ml of glacial acetic acid per liter of acetone used).

Buffer. Partially neutralize a $0.100 M$ solution of monochloroacetic acid with sodium hydroxide to give a solution of pH 3.0. Make the solution $0.10 M$ in sodium perchlorate to provide a constant ionic strength.

Perchloric acid. Partially neutralize a $0.10 M$ solution with sodium hydroxide to give a solution of pH 3.0.

Procedure

Add about 10 ml of 70–72% perchloric acid to the sample or to the standard rhodium solution. Fume to 1 or 2 ml to remove halides or nitrates. Add water to 80–90% of the final volume, adjust to pH 3.0, then dilute to the mark in a volumetric flask. Transfer 5 ml of the monochloroacetate buffer to a 50-ml Erlenmeyer flask, and add the pH 3.0 perchloric acid in such volume that the sample plus the perchloric acid is 10.0 ml. Add the sample solution (not more than 10 ml) and 1.0 ml of the "*sym*-diphenylcarbazone" solution; wash down the flask wall with acetone. Place the flask in a boiling water bath for 30 min, cool, and transfer the contents to a 50-ml volumetric flask. Rinse 2 or 3 times with methanol. Wash down the reaction flask with 5 ml of N,N-dimethylformamide; tilt the flask and rotate it so as to wet the wall with the solvent. Transfer the rinsings to the volumetric flask, and again rinse with small portions of methanol. Cool the mixture to room temperature, dilute with methanol to the mark, and measure the absorbance at 565 nm against a reagent blank prepared in the same way as the sample solution.

Tin(II) Chloride Methods

In experienced hands, tin(II) remains one of the most useful reagents for ascertaining the presence of a number of platinum metals including rhodium. Its application to the quantitative estimation of rhodium is only now being realized. Various authors have proposed its use and have suggested procedures for specific purposes. Sandell (150) stated that either the red color, developed in $2 M$ hydrochloric acid, or the yellow form, occurring in less acidic media, could be used for colorimetric purposes. The red form is less sensitive than the yellow complex, the former being $0.026 \mu g \ cm^{-2}$ of rhodium measured at 479 nm, and the latter $0.0084 \mu g \ cm^{-2}$, measured at 430 nm. The red complex is generally preferred because of its color stability; the yellow color forms slowly, and reproducible results can be obtained by an accurate measurement of the time interval of color development. The red and yellow colors are reversibly developed, the latter being converted to the red by the addition of an alkali metal chloride or more usually by the addition of hydrochloric acid. The yellow form appears on diluting the red solution with water. Some difference of opinion exists concerning the quantitative aspects of this change. Ayres *et al.* (198) stated that the full color intensity was attained in 3–5 min near the boiling point, and that when the amount of hydrochloric acid was varied from 1 to 40 ml of concentrated acid per

100 ml of final solution, the absorbance readings were constant within the limit of the accuracy of the measurements.

Stein (150) found that with 150 μg of rhodium heating for 15 min gave a color intensity about 4% less than that obtained by heating for 30 min; dilution to a final hydrochloric acid concentration of 1 M gave a solution whose absorbance was not constant after 1 hr, and which at the end of that time was a little less than that of a 2 M hydrochloric acid solution.

Maynes and McBryde (199) found that when samples of rhodium solutions were heated for 40 min, the absorbance showed a slow drift during the 24-hr period after removing the samples from the water bath. When the heating interval was 1 hr the absorbances were stable after 30 min for at least 24 hr. These authors diluted their samples with 2 M hydrochloric acid. Anyway, the present authors have had good success by heating for 1 hr and a final dilution with 2 M hydrochloric acid.

Some difficulties have also been experienced in the application of the yellow form to colorimetric work. McBryde and Yoe (157) failed to obtain satisfactory results with the yellow form, but were successful with the red form in 2 M hydrochloric acid, making their absorbance measurements at 460 nm.

Some slight differences in the applicable range have been recorded. Ayres *et al.* (198) found the optimum concentration range for measurement in 1-cm cells was about 4–20 ppm of rhodium. One may expect that these limits will vary with variations in specific procedures. Maynes and McBryde (199) reported that Beer's law applied up to 40 ppm of rhodium.

Absorbance measurements for the red form are made at 473 or 475 nm (green or blue filter, 480 nm); there is also a strong absorbance at about 330 nm. Near 330 nm, the absorbance is sensitive to small variations in the concentration of tin(II) chloride, which is not so at 475 nm. For the yellow form, measurements can be made with a blue filter (435 nm). As would be expected, there is considerable interference from other platinum metals and their removal is necessary.

Of the associated base metals, large proportions of nickel can be tolerated. This, together with the favorable tolerance for iridium, are good features of the tin(II) chloride method. Iridium forms brown chloroiridite, which does not absorb appreciably at 475 nm, and about 20 times as much iridium as rhodium can be present. Maynes and McBryde (199) provide a procedure for the spectrophotometric determination of iridium and then rhodium. These authors were unable similarly to determine rhodium then iridium; they suggest that 1.3 mg of rhodium is the maximum amount of rhodium for which the tin(II) chloride method is recommended when iridium is absent; in the presence of up to 5 mg of iridium, the applicable amount of rhodium is slightly decreased.

Up to about 50 ppm of copper and iron may be present; gold, silver, tellurium, and selenium must, of course, be absent. Chromium also interferes. The red rhodium complex can be extracted from a 4.4–5.2 M hydrochloric acid solution by amyl acetate to form a yellow, unstable solution. At the present time, no satisfactory extraction method has been recorded.

Ayres *et al.* (198) recorded a procedure for the simultaneous determination of rhodium and platinum, measurements being made at 475 and 403 nm, or, in order to reduce the absorbance of platinum in the presence of rhodium at 560 or 403 nm. Determinations such as these have limited, albeit useful applications, and are subject to large percentage errors, e.g., when there is a preponderance of one constituent as is usually so with rhodium and platinum.

This simultaneous determination was used by Smith (199a) to determine rhodium and iridium in plutonium with no prior separations of these three metals. Under the conditions used, plutonium was reduced by tin(II) chloride to plutonium(III), which has an absorption peak at 665 nm, at which wavelength the absorbances of platinum and rhodium are negligible. For the latter two metals, the absorbance was measured at 399 and 470 nm respectively. The absorbance of plutonium at these wavelengths is directly related to the absorbance of plutonium at 665 nm so that a correction may be applied by subtracting the absorbance of plutonium at 470 and 399 nm. The method is applicable over the range 5–14 ppm of the two platinum metals.

The tin(II) chloride method was applied by Gardner and Hues (200) to the determination of rhodium in uranium. No determinations were made with less than 0.02 mg of rhodium or with more than 200 mg of uranium. A wavelength of 520 nm was used rather than the customary 475 nm, because at the latter, the absorbance of uranium was significant, and also because conditions of time, temperature and acid strength, which were not critical for rhodium, proved to be critical for uranium. At 520 nm, reliable blanks were obtained for even large amounts of uranium. The data obtained for approximately 1% rhodium alloys agreed well with gravimetric results.

Tin(II) chloride was also used by Karttunen and Evans (201) for the determination of rhodium in uranium–base metals fission alloys. The rhodium was isolated by a prior solvent extraction and treatment on a cation exchange resin. The method of isolating rhodium from the fission products and the subsequent use of the tin(II) chloride methods for rhodium concentrations as low as 1 ppm were discussed by Cheneley *et al.* (202).

Various efforts have been made to increase the sensitivity resulting from the formation of the tin(II) rhodium complexes. Markham (203) found that tin(II) chloride gives a more intense color in the presence of iodide than in the presence of chloride alone. The sensitivity of this modification is 0.009 μg cm^{-2} at 435 nm as compared to 0.026 μg cm^{-2} at 475 nm for the red form. The method involves relatively close control of the conditions for color development.

Kalinin and Yakovleva (204) recorded a tin(II) chloride method with the claim that it was ten times higher than the older method with tin(II) chloride. The system is rhodium(III)–hydrochloric acid–water–perchloric acid–tin(II) chloride.

Beer's law is obeyed over the range 0.01–0.5 μg rhodium per milliliter, in the presence of 0.064 M tin(II) chloride. Maximum absorbance occurs at 330 nm. Optimum conditions for complex formation is 0.05 M hydrochloric acid in 6 M perchloric acid and a Sn/Rh ratio of $\geq$5000:1.

Kalinin *et al.* (204) recorded optimum conditions for the extraction-separation of trace amounts of rhodium as its complex with tin(II) chloride from hydrochloric acid media with butylorthophosphate. Rhodium can be re-extracted from the organic phase with 0.25–0.5 M hydrochloric acid after dilution with benzene.

Tin(II) Bromide and Tin(II) Chloride Methods

Berman and Ironside (205) were able to increase materially the sensitivity of the tin(II) method through the use of tin(II) bromide. The sensitivity of the method is 0.0035 μg cm^{-2}. The yellow solutions obey Beer's law over the range 0.8–1.6 ppm of rhodium, but satisfactory results are obtained over the range of 0.4–4 ppm. Like tin(II) chloride, the composition of the bromide complex is unknown. The absorption peak is at 427 nm; in the presence of chloride a second peak at 407 nm may appear. To avoid the presence of the latter, the tin(II) bromide is prepared by dissolving tin(II) oxide in concentrated hydrobromic acid. The instability of the colored species and the photosensitivity of the color development can be avoided by the addition of perchloric acid before adding the reagent. Occasionally, a precipitate may appear on addition of the reagent, but this dissolves when diluted and does not affect the absorbance. The color remains stable for at least 3 hr. Although hydrobromic acid additional to that required by the procedure produces an increase in absorbance, the time required to achieve maximum absorbance also increases.

The authors' data on the effect of sulfuric acid are well worth noting.

A very appreciable difference was found in the degree of interference from added sulfuric acid and a solution of rhodium fumed in the presence of sulfuric acid. One cannot overemphasize the possibility that interferences determined by the mere addition of constituents are not necessarily applicable to a specific dissolution procedure. The tin(II) bromide method was made applicable to fumed solutions by dilution with water, the addition of concentrated hydrobromic acid, and boiling vigorously for 5 min.

The method usually requires the separation of rhodium from other platinum metals and associated base metals. Iridium produces the least interference, and a fair estimation of rhodium can be made even when equivalent quantities of iridium are present. This procedure is perhaps the most generally satisfactory of all rhodium methods so far recorded.

PROCEDURE 40 (205)   TIN(II) CHLORIDE METHOD

Reagents

Tin(II) chloride solution.   Dissolve 100 g of tin(II) chloride in 160 ml of concentrated hydrochloric acid. When the solution is clear, dilute to about 400 ml, filter, and make up to 1 liter.

Standard rhodium solution.   Dissolve about 2.5 g of sodium rhodium chloride (21.29% rhodium) in water, add 10 ml of concentrated hydrochloric acid, filter, and dilute to 1 liter. Standardize with 2-mercaptobenzothiazole (Procedure 59).

Procedure

To a 25-ml volumetric flask add an aliquot of rhodium solution containing from 4 to 100 μg of rhodium diluted with 5 ml of water. Add 10 ml of tin(II) chloride, and heat for 1 hr in a boiling water bath. Cool to room temperature, then dilute with 2 M hydrochloric acid to volume. Measure the absorbance at 470 nm in a 5-cm cell against a reagent blank.

Note.   For amounts of rhodium of the order of 100–500 μg, use the same procedure except for a 50-ml flask and a 1-cm cell.

PROCEDURE 41 (205)

Reagents

Tin(II) bromide solution.   Dissolve tin(II) oxide in 40% redistilled hydrobromic acid to produce a 20% (w/v) solution.

Standard rhodium solution.   Dissolve about 2.5 g of sodium rhodium chloride (21.29% rhodium) in water, add 10 ml of concentrated hydrochloric acid, filter, and dilute to 1 liter. Standardize with one of the gravimetric reagents.

Procedure (If No Fuming with Sulfuric Acid Has Been Done)

Transfer the rhodium solution to a 25-ml volumetric flask, and dilute to about 5 ml with water. Add 10 ml of concentrated perchloric acid, followed by 2 ml of the tin(II) bromide solution. Dilute to volume with water. After 20 min, measure the absorbance at 427 nm against a reagent blank.

Procedure (After Fuming with Sulfuric Acid)

This method is used for fumed rhodium solutions. Cool the fumed rhodium-containing solution, containing 1 or 2 ml of sulfuric acid. Add 5 ml of water and 0.5 ml of concentrated hydrobromic acid. Boil the solution vigorously for at least 5 min, and allow to cool. Add 8 ml of concentrated perchloric acid followed by 1 ml of the tin(II) bromide solution. Transfer to a 25-ml volumetric flask, and dilute to volume. After 8 min, measure the absorbance at 427 nm against a reagent blank containing the same quantity of sulfuric acid as the sample. The mean deviation of the results should not exceed 0.2%.

IRIDIUM

Leuco-Crystal Violet Method

One of the most sensitive reagents for iridium, leuco-crystal violet, was used by Ayres and Bolleter (206) for the determination of iridium at concentrations of 0.5–4 ppm of metal. Ewen and Cook (207) modified the method and measured absorbance in a 2-cm cell for 0.2–1.2 ppm. The minimum amount for iridium in the sample solution or aliquot portion should be 2 μg. The modified method involved oxidation to the yellow product of leuco-crystal violet by hydrogen peroxide rather than fuming a solution containing nitric, phosphoric, and perchloric acid solution, thus increasing the precision of the method. The method is applicable to the determination of iridium in 5 N hydrochloric acid solutions after the iridium has been separated from rhodium, palladium, platinum, silver, and gold. Thus it can be used for solutions of iridium separated by chromatography (described in Part II). Rhodium, platinum, palladium, and silver interfere when they are present in ratios higher than 10-to-1, 50-to-1, and 2.5-to-1, respectively. Gold interferes seriously. The method is applicable in the presence of up to 0.005% of concentrated sulfuric acid. The useful concentration range for iridium on a calibration curve for a 2-cm cell is 0.2–1.2 ppm. Hence the minimum amount of iridium in the sample solution or aliquot portion of the solution taken for analysis should be 2 μg.

PROCEDURE 42 (206, 207)

Reagents

Standard iridium solution. Prepare by dissolving in 5 N hydrochloric acid a precipitate of hydrated iridium dioxide (Procedure 62) and standardizing by Procedure 61. Take an aliquot to provide 10 μg per one milliliter of 5 N hydrochloric acid.

Hydrogen peroxide solution. Dilute 1 ml of 30% (v/v) hydrogen peroxide to 100 ml. Prepare fresh solution daily.

Leuco-crystal violet solution (0.1%). Dissolve 50 mg of leuco-crystal violet in 50 ml of 10% phosphoric acid solution. Prepare fresh solution weekly.

Sodium hydroxide solution (6 N). Dissolve 60 g of sodium hydroxide in 250 ml of water.

Acetate buffer (pH value of 4.2). To 100 ml of the 6 N sodium hydroxide solution add 175 ml of glacial acetic acid (sp gr 1.059).

Hydrochloric acid (5 N). Dilute 41.7 ml of concentrated hydrochloric acid (12 N) to 100 ml.

Procedure

All the steps of the procedure must be carried out as quickly as possible, and under no circumstances must the solutions be left to stand overnight.

Ensure that the laboratory atmosphere is free from hydrogen sulfide and sulfur dioxide fumes. Transfer an appropriate aliquot portion of the sample solution to a 10-ml beaker (see Note 2). Add 1 ml of the 1% hydrogen peroxide solution, and evaporate to about 0.5 ml on a steam bath. Cool the solution, and add another 1 ml of hydrogen peroxide solution. Evaporate the solution on a steam bath to about 0.2 ml. Repeat once. Do not evaporate the solution to dryness (see Note 3). Cool, and add 1 ml of the leuco-crystal violet solution and 5 ml of the acetate buffer solution. Transfer to a 10-ml volumetric flask, and dilute to volume with water. Measure the absorbance of the solution in a 2-cm cell at a wavelength of 590 nm, against that of a reagent blank.

Transfer aliquot portions of the dilute standard solutions, containing 2, 4, 6, 8, and 10 μg of iridium, into five 10-ml beakers. Proceed as in the preceding.

Notes. 1.   The normality of the acid solution for the determination of iridium is not critical. However, as the separation of the noble metals can be carried out in a 5 N hydrochloric acid solution for convenience this medium was also used for the iridium determination.

2.   If the aliquot portion of the sample is greater than 3 ml, evaporate on a hot plate to about 2.5 ml and then proceed as above.

3.   The volumes are critical because the final pH of the solution is affected by the amount of acid left in solution, and 5 ml of buffer solution is the maximum amount that can be added to the flask.

RUTHENIUM

Dithiooxamide Method

Dithiooxamide (rubeanic acid) may be used for the determination of $0.60 - 5.9$ ppm of ruthenium.

Yaffe and Voight (208) determined the composition of the colored constituents in perchloric acid solution. Ruthenium(III) or (IV) produced $[Ru(SC(NH)(CSNH_2)]_2^+$ and $[Ru(SC(NH)CSNH_2)_3]^0$. The reaction in 1.5 M hydrochloric acid were studied by Kolarik and Konecny (209) who found complexes with a ruthenium to reagent ratio of 1:3, 1:2, and 1:1. It is stated that ruthenium(IV) is reduced by a mole of reagent to form ruthenium(III), which then produces a brown constituent with a reagent to ruthenium ratio of 1:1. This intermediate product, with a maximum absorption at 470 nm, is converted to a blue-green species and then to the blue product with a ratio of 3 of reagent to 1 of ruthenium(III). These blue reaction products are also produced by thiourea (210) under similar conditions, but the latter reaction is considerably less sensitive.

Ayres and Young (211) made a spectrophotometric study of this method as applied to the ruthenium chloro complexes. Measurements are made at 650 nm and the solution obeys Beer's law. The development of the color is accomplished by heating for 20 min at 85°C. A high concentration of acid and a large amount of ethanol are required to prevent the formation of precipitates and to give a stable color. The method tolerates small amounts of associated base metals and platinum metals except osmium. It should be noted again that the ratios of other platinum metals and of base metals to ruthenium present in most natural and synthetic products are considerably beyond the tolerances permitted in this method. Furthermore the degree of interference can be estimated accurately only when the mixed constituents have been subjected to the full chemical treatment. For these reasons and because osmium interferes, both the thiourea and the dithiooxamide methods will find application generally after separations have been made; and because one cannot, in general, follow the quantitative progress of these distillations, such separations will inevitably be "sharp."

Although it is claimed, with some dissension (212), that dithiooxamide is a more sensitive reagent than thiourea, the latter is preferred by the present authors, largely because of familiarity through long continued use of the procedure. Within the applicable concentration ranges of these two reagents, however, the advantage of superior sensitivity to both osmium and ruthenium is reduced by the control of composition through the concentrating processes of distillation and to a lesser degree by selective solvent extraction. For those who approach a ruthenium determination for the first time, the dithiooxamide method is recommended.

Rubeanic acid has been used also to determine ruthenium in the presence of large proportions of uranium (213). When the proportions of uranium to ruthenium exceed 120 to 1 the method for $10-50$ μg of ruthenium involves the addition of 1 ml of concentrated hydrochloric acid and 1–2 ml of a 0.1% ethanolic rubeanic acid solution. The mixture is heated for 5 min in a boiling water bath, cooled, and extracted with three 5-ml portions of isoamyl alcohol. The extracts are collected in a 25-ml flask, made up to mark with isoamyl alcohol, and the absorbance measured at 650 nm. The colored constituent in the extract decomposes on standing. The method can be applied to solutions whose ruthenium to uranium ratio is 1 to 500.

PROCEDURE 43 (211)

Transfer to a 25-ml Erlenmeyer flask the hydrochloric acid solution containing 0.3–8 ppm of ruthenium. Add 40 ml of 50% (v/v) concentrated hydrochloric acid–95% ethanol solution and 15 ml of a 0.2% solution of dithiooxamide in glacial acetic acid. Heat on a water bath at 85°C for 30 min, cool, transfer to a 100-ml flask by thorough washing with a 50% (v/v) 6 M hydrochloric acid–ethanol solution, and make up to the mark with the same solution. Measure the absorbance at 650 nm against a reagent blank.

Thiourea Method

For the determination of about 2–15 ppm ruthenium, thiourea has proved to be a satisfactory reagent. Various authors have made contributions to this method. A qualitative study of the reaction of thiourea and its derivatives was made by Steiger (214). Yaffe and Voight (215) stated that thiourea reduced ruthenium(IV) to ruthenium(III), which subsequently formed the two aforementioned blue-green dissolved constituents $[Ru(SC(NH)NH_2)]_2^+$ and $[Ru(SC(NH)NH_2)_3]^0$.

Pilipenko and Sereda (216) found that the maximum formation of the complex occurred in 6 M hydrochloric acid, and that the complex-forming reaction was

$$[Ru(SC(NH_2))_2]^{3+} + 2SC(NH_2)_2 \leftrightharpoons \text{blue } [Ru(SC(NH)NH_2)_3]^0$$

De Ford (217) used thiourea for colorimetric determinations, and introduced new features into the unusual procedure for separation. Distillations were made from solutions containing perchloric acid and sodium bismuthate. Ruthenium tetroxide was collected in a centrifuge tube containing a solution of sodium hydroxide. Ethanol was added to produce the hydrated oxide of ruthenium, which was subsequently collected by magnesium hydroxide. The mixture was isolated and then treated with hydrochloric acid and potassium iodide. The latter reagent prevented the oxidation of thiourea by ruthenium in its higher valency states. The amount of thiourea added could be varied by as much as ±20%. The optimum acid concentration was 4 N, although considerable excess could be added without changing the absorbance of the final solution. For good reproducibility the time of heating had to be controlled accurately.

Thiourea was used by Bergstresser (218) for the determination of ruthenium in plutonium. The method of separation was similar to that of De Ford (217). The tetroxide was distilled from a perchloric acid medium and collected in a 3 N sodium hydroxide solution from which it was precipitated and collected by magnesium hydroxide as a carrier. Photometric measurements were made at 620 nm of hydrochloric solutions of the mixed precipitate. The standard deviation of a single measurement was 2 μg in the range 40–400 μg for samples containing as much as 100 mg of plutonium.

Ayres and Young (219) used thiourea to determine 1.9–ppm of ruthenium. In order to produce quickly a stable color system, they recommend an initially 6 M hydrochloric acid solution, a large excess of thiourea, the presence of 50% by volume of ethanol, heating for 10 min at 85°C, and a final solution about 4 N in acid. Measurements are made on a band at 620 nm sufficiently broad to permit the use of filters. Although a sharp absorption band occurs at 305 nm, the former is preferable from the standpoint of avoiding interference from associated metals.

The tolerance of the method for iridium, rhodium, platinum, and nickel provides advantages when appropriate alloys or reasonably pure solutions of ruthenium are to be analyzed. The proportions of associated metals which are tolerated, however, scarcely allow for a direct determination. Although platinum can be tolerated, palladium interferes, and because these two metals are generally associated in natural occurrences, the isolation of ruthenium is generally a requirement. Furthermore, thiourea is applied to the usual halide complexes; it is not applicable directly to alkali-receiving liquids and it is sensitive to the presence of nitroso complexes. The latter must be fumed, etc., and under these conditions distillation becomes a requirement.

When distillations are required some advantages may result from the use of a thiourea solution as a collecting liquid for the tetroxide. Thus Westland and Beamish (220) distilled from perchloric acid and collected in a solution of thiourea, ethanol, and hydrochloric acid and heated at 85°C.

Used by an experienced operator, this method can be rapid and effective. Both temperatures and air flow must be regulated to prevent the formation of hydrated oxides of ruthenium in the tubes leading to the receivers. Some difficulty may also be encountered by deposits of sulfur formed by the presence of distilled oxidant. The method is not suitable for oxidations which release gaseous halogens. The following spectrophotometric method, recorded by Ayres and Young (219) is recommended, and can be applied to the hydrochloric acid–ethanol receiving liquids.

PROCEDURE 44 (219)

To a few milliliters of the ruthenium chloride solution, containing from 50 to 400 μg of ruthenium, add 5 ml of concentrated hydrochloric acid, 5 ml of 95% ethanol, and 1 ml of a 10% aqueous thiourea solution. Heat for 10 min on a water bath at 85°C. Cool and make up to a 25.0 ml with a 50% (v/v) 6 M hydrochloric acid–95% ethanol solution. Measure the transmittance at 620 nm against blanks containing the same concentration of reagents.

PROCEDURE 45 (220)

Transfer to the distillation flask (Fig. 3.3) the solution containing ruthenium within the appropriate concentration range. Add to the water trap 20 ml of a 1% potassium permanganate solution. To 15 ml of a 50% (v/v) concentrated hydrochloric acid–95% ethanol solution in the first receiver add 8 ml of a 5% thiourea solution in the 50% acid–ethanol solution. To 5 ml of the 50% acid–ethanol solution in the second receiver add 2 ml of the same 5% thiourea solution. Add to the ruthenium solution in the distillation flask 20 ml of 70–72% perchloric acid, and heat to fumes. Add a further 5 ml of the perchloric acid, and continue heating. Draw the vapors through the permanganate trap. Heat the pot and trap liquids to fuming and boiling respectively for about 15 min. Discontinue the air current, and place water baths, initially at 85°C, about the two thiourea receivers. The blue color forms immediately on contact with the distillate and reaches a maximum in about 15 min. Transfer the receiving solutions to a 50-ml volumetric flask, thoroughly rinse the receivers, and make up the solution to the mark with the 50% hydrochloric acid–ethanol solution. Care must be exercised here to remove all of the receiving liquid, and this is best accomplished by successive washings with small amounts of water. Filter the solution

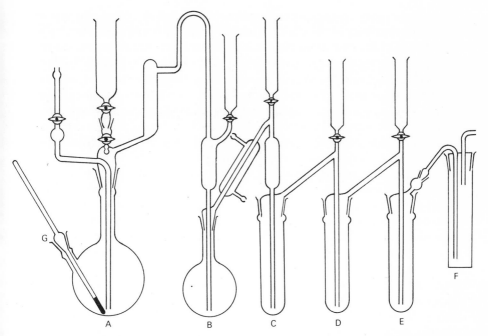

FIGURE 3.3. This apparatus is similar to Fig. 7.1 but is modified to allow for smaller flasks and receivers and to provide a direct line from the distillation flask to the trap when the condenser is removed. A thermometer is inserted through a ground glass joint sealed into either the wall of the distallation flask A or into the ground glass stopper for this flask. A is a 500-ml round-bottomed flask. B is a 125-ml round-bottomed flask. C,D, and E are 50-ml tubes. F is a bubbler containing a thiourea solution to detect incomplete retention by the receiving liquid, G is a thermometer inserted through a ground glass joint.

through an 11-cm Whatman No. 42 filter paper into the absorption cell, and measure its absorbance at 620 nm against a blank of the 50% ethanol–hydrochloric acid solution.

Note. In the presence of osmium, the latter together with ruthenium is distilled by oxidation with perchloric acid, collecting both volatile oxides in a hydrogen peroxide solution. Osmium is selectively distilled from the peroxide receiving liquid to which sulfuric acid is added. The ruthenium in the pot liquid is then removed by oxidation with perchloric acid as previously described or by bromic acid (see Part II).

OSMIUM

Thiourea Method

Despite a relative lack of sensitivity, for which various values have been reported, thiourea remains one of the most useful and best described spectrophotometric reagents for osmium. In this connection one must keep in mind that where distillations or extractions are required, any lack of sensitivity can be compensated to a good degree by the ease of concentrating the osmium content.

This use of thiourea was first proposed by Tschugaeff (221), who described the reaction as specific and sensitive to 0.1 ppm of osmium. The composition of the colored complex in the solid state was given as $Os(NH_2-CS-NH_2)_6Cl_3 \cdot H_2O$. Sulfur dioxide was used to eliminate the interference of nitric acid, which reacted with thiourea, and under the new conditions the sensitivity was recorded as 0.2 ppm. The color developed quickly and was stable. Recently, it has been shown that the red dissolved constituent formed when osmium(VIII) oxide reacts with thiourea in an acidic medium is best represented as the triply charged ion $[Os(NH_2CSNH_2)_6]^{3+}$.

Sandell (222) used thiourea for the colorimetric determination of traces of osmium in meteorites. The tetroxide was distilled and collected in sulfur dioxide–hydrochloric acid. It was stated that the evaporation of the distillate resulted in losses of osmium, and a procedure was devised that eliminated the evaporations (see the following). It was also observed that hexachlorosmate required a prior treatment to produce the colored constituent, consequently in the preparation of standards, the tetroxide was required. The system was found to obey Beer's law, and could be used over the range 1.3 ppm. Because of the yellow color of the thiourea–sulfur dioxide solution, visual methods were less satisfactory than photometric methods.

Allan and Beamish (223) used sulfur dioxide–hydrochloric acid receiving liquids, and found that the rose-colored complex was formed only with fresh distillates. The explanation for this is not known, but it probably involves the slow formation of stable sulfite or sulfate complexes. Sandell's data are not inconsistent with this suggestion, and the explanation for his observation that evaporation of osmium–sulfur dioxide–hydrochloric acid distillates resulted in losses of osmium may be found in the presence of these complexes and not in losses of osmium by volatilization.

Difficulties arising out of the presence of sulfur dioxide resulted in the conclusion by Allan and Beamish that the more satisfactory receiving liquid contained only thiourea and hydrochloric acid. Complete evaporations were then unnecessary. These authors used the thiourea method for the determination of the losses of osmium by fire assay. A narrow band filter centered at 480 nm was used, and the optimum concentration range was 0.4–5 ppm. It was observed that the weight of thiourea required for a fixed amount of osmium depended on the volume of the solution. For solutions of osmium of 5 mg/l, overnight standing was advisable to attain the maximum color density.

The hexabromosmate ion may also be used for the thiourea reactions (224). Ayres and Wells (225) made a spectrophotometric study of the thiourea–osmium system, and recorded a sharp transmittance minimum at 480 nm and a lesser applicable band at 540 nm. For the former wavelength, the optimum concentration is 5–50 ppm. Of the platinum metals, only palladium and ruthenium interfere. The latter, of course, is the more serious, and makes necessary the separation of ruthenium.

An acceptable separation from ruthenium was described by Sauerbrunn and Sandell (226). By this method, ruthenium is maintained in the reduced state by the addition of iron(II) sulfate, while osmium is oxidized by nitric acid and subsequently extracted into chloroform. The latter extract is treated with a sulfuric acid solution of thiourea, and the color measured at 480 nm. The method is suitable for 5 – 100 μg of osmium. Chloride interferes through the formation of chlorosmate, which resists oxidation to the octavalent state. When the presence of chloride is unavoidable, it can be removed by gentle fuming with sulfuric acid in the presence of iron(II) sulfate. Acceptable results are obtained in the presence of large proportions of iron(II), copper(II), palladium(II), and platinum(IV).

Various substituted thioureas and thiosemicarbazides have been proposed as colorimetric reagents for osmium. Geilmann and Neeb (227) proposed *o-di(o-tolyl)thiourea*. Steiger (228) recorded the reactions of some 21 compounds of the thiourea type, and concluded that color formation with osmium and ruthenium depended upon the C=S group acting in conjunction with one or two —NH$_2$ or —NHR groups. Bardodej (229) also considered C=S as a functional group in substituted thiourea reactions with osmium and ruthenium. Some sixty substituted thioureas were examined by Yoe and Overholser (230) with a view to securing greater sensitivity in color reactions, and it was concluded that, in general, these "offer no advantage to justify their use in place of thiourea." In the present authors' opinion this remains true.

The thiourea reaction is not sensitive to acidity or reagent concentration. Maximum color is developed in 5 min at room temperature for osmium tetroxide solutions in 2 – 4 M hydrochloric acid, and similarly in sulfuric acid. Indeed the latter acid is a preferable medium. Hydrochloric acid solutions of chlorosmates should be avoided because color development is quite slow; the rate may be increased by the addition of tin(II) chloride.

PROCEDURE 46 (150)

Apparatus

The distillation apparatus is shown in Fig. 3.4. In this apparatus the thistle tube for addition of reagent solution is fused into a ground-glass connection, so that it is easy to remove any insoluble material from the flask at the end of the distillation and examine it for osmium. The distillation flask may have a volume of 250–500 ml. The two receiving tubes are 20 and 30 mm in diam, respectively.

Reagents

Osmium tetroxide solution. A 0.005% solution in 0.1 N sulfuric acid. Prepare this solution by diluting a stronger one, which may be obtained in the following manner. Make a number of scratches with a file on a 0.5-g ampoule of osmium tetroxide, and weigh the

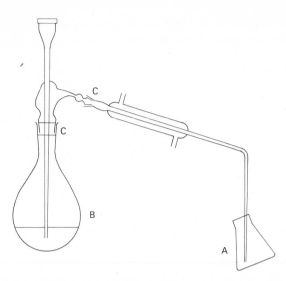

FIGURE 3.4.   A is a 100-ml wide-mouth Pyrex Erlenmeyer flask. B is a 500-ml Pyrex round-bottomed flask. C are ground glass joints. See Procedure 46 for details.

ampoule. Drop the ampoule into a 200-ml glass-stoppered bottle containing about 50 ml of water. Break the ampoule by shaking the bottle, and when the osmium tetroxide has dissolved, decant off most of the supernatant liquid into a volumetric flask (e.g., 250-ml). Rinse the bottle well with successive portions of water and transfer these to the volumetric flask, taking care to leave all the glass fragments in the bottle. Then transfer the glass fragments to a weighed filter crucible, and obtain the weight of the whole after drying. The weight of the osmium tetroxide used to prepare the stronger solution is thus obtained by difference.

Potassium permanganate solution.   This should be a 5% solution.

Thiourea solution.   This should be a 10% aqueous solution.

Sulfuric acid–thiourea solution.
(1) Five percent thiourea solution in 2 N sulfuric acid.   This is to be used when osmium tetroxide is separated by distillation.
(2) One percent thiourea solution in 1 N sulfuric acid.   This is to be used when osmium tetroxide is separated by extraction (Procedure 47). Both solutions are best prepared fresh daily.

Procedure

Add 0.5 ml of the thiourea solution to about 10 ml of sample solution, $0.5-4\ N$ in sulfuric acid, which may contain $10-1000\ \mu$g of osmium. If osmium is present as chlor- or bromosmate, add also 0.10 ml of a 10% tin(II) chloride solution in 1:4 hydrochloric acid (the hydrochloric acid concentration of the sample solution should not exceed $1\ M$). In the latter instance, heat the mixed solution on a water bath for 15 min; otherwise (with osmium present as the tetroxide) simply allow the solution to stand at room temperature for 10 min. Cool to room temperature, and dilute with water to 25 ml, or to a smaller volume when only small amounts of osmium are present. Determine the transmittance at 480 (or 540) nm or with a green filter against a reagent blank.

Procedure When Osmium Is Separated by Distillation

The sample solution should have a volume such that it is ready for distillation, after the addition of nitric acid and permanganate, the total volume is less than 50 or 60 ml. Chlorides must be absent, and if a permanganate oxidation is necessary the solution should be about 1 N in sulfuric acid.

Transfer the solution to the distillation flask (see Fig. 3.4), and if iron(II) or other reducing substances are present, add some potassium permanganate solution until a 1-drop excess is present as indicated by the color change; avoid getting permanganate on the neck of the flask. Next add approximately 50 mg of ammonium iron(II) sulfate hexahydrate to destroy the permanganate and higher oxides of manganese. The volume of the solution at this point should be 35–40 ml.

Add a few small grains of pumice, connect the flask to the condenser, and heat the solution slowly to near the boiling point to make certain that higher manganese oxides have been brought completely into solution. Dip the end of the condenser into 10 ml of the sulfuric acid–thiourea solution contained in a large vial or test tube marked to indicate 20 ml. Add 15 ml of concentrated nitric acid through the inlet tube of the flash, and distill at such a rate that 10 ml of distillate are collected in 10–15 min. Transfer the distillate to a 25-ml volumetric flask, rinsing the condenser and receiver with a few milliliters of water, and make up to the mark with water. Determine the transmittance of the solution at 480 or 540 nm against a blank carried through the complete procedure. In constructing the standard curve, add 0.25 and 0.5 μg of osmium tetroxide to distillates obtained from osmium-free nitric acid mixtures as described above.

Note. If the amount of osmium is likely to be less than 10 μg, use 5 ml of the sulfuric acid–thiourea solution contained in a 25-ml graduated cylinder for collecting 10 ml of the distillate. Read the volume of the solution in the graduate (which has been checked for accuracy), and determine the transmittance as described above.

The Isolation of Osmium Tetroxide by Extraction

PROCEDURE 47 (150)

The volume of the chloride-free solution should be such that when ready for extraction, it does not exceed about 50 ml. Preferably more than 5 μg of osmium should be present. If osmium is not in the octavalent form (e.g., when iron(II) salts are present), add 5 ml of 6 N sulfuric acid and 2.5 ml of 85% phosphoric acid to be approximately 25 ml of solution. Then add dropwise a solution of potassium permanganate until a pink color appears. (If much iron(II) is present, the potassium permanganate may be added as a finely ground solid to keep the volume of the solution as small as possible.) Discharge the permanganate color with 25 mg of ammonium iron(II) sulfate, added as solid or in solution. If the color is not completely discharged, add another similar amount of the iron(II) salt. The solution should be free from turbidity or insoluble material. Do not allow it to become warm during the oxidation, lest some osmium tetroxide by volatilized. Transfer the solution to a separatory funnel, and add 15 ml of concentrated nitric acid (or enough to make its concentration of 5 M). Cool the solution to 20–25°C, add 10 ml of reagent-grade chloroform (or carbon tetrachloride), and shake vigorously for 2 min. Allow the phases to separate, and deliver the chloroform layer into a dry glass-stoppered bottle or flask. Repeat the extraction with two or three 10-ml portions of chloroform, depending on the recovery desired.

To the combined chloroform extracts, add exactly 5 ml of the 1% thiourea solution in 1 N sulfuric acid, and shake vigorously for 5 min (a mechanical shaker operating at about 3 counts/sec is recommended). Separate the phases, and transfer the clear aqueous layer to a 1-cm absorption cell; obtain the transmittance at 480 nm in a spectrophotometer or filter photometer. If the red color is strong, dilute the solution with water to an appropriate volume before reading. Establish the standard curve by treating known amounts of osmium tetroxide in water with thiourea; the final concentration of the latter may be about 1% in 1 N sulfuric acid, but the concentrations are not critical.

1,4,6,8-Naphthylaminotrisulfonic Acid Method

For the range determination of 0.1–8 ppm of osmium, 1,4,6,8-naphthylaminotrisulfonic acid was used by Steele and Yoe (231) to produce a water-soluble, purple complex which changes to violet at pH 1, light blue at pH 6, and pale green at pH 8. Maximum absorbance at 555 nm is attained between pH 1.0 and 1.5. The sensitivity of the reaction is 0.0068 μg cm^{-2}. At room temperature, the color is stable for weeks; excess of reagent is without measurable effect, but there is some decomposition and a resulting change in absorbance after a month's standing. Maximum color development requires about 1 hr at 35°C. Because the associated platinum and base metals interfere, a separation is required, and osmium tetroxide is distilled from a nitric acid medium and collected in a potassium hydroxide solution.

PROCEDURE 48 (231)

Reagents

Reagent solution. Recrystallize 1,4,6,8-naphthylaminotrisulfonic acid from the minimum amount of warm water by the addition of isopropanol. Dissolve the product in distilled water to produce a 0.02% solution.

Buffer solution. Mix 97 ml of 0.2 M hydrochloric acid and 50 ml of a 0.2 M potassium chloride solution and dilute to 200 ml. The pH is 1.0.

Procedure

Transfer the sample to the distillation apparatus (Fig. 3.5), and regulate to ensure a continual flow of air. Add 6 *M* nitric acid if the sample is solid and can be thus dissolved. Resistant samples may require chlorination; see Procedure (115). A total of 25 ml of nitric acid is usually sufficient. Heat slowly to boiling, and boil for 30 min, distilling the tetroxide into 15–20 ml of a 0.05 *N* potassium hydroxide solution. For about 0.1 mg of osmium, 5 ml of the potassium hydroxide solution is sufficient. Transfer the distillate to a 25-ml volumetric flask, and add 5 ml of the 0.02% aqueous reagent solution. Add 0.1 *M* hydrochloric acid dropwise to a pH of 2.5–3. Warm in a water bath for 45 min, add 10 ml of the pH 1.0 buffer, and dilute to the mark with distilled water. Measure the absorbance at 555 nm, and compare with a standard curve similarly prepared, including the distillation.

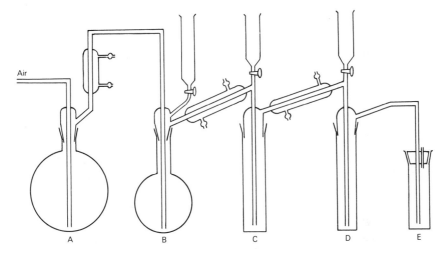

FIGURE 3.5.   A is a 500-ml flask, B is a 125-ml flask, and C and D are 40-ml flasks. E is a thiourea bubbler (see also Fig. 3.3).

Equipment in Fig. 3.3 may be substituted. If acid is to be added to the distillation flask after the dissolution of the osmium, the flask illustrated in Fig. 3.3 must be used.

CHAPTER 4

GRAVIMETRIC DETERMINATIONS

INTRODUCTION

The character of the natural sources of platinum metals and the requirements of the analytical laboratories involved in the determination of these metals have encouraged the development of the more rapid empirical analytical methods and the rejection of the relatively slow, but more accurate, precipitation procedures. Nevertheless, the gravimetric methods not only retain their status as the final arbiter of quantitative composition, but also provide a pool from which refining processes may be improved. Up to the present time, many of the large scale separational methods involved some type of selective precipitation and, of course, in most instances the isolation of each of the platinum metals requires an adoption of some gravimetric method. Although the separational value of precipitation methods will eventually give way to such techniques as ion-exchange separation, it is unlikely in the foreseeable future that either the platinum metals industries or the research analyst will be able to dispense with the good gravimetric methods. It is unfortunate that there are few available methods, either gravimetric or empirical, for the rarer platinum metals, while for palladium the surfeit of analytical methods is so great that even the specialists in this field are quite unable to choose with certainty the most generally useful procedure.

GOLD

In general the quantitative recovery of metallic gold requires exacting techniques. To avoid any loss of the fine, but heavy particles of gold, one must recognize the effects of acidity, rate of addition of precipitant, period of digestion, and particularly the tendency for these particles to adhere tenaciously to the beaker wall. The techniques recommended below include the initial addition of the precipitant by means of a fine capillary and the final transference of gold in the light of a flashlight placed upright beneath the beaker, assisted further by a small piece of ashless paper moved about by a stirring rod drawn to a fine stem that terminates in a small glass globule. Alternatively, one may use a filtering feather sealed directly into a glass rod. As one would guess from the pertinent reduction potentials, the tendency toward the formation of fine particles will vary with the reagents used.

The optimum heating range for precipitated gold has been the subject of various investigations. Ishii (232) recommended temperatures above 230°C. Kiba and Ikeda (233) used 700°C. Champ *et al.* (234) recorded useful data which provide some explanation for the variety of optimum ignition temperatures. As one would expect, occluded organic reductant will have some influence, but a hitherto unknown phenomenon was observed by Duval in that ''gold retains an amount of oxygen by adsorption, which varies according to the degree of subdivision, the nature of the precipitating reagent, and the environmental atmosphere during heating. When this phenomenon takes place the gain in weight is about 1%. It is reversible, and if the cooling curve is identical with the heating curve, exactly the same amount of oxygen is released as was taken up on heating.'' This characteristic is made clearly evident by heating gold precipitated by pyrogallol.

The most generally used quantitative methods for gold involves a reduction to the metal. As one would expect from the reduction potentials of gold salts, this precipitation can be accomplished with many reagents. Thus standard solutions of gold exposed to a laboratory atmosphere may be affected by the constituents of dust. The most effective quantitative reductants are organic compounds. Among those for which procedures have been recorded are oxalic acid, dimethylglyoxime, formic acid, hydroquinone, and a variety of aminophenol derivatives.

To a greater or lesser degree each of these and other reductants are effective only in the absence of dissolved nitrogen oxides. These are usually present when the material to be analyzed is initially in the solid form as the metal or an oxide which requires aqua regia for dissolution. Various techniques have been proposed for the removal of nitric acid by hydrochloric acid. The problem is made difficult by the ease of reduction of gold salts, and to avoid this sodium chloride is frequently added to produce sodium chloroaurate. With this method a few milligrams of sodium chloride are added to the evaporating solution, the final residue being treated repeatedly with hydrochloric acid and intermittently evaporated. A common error in this method is to add unnecessarily large volumes of hydrochloric acid. The removal of nitrogen oxides can be accomplished over steam within a few minutes.

When the presence of sodium chloride is objectionable, two methods of removing nitrogen oxides can be used. When great care is exercised, the above evaporations with hydrochloric acid can be accomplished in the absence of salt. In this instance the wet residue is held over the steam until dryness appears at the circumference of the layer of wet gold salt. With experience this layer will become dry after the removal of the beaker from the steam bath, or removal to a distance of a few inches from the metal surface of the steam bath. Three repetitions of this treatment normally are sufficient except where small amounts of gold are to be reduced by oxalic acid.

The preferable alternative method has been used by Pollard (235). He dissolves small amounts of gold in 2 drops of nitric acid and 6 drops of hydrochloric acid in a porcelain crucible that is then placed in a 150-ml beaker. The latter is closed by a stopper through which passes two tubes, one adjusted near the level of the crucible and the second to be attached to a suction pump with the opening close to the tope of the beaker. The first tube should be connected to a purified air supply. This air impinging on the surface of the aqua regia quickly removes the oxides of nitrogen. This method was used by Pollard for titrimetric applications, but insufficient data are available to recommend its general use for gravimetric purposes. The method is useful, however, for reagents such as hydroquinone that are relatively insensitive to the presence of nitric acid.

Oxalic Acid

Oxalic acid is one of the most frequently used reductants for gold. In 1841 Kemp (236) used this reagent to separate gold from platinum, and over a century later Magdalena (237) recorded directions for this precipitation. Malowan (238) recommended the addition of morpholine oxalate to the oxalic acid solution to produce an easily handled precipitate. With oxalic acid there is the advantage that excess reagent presents no problem and the reagent is stable and readily obtained in a pure state. On the other hand, a successful precipitation is peculiarly dependent upon the prior history of the gold solution. Samples that have been treated with aqua regia and subsequently with hydrochloric acid yield exceedingly fine precipitates, sometimes with a marked tendency to adhere tenaciously to the beaker wall; and not infrequently the precipitation is incomplete. These difficulties become much less evident when oxalic acid is used to precipitate gold from solutions of its pure salts or when the amounts of gold to be precipitated are in excess of a few milligrams. The exact cause of this difference in behavior is not clear. Anyway, this method usually required extended digestion, and the following procedure is recommended only for pure gold salt solutions or for amounts of gold exceeding 10 mg if the gold has been previously treated with aqua regia and then evaporated in the presence of hydrochloric acid.

PROCEDURE 49 (239)

Dilute the gold solution, previously treated to remove all nitrogen oxides, to 100 ml with distilled water. Add 2 ml of concentrated hydrochloric acid and 5 drops of concentrated sulfuric acid. Add 2 or 3 ml of a 10% oxalic acid solution and place on a steam bath overnight. Transfer the supernatant liquid to a 7-cm filter paper or an A2 filtering crucible. Wash the residue from the beaker with a fine stream of distilled water. Thoroughly clean the wall of the beaker with a small piece of filter paper, with intermittent washing and transference of the residue. This technique is particularly essential with oxalic acid as the precipitant. Transfer the paper and residue to a porcelain crucible and ignite to the metal. Cool the crucible for 15 min in air, then 20 min in a closed container. Allow the crucible to stand on the balance pan for 20 min, then weigh.

Hydroquinone

Among the reagents of this type hydroquinone has proved to be generally applicable. It is an effective reductant of gold for the analysis of fire assay beads (240). As compared with the older reducing reagents, there is less interference from platinum and palladium. The latter metal may be removed directly from the filtrate with dimethylglyoxime in which instance the subsequent determination of platinum requires the destruction of organic matter. A precipitation by zinc, however, may be applied in the presence of organic matter. Compared to oxalic acid the interference of nitrogen oxides is less critical with hydroquinone. The removal of these oxides is recommended, however. Precipitation by hydroquinone can be accomplished in the cold, in which instance about 2 hr contact is required.

PROCEDURE 50 (240)

Dissolve the gold sample in the minimum volume of aqua regia. Add a few milligrams of sodium chloride, and evaporate to dryness on a steam bath, taking care to avoid baking. Add a few drops of concentrated hydrochloric acid, and evaporate again. Repeat this thrice. Add 15 ml of water, and filter through a 7-cm filter paper. Wash with water to a volume of about 50 ml. For highest accuracy, burn the paper, and treat the ash with aqua regia and sodium chloride. Evaporate the solution thrice in the presence of hydrochloric acid, then dilute with a few milliliters of water and filter into the original gold solution. Heat the filtrate to boiling, and add by a capillary tube a few drops of a 5% aqueous hydroquinone solution (see Note). With continued boiling add 3 ml of the hydroquinone solution per 25 mg of gold. Continue heating for 20 min, cool, and filter through a porous-bottomed A2 filtering crucible. Alternatively use a 7-cm filter paper. Wash the precipitate with 100 ml of hot water. Take precautions at this point to see that all of the gold is collected. Fine particles remaining in the bottom of the beaker can best be observed by means of a flashlight. Use small pieces of ashless paper to remove any small particles of gold that adhere to the beaker wall. Perhaps the most persistent cause of imprecision and inaccuracy is the failure to observe the very fine particles of gold remaining in the beaker! Ignite carefully and weigh as metallic gold.

Note. This solution retains its reducing efficiency for as long as a week.

Procedures have been recorded for the application of hydroquinone to the analysis of alloys of copper, nickel, zinc, and gold (240), for the separation of gold from tellurium and selenium (241), and for the analysis of mill cyanide solutions (241).

Sulfur Dioxide

Sulfur dioxide should not be used in the presence of selenium, tellurium, and lead with which gold is frequently associated. With appreciable amounts of platinum and palladium there is significant coprecipitation, and a reprecipitation is necessary for which oxalic acid is suitable if the amounts of gold are of the order of 10 mg or more; for smaller amounts reprecipitation by hydroquinone is recommended. In general sulfur dioxide is the preferred reagent for large amounts of gold because hydroquinone adds an organic reaction product to the filtrate. Sodium sulfite may be used as the source of the sulfur dioxide, but the addition of alkaline salts may be objectionable in subsequent treatments of the filtrate.

PROCEDURE 51

To 100 ml of the gold solution, free from nitric acid and containing 500 mg or less of the metal, add 3 ml of concentrated hydrochloric acid. Add 25 ml of a saturated sulfur dioxide solution, and digest on a steam bath for 1 hr. Add a further 5 ml of sulfur dioxide solution or sufficient to provide a strong odor of the gas (see Note). Set aside to cool. Transfer the supernatant liquid to a porous porcelain filtering crucible or to an ashless filter paper. Wash the precipitate by decantation with 0.1 M hydrochloric acid, and then transfer the precipitate with a thin stream of the acidic wash solution. Remove the gold adhering to the beaker wall by repeated wiping with a small piece of filter paper. Finally, wash to the disappearance of the thin line of gold precipitate that may be observed in the tilted beaker by means of a strong light. Ignite, cool, and weigh. If reprecipitation is required, dissolve the ignited residue in aqua regia, and treat as described above with hydroquinone or with oxalic acid.

Note. Hecht and Lamac-Brunner (242) used a stream of sulfur dioxide gas passed over the surface of the gold solution while the latter was shaken.

PALLADIUM

Dimethylglyoxime*

Undoubtedly the most important type of precipitant for palladium is the oxime family. The applications of dimethylglyoxime to quantitative analysis were first recorded by Duparc (243) and Wünder and Thuringer (244) who found that the reagent quantitatively precipitated palladium and nickel, the former from acidic media and the latter in alkaline media. The precipitation was carried out in the cold, and the reagent was dissolved in dilute hydrochloric acid. The palladium complex was ignited to the metal. In one method (245) dimethylglyoxime was used for the precipitation of gold and palladium, heating the solution in the presence of excess of reagent to precipitate gold. Gutbier and Fellner (246) used dimethylglyoxime to separate palladium from tin. Davis (247) thus separated palladium from platinum with no contamination by the latter. Gol'Braikh (248) used dimethylglyoxime to determine small amounts of palladium in the presence of large amounts of nickel. Zschiegner (249) used sodium nitrite to separate gold, silver, and base metals from palladium, with the latter then being determined by the direct weighing of the palladium–dimethylglyoxime complex. Holzer (250), however, was unable to recommend direct weighing because the complete removal of excess of reagent resulted in some loss of precipitate. Whereas the use of a large excess of dimethylglyoxime will undoubtedly introduce errors, the amount of reagent can be so regulated that the direct weighing of the precipitate is in general the more acceptable procedure, and the metal is thus seldom used as a weighing form.

The optimum acidity for the precipitation of the palladium–dimethylglyoxime complex has been a subject of dispute. A $2\,M$ solution was considered optimum by Peshkova and Shlenskaya (251). Morachevskii *et al.* (252) state that concentration of nitric, hydrochloric, and perchloric acids higher than $1.5\,M$ reduce the completeness of precipitation; a similar effect occurs with $2N$ sulfuric acid. Precipitation in near neutral solutions was advocated by Filott *et al.* (253). These

*Use of reaction of the induced precipitation of nickel dimethylglyoxime for determination of palladium microconcentration (242a).

authors claimed a higher selectivity for palladium at pH 5.5, and they avoided interferences from associated cations by the additions of EDTA. Under these conditions palladium could be precipitated in the presence of ammonium, which in basic media is an interfering constituent.

The effect of concentrated hydrochloric acid on the precipitation of the palladium–dimethylglyoxime complex has been discussed by Sharpe and Wakefield (254) who have identified certain of the products. Although there has been no agreement as to the optimum acidity for gravimetric work, the present authors have been able to secure good selectivity and quantitative recovery in 0.2 M hydrochloric acid solutions.

The safe heating range for the dimethylglyoxime complex was recorded as 45–171°C by Champ *et al.* (255), and as 100–200°C by Tashiro (256). The efficiency of this gravimetric method as regards both the determination of palladium and its separation from other platinum metals was discussed by Ayres and Berg (257). They found solubility losses to be insignificant except where small amounts of palladium were to be determined. Separations from platinum, rhodium, and iridium were effective except when large proportions of palladium were precipitated.

A method for the homogeneous precipitation of palladium dimethylglyoximate was developed by Gordon and co-workers (258). To 3% hydrochloric acid were added excesses of biacetyl and hydroxylammonium chloride. The procedure was developed for a maximum of 25 mg of palladium, and a standing period of at least 16 hr was required. Although there was improvement in the case of filtration, the method offered no significant advantages over the conventional method.

Concerning the question of the contamination of the palladium precipitate it has been the authors' experience that the character of the prior treatment of the platinum metal will somewhat determine the degree of coprecipitation. Undoubtedly, dissolution processes in hot sulfuric acid encourage the simultaneous precipitation and perhaps the coprecipitation of the platinum–dimethylglyoxime complex (259). Boiling hydrochloric acid solutions of the mixed platinum and palladium salts will also encourage the coprecipitation of the blue-bronze, anisotropic platinum dimethylglyoximate crystals. Serious instances of contamination are usually detected by a green tint in the yellow palladium complex (260). A rather peculiar interference with the precipitation of the palladium–dimethylglyoxime complex is encountered in filtrates resulting from the removal of gold by hydroquinone (261). In this medium a certain threshold value of the palladium concentration is required before palladium can be precipitated directly (see the following).

The problem of dissolving an impure dimethylglyoximate is not a simple one. The usual process of dissolving the precipitate in aqua regia and reprecipitating is not acceptable. Gilchrist (262) has shown that the process results in small but significant losses than can be recovered from the filtrate by fuming with sulfuric acid. With large amounts of palladium fuming with sulfuric may result in the appearance of a palladium oxide resistant even to aqua regia. On the other hand, ignition processes frequently involve losses by volatilization of the metal–dimethylglyoxime complex or one of its ignition products. Not infrequently it is advisable to ensure purification of the initial palladium dimethylglyoximate precipitate, and where this is not desirable one should arrange prior treatment so as to avoid extensive contamination.

The authors' experience is in agreement with Ayres's findings (257) concerning the solubility losses. With gram amounts of palladium, the losses need not be significant even though the presence of palladium in the filtrate can be detected. With small amounts of the metal there is the mechanical difficulty of collecting the finely divided precipitate. This process is, of course, assisted by a comparable reduction in the volume of the liquid. The fact that the palladium can be practically completely converted to the dimethylglyoxime complex is attested by its application, through solvent extraction, to submicrodeterminations.

Although ignition of the palladium complex is not now generally used for gravimetric purposes, there are some, including the present authors, who prefer the method. The difficulty associated with the danger of volatilization can be avoided by initial heating with a Bunsen burner. This permits a greater manual control than muffle ignition. A technique, used with success by the authors, involves wrapping the complex and filter paper in a second wet paper and igniting, without preliminary slow drying, in the full heat of the Meker burner. It may be that by this process the complex is dissociated before there is an opportunity for volatilization. However unorthodox the method may be, it is an effective method of achieving accuracy with this rather difficult ignition.

Undoubtedly, an attempt by the analytical chemist, inexperienced in the field of platinum metals, to select the most efficient of the various oxime reagents for palladium must be an unrewarding experience. The multiplicity of claims for superiority, the lack in almost every instance of data concerned with dissolution processes required in practice, and the absence of properly determined degrees of interference make the tasks an almost impossible one. In the present authors' opinion, there is little to choose between the methods. Dimethylglyoxime remains the preferred reagent for most purposes; it is readily available, inexpensive, sufficiently sensitive, and selective, and the precipitate can be handled with ease.

Concerning the question of its solubility in water, little difficulty is experienced in removing excess of reagent from the palladium–dimethylglyoxime complex. In situations where water solubility really does become an important consideration, the authors have found no difficulty with the sodium salt, a reagent which seems to have been peculiarly neglected for the precipitation of palladium.

As stated above, platinum may under certain circumstances interfere with the palladium–dimethylglyoximate precipitation. With the normal course of dissolution by aqua regia and conversion to chloride, however, there is little tendency for the indigo-blue or bronze platinum complex to precipitate. Osmium, ruthenium, iridium, and rhodium salts dissolved in dilute hydrochloric acid or dilute sulfuric acid do not produce a precipitate on standing for several days in contact with an alcoholic dimethylglyoxime solution, so that any interference from these associated metals would only occur through some type of adsorption or occlusion.

It has been stated that iron(III) tends to reduce the completeness of precipitation (252) particularly in a nitric or perchloric acid medium, and this effect increases with decreasing acidity.

PROCEDURE 52

Add to the 0.25 M hydrochloric or nitric acid solution containing 5–25 mg of palladium an excess of an aqueous sodium dimethylglyoximate solution (about 8 mg of precipitant per milligram of palladium). The total volume should be about 100 ml for 5 mg and 200 ml for 25 mg of palladium. Stir, and place on a steam bath for 15 min, then allow to stand for 45 min, during which time the orange-yellow precipitate should coagulate and settle. Filter the supernatant liquid through a 3-ml capacity A2 porcelain filtering crucible. Transfer the coagulated precipitate by a glass rod drawn to a suitable diameter, then remove the final precipitate, and clean the beaker wall with distilled water and a rubber policeman. Wash with about 200 ml of water, dry at 110°C, and weigh. If a reprecipitation is required, place the filtering crucible and contents in a 200-ml beaker. Add to the crucible 2 ml of concentrated sulfuric acid and a few drops of concentrated nitric acid. Warm on a steam bath to oxidize the organic matter, add 10 ml of water, warm the solution, and remove the crucible with platinum-tipped tongs, washing well with water. Suck clean and wash the crucible, and transfer the resulting liquid to the original solution. Evaporate the solution to fumes, add a few drops of nitric acid, evaporate again to fumes, add a few milliters of water, and again evaporate to fumes. Adjust to 0.25 M in acid, filter the solution, and reprecipitate.

Note. The process of dissolving the precipitate can be assisted if the initial filtration is made through a 7-cm filter paper. Dissolution of the precipitate by aqua regia, dilution with water, and reprecipitation may be used when high accuracy is not required. It should be noted, however, that the sulfuric–nitric acid method for destroying organic matter will encourage the precipitation of platinum dimethylglyoximate. When platinum contaminates the initial precipitate of palladium dimethylglyoximate, the latter should be ignited, with the metal subsequently dissolved in aqua regia. It should also be noted that continued fuming with sulfuric acid in the presence of much palladium may produce a red-brown palladium deposit. This can be avoided by a careful fuming, and anyway it should redissolve upon the addition of water, and boiling.

IGNITION

Very little information has been directly recorded in the literature with respect to the danger of loss of metal by burning palladium dimethylglyoximate. Tacit recognition of this difficulty is made in the various suggestions, such as wrapping the precipitate and paper in a second paper before ignition.

The following two procedures are recommended:

1. Transfer the well-wrapped paper and precipitate to a porcelain crucible of suitable size, and place this crucible within a larger porcelain crucible. Ignite with a microburner or a small flame to produce a thin line of smoke. This should be accomplished simultaneously with the drying of the paper because complete drying before ignition invites sudden combustion and an almost certain loss of metal, even though the filter has been wrapped in a second paper. Upon ignition to an ash complete the combustion with a full flame.

2. Transfer the wet paper and its contents to a porcelain crucible placed within a second crucible, and immediately subject the crucible to the full flame from a 7-in. Meker burner. Continue until the char is ignited to ash. If this manipulation is properly made, there is no visible evidence of the emission of carbon

Nioxime

Nioxime is water soluble and produces with palladium a yellow compound that is somewhat more insoluble than the dimethylglyoximate. The minimum excess of reagent is 30% and the applicable range of concentration is about 5–30 mg of palladium. The selectivity is about the same as that of dimethylglyoxime, e.g., there is interference from gold, and under certain conditions from platinum.

PROCEDURE 53 (263)

Adjust the volume of the palladium solution to 200 ml. Add hydrochloric acid to a pH of one, and heat to about 60°C. Add dropwise a 30% excess of a 0.8% aqueous solution of nioxime (0.43 ml/mg of palladium), and digest with occasional stirring at 60°C for 30 min. Filter through an A2 filtering crucible, and wash with hot water. Dry at 110°C for 1 hr, and weigh.

PLATINUM

Hydrogen Sulfide

In general the most useful method for isolating platinum from the remaining platinum metals involves an oxidation by bromate to platinum(IV) followed by a hydrolysis with sodium hydrogen carbonate. By this procedure the filtrate containing platinum also contains dissolved sodium salts. Prior to the hydrogen sulfide treatment, platinum filtrates are evaporated with hydrochloric acid to destroy bromate with the result that appreciable amounts of sodium chloride are associated with the dissolved platinum.

The platinum metals belong to the acid sulfide group, and consequently sulfide is not a selective precipitant. For quantitative purposes precipitation as sulfide is applied extensively only to rhodium and platinum. For the latter metal it is one of the oldest recorded methods. It was used by Berzelius in 1826 (264). Over this long period little data have been recorded concerning the mechanism of the precipitation, and today we have little more than an empirical approach to the problem of an accurate recovery of platinum sulfide.

In 1896 Antony and Lucchesi (265) considered the precipitate obtained at 90°C to be pure PtS_2; the mixed brown precipitate obtained at room temperature yielded H_2S on heating until at 200°C PtS_2 was produced. Some two decades earlier, von Meyer (266) thought the precipitated sulfide to be a "loose" compound of platinum disulfide and hydrogen sulfide.

In 1950 Jackson and Beamish (267) provided some evidence for the latter hypothesis. These authors formed the sulfide by adding to hexachloroplatinic acid a saturated aqueous solution of hydrogen sulfide, 0.1 M in hydrochloric acid. The precipitate was washed consecutively with water, ethanol, pyridine, and ether, and was dried at 100°C. The composition of the precipitate was $PtS_2 \cdot H_2S$; when heated in nitrogen between 150 and 250°C, it evolved hydrogen sulfide. The thermogram of platinum sulfide published by Champ et al. (268) failed to indicate the above composition, although the procedure which was used to produce the sulfide was the usual "gassing procedure."

One may arrive at certain tentative opinions concerning the character of the precipitate by examining the polemical discussion provided by Feigl (269) and by Kolthoff and Pearson (270). To account for the contamination of copper sulfide by zinc sulfide, the latter rejected Feigl's suggestion that the mass law was inoperative and that the explanation lay in the production of coordinated compounds. Kolthoff provided evidence to support the view that solubility equilibria were involved and that the supersaturation of zinc sulfide was more or less slowly reduced, partly by the selective adsorption of hydrogen sulfide on the surface of the copper sulfide precipitate. The mechanism of this adsorption is not explained, but if there is an analogy with Jackson's data (267) one may entertain the conception that these associations of hydrogen sulfide are not always simple adsorption phenomena.

The early literature contains a variety of procedures for the precipitation of platinum sulfide (172, 271, 272), some of which are characterized by unacceptable techniques and unnecessarily complicated procedures (273, 274). An interesting example of the homogeneous precipitation of the sulfide, recorded some 30 years ago by Doht (275), involved the production of hydrogen sulfide by adding dihydrogen sodium phosphate and sulfurous acid to the hydrochloric acid solution of platinum. The reactions resulted in the formation of some sulfur.

The essential features of most of the early procedures are incorporated into three very similar standard methods (172, 271, 272). It has been the present authors' experience that the application of each of these procedures to solutions of platinum that had been evaporated in the presence of sodium chloride during their preparation yielded high results. It was found also that (a) the accuracy obtained with samples of hexachloroplatinic acid previously evaporated with sodium chloride depends on the concentrations of sodium chloride and hydrochloric acid; (b) accurate results are obtained in the presence of sodium chloride if there has been no evaporation or boiling with hexachloroplatinic acid.

As a result of a critical examination of these methods, Jackson (267) reported that the positive error could be avoided by the addition of ammonium chloride to sodium chloride solutions of platinum prior to the evaporation required for the removal of nitric acid. In addition, it was found that the requirement of prolonged gassing with hydrogen sulfide was unnecessary. Complete and easy recovery was effected by the addition of hydrochloric acid solutions saturated with hydrogen sulfide. Furthermore, contrary to some opinions, the positive error usually encountered after ignition was not due entirely to the retention of sulfur by the platinum; the contribution made toward this positive error by the presence of sodium chloride was probably the result of some type of coprecipitation of a complex platinum constituent, and its elimination by proper treatment with ammonium chloride was presumably a type of replacement phenomenon. Anyway, the authors were able, with the following procedure, to obtain accurate results.

PROCEDURE 54 (267)

To the aqua regia solution containing about 10 mg of platinum, add 1 g each of sodium chloride and ammonium chloride. Evaporate to dryness on a steam bath. Add 10 ml of distilled water, and repeat the evaporation to dryness. If maximum accuracy is required, dissolve this residue in water, filter, and again evaporate the filtrate to dryness. Add a solution made up of 20 ml of 1 N hydrochloric acid and 30 ml of water saturated with hydrogen sulfide, and place on the steam bath until the precipitate has settled. Alternatively, acidify the platinum solution, boil, and pass in hydrogen sulfide for 30 min. Cool the mixture, filter through a 7-cm Whatman No. 42 paper, and wash well with a 1% aqueous ammonium chloride solution. Fold the paper carefully, and transfer it to a minimum-sized porcelain crucible. Dry, and ignite carefully at 800°C for 1 hr. Cool in air, and leach with the 1% ammonium chloride solution. Filter the leaching liquid through a second small paper, and transfer the paper to the original crucible for a second ignition at 800°C for 1 hr. Cool the platinum residue in a desiccator over a saturated calcium nitrate solution, and weigh.

The Standardization of Platinum Solutions by Hydrogen Sulfide

Evaporate to dryness the aqua regia solution of platinum on a steam bath. Add a few milliliters of water, and again evaporate; repeat to remove oxides of nitrogen. Add 25 ml of water, filter, wash to 50 ml, and proceed as described above.

In the absence of sodium chloride the platinum sulfide method provides very accurate results.

Dimethylphenylbenzylammonium Chloride

This reagent was used by Ryan (276) to produce the orange-colored precipitate $[C_{15}H_{18}N]_2[PtBr_6]$. It was dried at 110°C and weighed. The results from platinic chloride solutions were low, but by conversion to the bromide the author obtained excellent results on samples ranging from about 1–10 mg. Separation from most of the other platinum metals was necessary, but the associated base metals such as Cu(II), Fe(III), Ni(II), and Cr(III) produced no interference; furthermore the presence of moderate amounts of lead was not detrimental.

The method can be applied usefully to the solution of platinum(IV) obtained by precipitating the remaining platinum metals as hydrated oxides. The reagent is superior to strychnine or brucine, which, while producing more insoluble chloroplatinates, provides poor precision and requires a more critical acidity for precipitation.

Westland and Westland (277) made a critical study of the conditions required for the precipitation of platinum using the above reagent. It was found that, whereas the heating technique used by Ryan (276) was suitable for 10 mg of platinum, samples as large as 20 mg introduce errors. Prolonged heating or drying at temperatures lower than 110°C failed to correct the errors. Furthermore, on applying the method recommended by Ryan to solutions of platinum subjected to prior fuming with sulfuric acid, it yielded lower results.* The revised method, recommended for micro or semimicro amounts only, is described in the following.

PROCEDURE 55 (276)

Reagent

Prepare the reagent by mixing equimolar amounts of benzylchloride and di-methylaniline and allowing the mixture to stand at room temperature until a crystalline mass is formed. Wash the solid with ether to remove unchanged benzylchloride and dimethylaniline. Dissolve the solid in ethanol and recrystallize by the addition of ethyl acetate to give a white crystalline product.

Procedure

For platinum filtrates obtained by hydrolytic separation, evaporate the filtrate to dry-ness, add a few milliliters of concentrated hydrochloric acid, and evaporate. Continue this

---

*Recently Ryan found that Westland's criticism resulted from the latter's use of an impure pre-cipitant (277a).

treatment to destroy the bromate used to secure the higher valencies of platinum metals prior to the hydrolytic precipitation. In the absence of bromate evaporate to remove hydrochloric acid in the presence of 0.5 g of sodium bromide. Add to either of the above evaporated residues a few drops of 6 M hydrobromic acid and 10 ml of water, and filter through a porous-bottomed filtering crucible. If there is a dark residue, dissolve this with a few drops of 1:3 aqua regia and filter into a separate container. Wash the solution from the crucible, and evaporate the aqua regia leach to dryness. Repeat the evaporation several times in the presence of a few drops of hydrobromic acid. Acidify slightly, dilute to a few milliliters, and add the filtrate to the original solution.

Treat the resulting solution with 48% hydrobromic acid to provide 4 ml of the acid per 10 mg of platinum present. The acidity may be adjusted to as high as 30%. Add water until the volume is 100 ml, and then add slowly, with stirring, a 5% filtered aqueous solution of dimethylphenylbenzylammonium chloride. This solution should be kept in a dark glass container and be freshly prepared each week. Although an excess of reagent is not detrimental, a volume of 5 ml, plus 1 ml for each 10 mg of platinum present, is adequate for complete precipitation. Allow the mixture to stand for 3 hr, and filter through a tared, fritted crucible. Wash with a 0.1% aqueous solution of the reagent, then with 3 ml of dioxane, followed by 4 ml of cyclohexane. Heat at 80°C to constant weight (0.5–1 hr).

THE TREATMENT OF SOLUTIONS FUMED WITH SULFURIC ACID

The conversion to bromoplatinate when platinum has been complexed with sulfate is more difficult than the corresponding conversion from chloroplatinate. The conversion proceeds rapidly, however, when the temperature exceeds the boiling point of hydrobromic acid. The following procedure is applicable to a sulfuric acid solution of platinum.

PROCEDURE 56

Evaporate the sulfuric acid solution to about 2 ml, cool, and add 1 ml of water and 10 drops of 48% hydrobromic acid. Heat to the appearance of hydrobromic acid fumes. Cool, and repeat the fuming after adding a few more drops of hydrobromic acid. The heating must not proceed to a point where hydrobromic acid fumes cease to be evolved. Filter the sample through a porous-bottomed crucible, and precipitate as described above.

Thiosalicylamide

Thiosalicylamide reacts with platinum(IV) to give two distinct types of complexes; in $1.0-6.0\,M$ hydrochloric acid medium, platinum(IV) forms a complex (A) with thiosalicylamide which corresponds to the formula $Pt]C_7H_6ONS)_2$, $2\text{-}C_7H_7ONS]Cl_2$, whereas in acetate medium (pH 4.5–6.0), the reagent reduces platinum(IV) to the bivalent state before the formation of the complex (B),

$Pt(C_7H_6ONS)_2$. Both complexes can be weighed directly in order to determine the metal.

PROCEDURE 57 (278)

Chemicals

Chloroplatinic acid (E. Merck) was dissolved in dilute hydrochloric acid, and the solution was standardized by precipitating the metal by the formic acid method. Weaker solutions were made by diluting the stock solution with distilled water. A 1% solution of thiosalicylamide in 20% ethanol was used for the precipitation of platinum. For spectrophotometric measurements a 0.01 M reagent solution in ethanol was used.

A different complex was obtained when platinum(II) was precipitated in hydrochloric acid medium. The complex (C) was yellowish white in color. Elemental analysis showed the formula to be $Pt(C_7H_7ONS)_4Cl_2$. This complex decomposed at 218°C.

Procedure

Dilute the platinum(IV) solution to 125–150 ml, adjust its acidity 2–3 M with respect to hydrochloric acid and warm to 60°C. Add 10–15 ml of 1% thiosalicylamide solution and digest the orange-yellow precipitate for 30–60 min on a hot water bath. Filter and wash the precipitate with hot 10% hydrochloric acid on a sintered glass (no. 4) crucible. Dry the precipitate at 110–120°C to constant weight, cool in a desiccator and weigh.

Effect of Diverse Ions

Platinum was precipitated and determined as described above in the presence of many other ions. In the determination of 5.70 mg of platinum, no interference was found from 500 mg of Zn^{2+}, Cd^{2+}, Zr(IV), or Th(IV); 250 mg of Mn^{2+}, Cr^{3+}, or Ga^{3+}; 200 mg of Ni^{2+}, Co^{2+}, Al^{3+}, or In^{3+}; 100 mg of Ti(IV), U(VI), or Mo(VI); 20 mg of Ir(IV), and 10 mg of Rh^{3+} or Ru^{3+}. Interferences of Os(VIII), Os(VI), Fe^{3+}, and V^{5+} were avoided by prior reduction (see the following). Palladium, copper, and mercury, however, interfered.

Separation of Platinum from Iron(III), Vanadium(V), Osmium(VI), and Osmium(VIII)

A solution containing platinum(IV) and the diverse ion was diluted to 100 ml. Freshly prepared saturated sulfur dioxide water (20 ml) was added to reduce the metal ions and the acidity was maintained at 2 M with respect to hydrochloric acid. The solution was warmed and platinum was precipitated with thiosalicylamide as before.

Precipitation from acetate medium. Platinum was completely precipitated with thiosalicylamide between pH 4.0 and 6.5; at lower pH incomplete precipitation occurred. Precipitation was complete when the supernate contained 0.04% (w/v) of thiosalicylamide in excess at pH 5.0.

Procedure

Dilute the platinum solution to 150 ml and adjust the pH to 5.0–6.5 with sodium acetate. Heat the mixture to 80°C and add 10–15 ml of ethanolic 1% thiosalicylamide solution with constant stirring. Digest the yellowish brown precipitate on a hot water-bath for 2 hr with occasional stirring. Filter, wash with water, and dry at 130–140°C to constant weight. Calculate the metal content on the basis that the precipitate (complex (B)) contains 39.05% of platinum.

Effect of Diverse Ions

When 17.1 mg of platinum(IV) was precipitated by the above method in the presence of tartrate as masking agent, there was no interference from 200 mg of Ni^{2+}, Co^{2+}, Mn^{2+}, or Al^{3+}; 150 mg of Zn^{2+}, W(VI) or Th(IV); 100 mg of Fe^{3+}, Ga^{3+}, In^{3+}, or U(VI); 50 mg of Tl^{3+} or V^{5+}; and 20 mg of Ir^{4+}. Interferences of copper and titanium were avoided by masking with EDTA and fluoride ions, respectively, at pH 5.5; up to 50 mg of copper(II) and 60 mg of titanium(IV) could then be tolerated. Rhodium, ruthenium, and osmium, however, interfered with the precipitation of the metal.

RHODIUM

Hydrogen Sulfide

Precipitation by hydrogen sulfide in acidic solutions remains one of the most useful methods for the determination of rhodium. Unfortunately, sulfuric acid interferes. This effect is said to be eliminated by heat treatment with sufficient hydrochloric acid, resulting in the conversion of the yellow sulfate to the pink chloro complex (262, 279). The efficiency of this commonly used method has been disputed, however, and a more satisfactory conversion to the pink solution is obtained by fuming to crystals in the presence of sodium chloride or preferably ammonium chloride (280). Presumably the sulfuric acid reacts with the added chloride providing sufficient hydrochloric acid to accomplish conversions to the chloride. The sodium salt may contaminate the rhodium metal, and for this reason the more volatile ammonium chloride is recommended. Under suitable conditions precipitation by hydrogen sulfide is a highly satisfactory process. The precipitate has very suitable physical characteristics and is easily coagulated and filtered. The rhodium sulfide thus prepared, however, is not easily purified, and the washed precipitate is generally ignited to the oxide and subsequently reduced to the metal.

PROCEDURE 58 (280)

If rhodium is present in a sulfuric acid solution, fume the latter to 5 ml. Add 50 ml of a 20% ammonium chloride solution, and evaporate until crystallization takes place. During this treatment the solution should change from yellow to pink. Add 200 ml of water. The solution should now have a pH of 0.9–1.1. If rhodium chloride is the initial dissolved constituent, adjust the volume to 200 ml, and add 0.5 ml of concentrated hydrochloric acid. Heat to boiling, and while boiling pass in a rapid stream of hydrogen sulfide until the precipitate is well coagulated and the supernatant liquid is clear. Transfer the precipitate to a Whatman No. 42 filter paper of a size appropriate to the weight of the precipitate. Wash with 0.1 N sulfuric acid, then with 0.1 M hydrochloric acid. Remove any adhering particles of sulfide from the beaker wall with a small piece of ashless paper that is then added to the filter. The ignition, reduction, and weighing techniques are as described for ruthenium (Procedure 64). When procelain crucibles of the order of 15–20 ml are used, cool in the constant humidity desiccator, (containing saturated calcium nitrate solution) for 20 min, with the same period of standing in the balance case before the first weighing. Prepare blanks similarly.

Note. The results from the hydrogen sulfide precipitation are generally slightly high; this error can be reduced somewhat by a very slow initial ignition. Recently a method has been reported using rhodium sulfide as a weighing form. Taimni and Salaria (281) precipitated $Rh_2S_3 \cdot 3H_2S$ from a solution made alkaline with an ammonia solution then treated with a large excess of sodium sulfide. Subsequently, large excesses of acetic acid and ammonium acetate were added, and the mixture was boiled. The precipitate was said to be purified by washing with organic solvents and drying in a vacuum desiccator. The results obtained indicate errors of less than 0.1% for 18–45 mg of rhodium. The various attempts by the senior author to apply the method according to the directions recommended resulted in positive errors of 10–20%. Furthermore, no modification of the method succeeded in effectively improving either the accuracy or the precision.

2-Mercaptobenzoxazole and 2-Mercaptobenzothiazole

Both 2-mercaptobenzoxazole and the corresponding thiazole were used by Haines and Ryan (282) to precipitate complexes of rhodium from solutions containing ammonium chloride. Complete precipitation in 0.05 M nitric acid occurs with the thiazole but low results are obtained in this medium with the oxazole. The latter forms a complex with rhodium containing three formula weights of precipitant to one atomic weight of rhodium. The accuracy and precision of this method are comparable to that obtained with thiobarbituric acid.

PROCEDURE 59 (282)

To the rhodium chloride solution containing 5–20 mg of rhodium, add 5 ml of a 1% aqueous ammonium chloride solution, and make up to about 50 ml with water. Heat to boiling, and add 3–4 ml of the reagent solution containing 1.5 g of 2-mercaptobenzoxazole or 2-mercaptobenzothiazole in 100 ml of glacial acetic acid. Continue the incipient boiling for at least 2 hr, cool, and filter through a 7-cm Whatman No. 42 filter paper. Wash well with 0.1 M acetic acid, ignite in air, reduce in hydrogen, cool in carbon dioxide, and weigh as rhodium metal. The technique of ignition, reduction, and weighing is as described in Procedure 64.

Hydrolytic Method

The hydrolytic precipitation of rhodium is often preferred to the sulfide method because, where reprecipitations are required, the oxide is more readily converted to the soluble chloride. The technique of the precipitation is simple, the physical characteristics of the oxide are advantageous, and the subsequent conversion to the metal is not difficult. For the frequently encountered small quantities of rhodium, however, the method encourages significant contamination. The near neutral solution required for the precipitation allows the simultaneous precipitation of a wide variety of impurities whose weight, although perhaps small, becomes significant in relation to a few milligrams of recovered rhodium. This positive error can, of course, be reduced somewhat by a reduction of the volume of the solution and the quantity of reagent, but this precaution does not always eliminate a high relative error with small weights of rhodium. With metals of this type, being insoluble in mineral acids, the elimination of these contaminants may appear to be a simple problem of selective dissolution of impurities. These leaching processes in general seldom effectively remove all of the impurity, and sometimes they remove some of the required precipitates. They are last resort processes, and often do not produce significantly improved accuracy or precision with small amounts of rhodium.

Several procedures involving the principle of leaching have been proposed. To determine rhodium in the presence of copper, Moser and Graber (283) precipitated both metals as their sulfides, oxidized then reduced them, and then removed copper with nitric acid. None of these methods is recommended except for approximate determinations.

The existing hydrolytic methods are the products of a large number of techniques. The early methods involved the use of both tervalent and quadrivalent rhodium. Moser and Graber (283) used an oxidizing hydrolysis using potassium bromate and potassium bromide. Reagents such as solutions of sodium hydrogen carbonate in bromine water probably resulted in the precipitation of the hydrated rhodium dioxide (284). A variety of carbonates would produce rhodium(III) oxide at the favorable pH of about 7 (285). There is evidence to suggest that basic media, in the absence of oxidizing constituents, encourage the formation of rhodium(III) even in the initial presence of quadrivalent rhodium. Anyway, from the analytical point of view, the green, hydrated rhodium dioxide has the more desirably physical properties, and all acceptable modifications of the hydrolytic method involve the use of an oxidant during the neutralization of the rhodium solution. Sodium bromate has become the accepted oxidant, and the technique for its addition and of the method of neutralization are somewhat critical. In the method by Moser and Graber (283), which used potassium bromate in a nearly neutral solution containing bromide, the completeness of precipitation could be indicated by the absence of a bromine odor. The following method, a modification of that recorded by Gilchrist (286), has been used successfully for the determination of 10–25 mg of either rhodium or iridium.

PROCEDURE 60 (286)

Heat the 150–200 ml of rhodium chloride solution to boiling, and add 10 ml of a 10% solution of sodium bromate. Add dropwise to the gently boiling solution a filtered 10% solution of sodium hydrogen carbonate until the pH is about 6, as indicated by bromocresol purple applied as an external indicator. This technique can be managed by dipping a slender stirring rod, drawn to a small glass bulb, into the solution to obtain a drop of the precipitating medium, then adding to this drop by means of a capillary a drop of the indicator. The color change can be observed easily around the glass bulb. Alternatively, if a pH meter is used, cool the solution before immersing the electrodes (see Note 1). At about pH 6 add another 5 ml of the bromate solution, and boil for 5 min while the dark suspension is coagulating. If necessary add the required amount of sodium hydrogen carbonate solution to reach a pH of about 7.5 as indicated by a pH meter or by using cresol red as an external indicator added to the stirring rod. Filter hot through a 7-cm Whatman No. 42 paper, and wash with 200 ml of a 1% ammonium chloride solution. Unless this washing is thoroughly carried out, the results will be excessively high and inconsistent. Clean the beaker with one or two small pieces of filter paper; use a third small piece of paper to clean the funnel area adjacent to the upper rim of the filter paper, because there is some tendency for the dioxide to creep (see Note 2).

Notes. 1. With considerable experience the operator should be able to arrive at the required end point through the physical appearance of the green rhodium dioxide assisted by the odor of bromine that is evolved as the solution becomes basic. Anyway, the optimum pH lies between 6 and 8. Thus 2 or 3 ml excess of the bicarbonate solution seems to have no adverse effects such as peptizing the precipitate.

2. Tear the three scraps of paper from the same filter paper to maintain a consistent blank. This method of cleaning is preferred to the use of the feather.

For the determination of small amounts of rhodium that have been isolated from a lead button and thus contaminated with boron, silica, aluminum, etc., the hydrolytic Procedure 60 may be used.

IRIDIUM

Although a number of organic reagents will form precipitates with iridium none of the resulting complexes is recommended as a weighing form and only one has been used for quantitative purposes.

2-Mercaptobenzothiazole

2-Mercaptobenzothiazole forms a bulky orange precipitate from solutions containing acetic acid and ammonium acetate. A 15-hr digestion period is required for its precipitation, and high concentrations of mineral acids must be avoided; their optimum concentration is 0.005–0.01 M. The method is suitable for small amounts of iridium, but amounts in excess of about 20 mg produce unmanageable quantities of precipitate. With these large quantities the hydrolytic precipitation is advantageous.

PROCEDURE 61 (287)

To a 5-ml sample of a sodium chloroiridate solution containing approximately 1 mg of iridium per milliliter add 10 ml of glacial acetic acid, 1 ml of a 20% ammonium acetate solution, and 25 ml of water. Heat to incipient boiling, and add 10 ml of a freshly prepared 1% solution of 2-mercaptobenzothiazole in 95% ethanol, together with two small glass beads to minimize bumping. (The precipitant is purified by crystallization from ethanol and has an melting point of 180.5–181.0°C.) Boil the mixture vigorously for 1 hr, during which time the iridium separates as a bulky orange precipitate. At the end of the hour (or before if the volume of the liquid becomes less than 20 ml) wash the cover glass and beaker wall with a hot solution that is 2% in ammonium acetate and 2% in acetic acid. The final volume of the liquid should be 50–70 ml. Allow the mixture to stand at least 24 hr on a steam bath, filter through a 7-cm Whatman No. 42 filter paper, and wash with 100 ml or more of the hot solution that is 2% in ammonium acetate and 2% in acetic acid. Partially dry the paper (a heat lamp is suitable), and transfer to a tared crucible. Char the paper in a muffle at 350°C, heat for 45 min at 650–700°C, cool, reduce in hydrogen, cool in nitrogen, and weigh. An increase in precision is obtained by cooling in a desiccator containing saturated calcium nitrate. Determine a blank with the full procedure. This should not exceed 0.1–0.2 mg.

Hydrolytic Method

During the analytical recovery of iridium from natural sources by fire assay, etc., appreciable amounts of impurities such as silica, aluminum, boron, etc., may be introduced. Under these circumstances the following hydrolytic procedure is recommended.

PROCEDURE 62

After the preparation of an iridium solution, e.g., as the effluent from a cation-exchanger (see Procedure 96), evaporate to dryness. To eliminate boron, add a few drops of hydrochloric acid and 10 ml of pure methanol, and evaporate slowly to dryness. Add a few drops of strong hydrochloric acid and 50 ml of water. Precipitate hydrolytically as described for rhodium in Procedure 60. Filter the hydrated iridium dioxide onto a 2-ml filter crucible, ignite, and reduce as described for rhodium. Chlorinate at 700°C, dissolve the residue in the crucible and in the tube with about 100 ml of 0.1 M hydrochloric acid, and filter. Evaporate the filtrate to dryness, add 5 ml of water, then 2 ml of a solution containing 70 mg of tartaric acid, and proceed as described for 2-mercaptobenzothiazole (Procedure 61).

Notes. **1.** Precipitation at this stage by 2-mercaptobenzothiazole has been found to be incomplete. The hydrolytic precipitation is made here only as an intermediate step because the accumulated blank may amount to 5 mg or more. Here, particularly, some difficulty may be experienced in achieving a good hydrolytic precipitation.

2. After a fire assay separation, the silica, boron, and aluminum are derived from an admixture of the slag in the collection button.

Iridium in Lead

Because the analytical recovery of iridium frequently involves an association with lead it should be noted that the 2-mercaptobenzothiazole method can be applied in the presence of limited amounts of lead. The efficiency with which this precipitation is accomplished is indicated by Table 4.1. The details of the separation are given in Procedure 61.

RUTHENIUM

Because no specific reagent is available for the determination of ruthenium, it is fortunate that this metal, along with osmium, may be isolated easily by the distillation of its volatile tetroxide. The most generally used distillation procedures involve (a) the treatment of a caustic solution of the metals with chlorine; (b) the selective removal of osmium by nitric acid followed by the oxidation of ruthenium by bromate; (c) the collective distillation of both metals into a hydrogen peroxide solution by oxidation with perchloric acid, and the subsequent separation of osmium by selective oxidation with hydrogen peroxide. For very small amounts of the metal, a condition almost invariably found in ores and concentrates, the authors prefer oxidation by perchloric acid. The use of nitric acid, which must be removed after the separation of osmium, sometimes encourages low values for osmium and ruthenium and introduces a greater number of technical difficulties into the determination of the remaining platinum metals. For each of these distillations, specific collecting liquids have been recommended, and the composition of these determine the treatment prior to precipitation and to a degree control the choice of precipitant.

Whereas caustic solutions are among the most efficient receivers for ruthenium tetroxide, there is the difficulty that almost all of the gravimetric reagents are

TABLE 4.1

Determination of Iridium in the Presence of Lead with 2-Mercaptobenzothiazole

Set no.	Iridium taken (mg)	Lead taken (mg)	No. of detns.	Average iridium recovery (mg)	Average deviation (mg)	Error of average (%)	Test of filtrates
1	5.02	—	4	5.02	−0.01	—	No color
2	10.03	—	3	10.03	−0.01	—	No color
3	15.05	—	4	15.03	−0.01	−0.1	No color
4	20.06	—	2	20.05	—	—	No color
5	25.08	—	1	25.04	—	−0.2	Faint color
6	—	1	.	−0.01	—	—	
7	4.95	—	2	4.95	—	—	No color
8	4.95	1	3	4.98	−0.02	+0.6	Faint color
9	4.95	2	2	4.96	—	+0.2	No color
10	4.95	3	4	4.93	−0.02	−0.4	
11	4.95	7	3	4.95	−0.04	—	Faint color

either applied in acidic media or the preparation for precipitation involves an approach to neutrality from the acidic side.

In addition to the tendency for ruthenium and most other platinum metals to pass through a colloidal stage when one approaches the optimum acidity from the basic side, there is the inconvenience of the accumulating salts in the subsequent precipitation.

Because of these facts, a receiving solution of hydrochloric acid and sulfur dioxide is frequently used, but in this instance prior evaporation to remove the acid is required. Recently (288), an aqueous solution of hydrochloric acid, without the addition of a reducing reagent, has been used successfully for microgram amounts of ruthenium, but in general this adaptation is inadvisable because the range of metal concentration over which the acid alone can be used has not been determined.

To avoid the necessity of complete evaporation, hydrobromic acid solutions have been used effectively as collectors. In this instance the preparation for precipitation requires only a sufficient evaporation to obtain the required acidity.

Hydrogen peroxide has proved to be a useful collector, requiring little or no evaporation prior to the addition of the reagent.

There exists for ruthenium the usual number of classical precipitants such as hydrogen sulfide, zinc, magnesium, ethanol, etc. In addition to these there are now available procedures for its gravimetric determination by thionalide and by hydrolysis to the hydrated oxide. In all instances the weighing form is the metal, and its preparation involves heating in air and hydrogen.

Thionalide

During the development of this method the quantitative recovery of ruthenium was confirmed directly by the examination of the filtrate by distillation and by radioactive tracer techniques (288). It was concluded that the thionalide precipitation was, under the recommended conditions, complete to less than 1 part in 10,000 on 6 mg of ruthenium. Since that time many hundreds of determinations of ruthenium by thionalide have been made with excellent precision and accuracy by the authors' colleagues. Recent literature, however, contains several references indicating that the determination of ruthenium may be attended by low results. These references have their origin in a statement (289) that "with semimicro quantities the results tend to be low—the maximum error amounting to about 10%." It would seem that this estimate is not based upon experience with the methods, but rather upon an indiscriminate use of values provided in the original publication (290) to indicate the adverse effects of certain dissolved constituents. Obviously the use of such exploratory data should eliminate any significance that may have been given to the above criticism. A second claim for ineffective precipitation by thionalide, however, has also been recorded (219). In this instance the authors rejected thionalide for the standardization of ruthenium solutions, and stated that "precipitation with thionalide using samples containing about 20 mg of ruthenium gave extremely poor precision resulting from the rather high solubility of the precipitate; ruthenium could always be detected both in the mother liquid and in the solution." Despite these adverse opinions thionalide is an excellent quantitative reagent, and the precision and accuracy are of a good order when the directions are carefully followed. It is well to realize, however, that thionalide is by no means a specific reagent and in the presence of many associated metals, ruthenium must first be isolated. It should be emphasized that, as with most methods for the platinum metals, the presence of nitrates and nitroso species provides serious interference. Some of the resulting nitroso compounds have been described by Zvyagintsev (291).

Quantitative precipitation by thionalide can be accomplished from practically all of the media used for receiving the tetroxide. Distillations in which the absorbing medium is a hot, acid-free solution of thionalide are not satisfactory because of mechanical difficulties. Whereas aqueous solutions of sodium hydroxide may be used, the subsequent preparation for precipitation is lengthy (290). A collecting medium of hydrogen peroxide is particularly suitable, because by a proper adjustment of the volume and acidity of the absorbant, quantitative precipitation can be accomplished directly without evaporation. One should, of course, filter the solution prior to the precipitation. The ruthenium–thionalide complex does not coagulate well if hydrogen peroxide containing acetanilide as a stabilizer is used, low results always being obtained. In all instances a distillate must be heated to boiling before the addition of thionalide. If this is not done the residual hydrogen peroxide reacts with the precipitant. The product of this reaction is a white unidentified material with a melting range of 178–183°C.

The thionalide–ruthenium precipitate contains one atomic weight of ruthenium to two formula weights of thionalide. All efforts to use the precipitate as a weighing form have failed. There is some evidence to suggest that, although the reagent is soluble in hot water, the required excess is not selectively removed by washing. Alternatively the high results obtained by direct weighing may be due to the presence of a partially decomposed ruthenium(III) complex containing hydroxide. This would agree with the customary trivalency of ruthenium in its complexes.

PROCEDURE 63 (290)

To 50 ml of ruthenium chloride solution, containing from 2 to 20 mg of ruthenium, add 0.6 ml of concentrated hydrochloric acid. If the hydrogen peroxide collecting liquid is to be used, wash the latter into a 150–ml beaker, similarly acidify, and then boil to destroy the residual hydrogen peroxide. Weigh out the thionalide in excess of the calculated amount, dissolve in about 3 ml of ethanol, and add to the ruthenium solution by means of a capillary tube, (see Note 1). Boil until the precipitate is well coagulated. (Insufficient boiling results in low values—1 hr should be sufficient.) Filter through a 7-cm Whatman No. 42 paper or an A2 porcelain filtering crucible. In the latter instance, ignite carefully in order to avoid losses by volatilization. If paper is used, place the latter together with the funnel under an infrared lamp to remove excess of moisture prior to folding and transferring to the crucible, a suitable one being the high form, 2-ml capacity, 5/0 Coors porcelain crucible. Fold the paper carefully to protect the precipitate, and transfer to the crucible. Place the latter in a muffle, heat slowly from the cold in order to char, and then ignite the organic matter. In the authors' opinion ignition is more effectively controlled with a micro gas burner. Transfer the crucible (several may be transferred at one time) to a porcelain boat, and place the boat in a Vycor tube, 750 × 25 mm (i.d.) heated by a suitable tube furnace (see Note 2). Allow hydrogen to pass through the cold tube at 2–3 bubbles per second, as indicated by a water flow meter placed in series at the exit of the tube. Increase the temperature to 600–700°C over about 30 min, and continue the reduction for about 20 min, although a much shorter time is sometimes sufficient. Allow the furnace to cool somewhat, and replace the stream of hydrogen with nitrogen. Cool somewhat, move the tube from the furnace, and place it on a metal sheet to cool, maintaining the nitrogen flow. Cool to room temperature, transfer the crucible to a silver-plated brass block in a constant humidity desiccator for 10 min, place it on a balance pan for 10 min, and then weigh it. A suitable humidity may be obtained by adding to the desiccator a saturated solution of calcium nitrate.

Notes. 1.   It is preferable to filter the ethanolic reagent solution; for a series of determinations a standard reagent solution can be prepared and suitable aliquots used for precipitation.

2.   The split type furnace allows visual observation, and is recommended.

Hydrolytic Method

The earliest hydrolytic method for ruthenium involved the absorption of ruthenium tetroxide in a dilute solution of potassium hydroxide containing some ethanol and the precipitation of ruthenium as an oxide by warming on a steam bath. This method encourages the adsorption of the alkali salt, and continued washing of the precipitate is generally ineffective for its complete removal. Although the evolution of these hydrolytic processes has been characterized by many contributions, a good degree of perfection has been reached through the researchers at the National Bureau of Standards. These advances involved the approach to neutralization from the acidic side, and thus more effectively avoided the occlusion of silica and the adsorption of alkali salts. The tendency toward contamination of precipitates formed in nearly neutral solution, however, is apparent here also, and becomes significant for very small amounts of ruthenium. A reduction in the weights and volumes of reagents in proportion to the smaller weight of ruthenium is not always feasible. Thus it would seem that a large proportion of the absolute error associated with the macrodetermination also applies to the microdetermination. The inaccuracies appear to be partly due to the inefficiency of the final leaching process. Where good accuracy is required the method is not recommended for determination of a few milligrams of ruthenium.

PROCEDURE 64 (292)   (MODIFIED)

Heat to boiling 100 ml of the ruthenium chloride solution containing 50–200 mg of ruthenium. If ruthenium has been distilled and then collected in a sulfur dioxide–hydrochloric acid solution, evaporate the acid distillate to a moist residue. Repeat this procedure three times with hydrochloric acid, then add 25 ml of water, boil to the complete dissolution of the ruthenium salt, filter, and wash with 0.1 M hydrochloric acid to about 150 ml. Heat the solution to boiling, and add a 10% solution of sodium bicarbonate to the appearance of a precipitate. This addition should be carried out slowly, and it is best accomplished by the use of a capillary tube or by a small bore pipet. Continue the heating and addition of the bicarbonate to a pH of 6. The optimum rate of neutralization is generally achieved only with some experience (see Note 1). Boil the solution at pH 6 to ensure coagulation. Filter the precipitate through an A2 porcelain filtering crucible and wash with 100 ml of a 1% aqueous solution of ammonium chloride. Then add about 50 mg of wet, solid ammonium chloride to prevent decrepitation during the ignition. Slowly remove the ammonium chloride by heating in an atmosphere of hydrogen, and continue the heating in this atmosphere for 20 min. Cool the crucible for 20 min in carbon dioxide, and weigh the residue as metal (see Note 2).

Notes. 1.   Too rapid an addition of the bicarbonate may result in local neutralization or in a final pH greater than 6. In both instances readjustment with an acid solution may result in low recoveries. The pH may be determined instrumentally or by the use of a suitable indicator. A 0.04% solution of bromocresol purple may be used, and it is effectively added from a capillary onto the stirring rod after stirring the hot solution. A convenient rod is made by drawing the standard glass rod to provide a narrow neck, terminated by a knob about the diameter of a pea. A little practice allows an easy detection of the yellow-to-blue change.

2. The results are persistently a little high, so it is sometimes advisable to wash the metal with hot water before weighing. This leaching process may lessen the positive error, but it does not eliminate it. For the larger amounts of ruthenium, i.e., 50–200 mg, the hydrated ruthenium(III) oxide can be filtered through a 7- or 9-cm Whatman No. 42 filter paper. After its saturation with ammonium chloride, the paper is transferred to a porcelain crucible and charred very slowly, preferably over a bunsen flame. With proper care with the rate of heating a thin column of fume can be maintained. Rapid burning will result in decrepitation and mechanical carry-over of some precipitate. After charring, the precipitate is strongly ignited and reduced as described above. For the larger amounts of ruthenium the hydrolytic precipitation provides very acceptable accuracy, and the precipitate is readily formed, filtered, and washed. In the presence of nitric acid the precipitation is not complete, and if the ruthenium solutions have been treated with nitrate or nitrite, care must be taken to remove these completely. This can be accomplished by repeated evaporations with hydrochloric acid.

OSMIUM

Hydrolytic Method

The characteristics of this method have been discussed above. Emphasis is given here to the necessity of a carefully made initial precipitation, particularly when small amounts of osmium are to be determined. One may experience with the hydrolytic precipitation of osmium a phenomenon that is recognized by analytical chemists of long experience, viz., difficulties incident to inexperience with a method may be overcome with no clear recognition of the nature of the technical improvement. Most assuredly, once accomplished successfully, the hydrolytic precipitation is one of the most satisfactory methods of analysis. There is little doubt that some of these difficulties arise from our ignorance of the identities of the dissolved platinum metals constituents.

As for ruthenium, no method for osmium should be expected to apply to all of the various solution compositions normally obtained during the distillation and collection of osmium tetroxide. The receiving solutions used for osmium are generally those used for ruthenium, and each type requires some variation in technique before the precipitant is added. The hydrochloric acid–sulfur dioxide solution is a useful receiving liquid, but all types of following analytical gravimetric determinations require the removal of sulfur dioxide prior to precipitation. Gilchrist (293) removed sulfur dioxide by evaporation and subsequent boiling with hydrochloric acid. Various authors, however, have obtained low results for osmium by this process; these errors, although small, become significant with milligram amounts of osmium. This error has been attributed by some (222) to the volatilization of osmium during evaporations, but other investigators (294) have found no loss of osmium during evaporations. Anyway it has been found that if the hydrochloric acid–sulfur dioxide distillates are allowed to stand for 12–16 hr at room temperature no loss of osmium occurs when the solution is concentrated by boiling. It is also a fact that while the addition of thiourea to fresh distillates results in the development of the intense rose color of the osmium–thiourea complex, distillates several hours old give a color of reduced intensity and distillates 24-hr old may yield no color.

These observations have received various explanations. It has been suggested that the fresh distillates contain some dissolved osmium tetroxide that is lost when the solutions are heated before the reduction of the tetroxide is complete. Geilmann and Neeb (294) have shown that the evaporation of the sulfur dioxide–hydrochloric receiving solution does not result in loss of osmium. It is not improbable that the change in the thiourea reaction results from changes in the identity of the dissolved osmium complex.

The hydrolytic method described below is recommended for amounts of osmium of the order of 15–50 mg. It is not recommended for a few milligrams of osmium when high accuracy is required. By an appropriate adjustment of volume and reagents, amounts of osmium as large as 200–300 mg may be similarly determined. It may occur to the analytical chemist that the hydrolytic precipitation, with its relatively great advantage in the favorable physical characteristics of the precipitate, can be applied to even very small weights of metal, despite the significant accumulation of impurities. The application to milligram amounts would merely require the determination of a true blank by a final oxidation to remove the osmium as the tetroxide. Some data have been recorded which bear upon this problem (225). Spectrograms of these residues remaining after volatilization of the osmium revealed the presence of iron, silica, and osmium. The oxidation of the residues at high temperature failed to remove all of the osmium. The treatment of these residues with nitric acid was also ineffective, although fusions with sodium carbonate released the osmium. The fact that osmium is sometimes retained even after ignition scarcely encourages the indiscriminate use of the "true blank" method to achieve improved accuracy.

PROCEDURE 65 (225)

To 50–75 ml of the boiling dilute hydrochloric acid solution of the osmium salt or the hydrochloric acid–sulfur dioxide distillate, treated as previously described to remove sulfur dioxide, add, initially dropwise, a 10% solution of sodium hydrogen carbonate. After the appearance and coagulation of the hydrated oxide of osmium, continue the addition of the hydrogen carbonate solution to a pH of 4, as indicated by adding a drop of

0.04% bromophenol blue indicator solution to the stirring rod as described for ruthenium in Procedure 64. Boil to complete coagulation. Add 10 ml of filtered 95% ethanol to assist the coagulation. Place on a steam bath for at least 2 hr. Decant the supernatant liquid through a weighed porous A2 porcelain crucible. Add to the precipitate in the beaker 25 ml of a 1% ammonium chloride solution and 10 ml of 95% ethanol. Return the beaker to the steam bath for 15 min, and then decant as before. Wash the precipitate 4 times with the 1% ammonium chloride solution, transferring each time to the crucible. Complete the transfer of the oxide with the assistance of a filtering feather, wash with a few milliliters of ethanol, and then cover with recrystallized ammonium chloride. Place the crucible in a quartz ignition tube through which hydrogen is allowed to flow. After 5 min, heat with a Meker burner until the ammonium chloride is volatilized. Ignite at full heat for 1 hr, cool in hydrogen for 5 min, and then in nitrogen for 15 min. Place in a desiccator with a saturated solution of calcium nitrate for 10 min, and weigh after standing in the balance case for 20 min. Obtain the blank by igniting the osmium metal in air and repeating the heating and the weighing technique previously described. Table 4.2 indicates the effectiveness of the determination when it is applied to 5 mg of osmium. Because the blanks are not proportional to the weight of osmium metal, but are dependent largely upon the amount of reagents and the volume used, the accuracy obtained with large samples can be very acceptable.

Thionalide

This precipitant, whose application to ruthenium was previously described, is the most generally applicable of the three known organic reagents for osmium.

TABLE 4.2
Hydrolytic Determinations of Osmium[a]

Sample no.	Osmium taken (as ammonium bromosmate) (mg)	Total wt of reduced precipitate (mg)	Wt of residue after volatilization of osmium tetroxide (mg)	Osmium volatilized (mg)	Difference between osmium taken and volatilized (wt %)
1	4.893	5.441	0.716	4.725	−3.5
2	5.193	5.686	0.594	5.092	−2.0
3	5.051	5.638	0.704	4.934	−2.3
4	4.920	5.306	0.485	4.821	−2.0
5	5.236	6.129	1.372	4.757	−9.2
6	5.321	6.252	1.042	5.210	−2.1
7	5.219	5.791	0.745	5.046	−3.3
8	5.337	6.057	0.986	5.072	−5.0

[a]Osmium dissolved in sulfur dioxide–hydrochloric acid solutions after being distilled from ammonium bromosmate solutions by nitric acid (102).

Unlike strychnine and 2-phenylbenzothiazole, thionalide can be used for hydrochloric acid–sulfur dioxide distillates of osmium tetroxide. It is not recommended for hydrogen peroxide–sulfuric acid distillations, however, because the coagulation of the precipitate is incomplete. The cause of this is not known with certainty, and its circumvention would be advantageous. It may be noted that the blank residue from the thionalide precipitation obtained in the final volatilization of osmium may contain magnesium; the latter is present through its use as a reducing reagent for the preparation of thionalide. When thionalide cannot be readily purchased, one may prepare the reagent according to a recipe prepared by the senior author (295). The thionalide complex contains one atomic weight of osmium to three formula weights of thionalide, in contrast to ruthenium, which has one of ruthenium to two of thionalide.

PROCEDURE 66 (296)

Add to a 100-ml beaker either a sample of ammonium bromo- or chlorosmate containing 2–30 mg of osmium or the osmium distillate in a hydrobromic or hydrochloric acid–sulfur dioxide solution. If the latter is used the sulfur dioxide must be removed by the usual repeated evaporations with hydrochloric acid. Allow the distillates collected by the sulfur dioxide–hydrochloric acid solutions to stand overnight before evaporating them. Adjust to make the 50 ml of solution $0.5 M$ in acid (see Note).

Add an excess of thionalide dissolved in 15 ml of ethanol. The method of this addition will contribute to the success of the determination. Gently boil the solution and initially add the thionalide effectively by means of a capillary tube, or anyway in small drops, over about 30 min. Alternatively, a buret with the tip extending into a hole bored in a borosilicate watch glass is convenient for the slow addition. Boil the contents of the beaker for 2 hr to give a well-coagulated precipitate and a clear supernatant liquid. Heat on a steam bath for another hour, keeping the volume constant by adding water when necessary. Decant the supernatant liquid through an A2 porcelain crucible previously heated in hydrogen to constant weight. Use the technique of filtration, ignition, and weighing as described in Procedure 64.

Note.   This solution need not be filtered prior to precipitation because true blanks may be obtained by direct ignition in air of the osmium metal.

CHAPTER 5

VOLUMETRIC DETERMINATIONS

INTRODUCTION

There is a marked deficiency of acceptable volumetric methods for most of the platinum metals. Furthermore, many of the proposed volumetric methods can be considered as useful only when applied to the isolated platinum metal constituent under carefully controlled conditions. As in the spectrophotometric methods, the analytical researcher seems not always to have realized the complexity of the equilibria existing in mineral acid solutions of these metals. This is made evident by the general failure to deal adequately with methods of dissolution prior to the application of the volumetric methods. These, like colorimetric methods, are sometimes peculiarly sensitive to changes in the identity of the dissolved metal constituent. This is particularly true of some of the proposed potentiometric titrations, where one may encounter almost totally different valence states in various acid concentrations. Indeed, there are recorded volumetric methods that claim quantitative application in weak acids for one valence, and for a second valence in strong acids. One cannot avoid concern about procedures whose successful application is dependent upon either the enforcement of a practically complete equilibrium through control of acidity, prior oxidation, etc., or upon the development of an equilibrium mixture of constant composition.

Regarding the general status of volumetric methods for the platinum metals, there is not only a paucity of acceptable methods for all of these metals but for some no really acceptable volumetric method is available.

GOLD

Most of the volumetric procedures for gold require a precipitation of the metal or gold(I) salt. In all but the iodometric gold methods, the end points are determined by potentiometric techniques or by back-titrations against excess of reductant. As would be expected from the reduction potentials of gold salts, there is available a wide variety of reducing agents. Among the most suitable are hydroquinone, iron(II), arsenite, and ascorbic acid. Others of use for specific purposes, but not recommended for general use, are titanium(III), tin(II), chromium(II), copper(I), hydrazine salts, sulfur dioxide, etc.

Potassium iodide is a notable exception to the general type of reductant involved in gold titrations. One of the earliest reagents, it was used in 1898 by Peterson (297), who suggested that the reaction resulted in the stoichiometric formation of gold (I) iodide and free iodine, the former reacting with sodium thiosulfate to produce $NaAuS_2O_3$. Thus Peterson believed that the quantitative relationship was one atomic weight of gold to three formula weights of sodium thiosulfate. Whether or not the above gold complex is formed, there can be little doubt that the proposed equation is unacceptable. Anyway, Gooch and Morley (298) indicated that the ratio of gold to iodine is 1:2 and found "no evidence whatever" for the reaction to form $NaAuS_2O_3$. These authors used amounts of gold of the order of 1 mg or less, and recorded that the reaction was influenced by the "volume of the solution, the mass of iodine present, and the time of action." To avoid the presence of foreign oxidizing substances from the dissolution of gold by chlorine water or aqua regia, the solution was twice treated with ammonia in excess, boiled, and acidified with hydrochloric acid. The titration procedure involved the addition of an excess of potassium iodide just sufficient to dissolve the gold iodide. The liberated iodine was titrated with a standard sodium thiosulfate solution to a clear solution in the presence of starch, and then standard iodine was added until the rose-colored end point was reached with a subsequent back-titration with sodium thiosulfate. Excellent results were obtained with 0.01 N or 0.001 N solutions of the titrant, and in general the work of these authors reaches a high standard. Scott (279) recorded the above method for small amounts of gold, but, without explanation, recommended the determination of an experimental gold factor for the standard thiosulfate solution.

For larger amounts of gold the Gooch and Morley method has been found inapplicable. Rupp (299) ascribed the difficulty to the instability of gold(I) iodide, which slowly dissociated into gold and iodine. Herschlag (300) rejected the suggestion that the instability of gold(I) iodide was the main cause of error, and without sufficient experimental evidence claimed that the error was caused "by a side reaction involving the aurous iodide complex and thiosulfate." This statement was made because Lenher (301) was able to obtain accurate results on samples of gold ranging from 0.0395 to 0.7301 g by a titration of released iodine with a standard sulfurous acid solution, in which the back-titration with iodine used by Gooch and Morley was not required.

Vanino and Hartwagner (302) rejected Peterson's method because the calculations based on the assumption of one formula weight of gold to three of sodium thiosulfate produced high experimental results, and they rejected Gooch and Morley's method because of inaccuracy, which became more evident when much larger amounts of gold were used. The proposed method, used for 3–12 g of gold, involved the addition to the solution of $HAuCl_4$ of potassium iodate and an excess of potassium iodide to form a yellowish-white precipitate that was dissolved in excess of potassium iodide. The following reactions expressed the quantitative relationships:

(i) $6HAuCl_4 + 18KI \rightarrow 6AuI + 12I + 18KCl + 6HCl$
(ii) $6HCl + 5KI + KIO_3 \rightarrow 6I + 6KCl + 3H_2O$
(iii) $18I + 18Na_2S_2O_3 \rightarrow 18NaI + 9Na_2S_4O_6$

The results obtained with pure $HAuCl_4$ solutions were an improvement on those from older methods, but the method will have a very limited application because acids other than the combined hydrogen chloride are inadmissible. Unfortunately these authors' data (302) included no reference to the preparation of gold solutions by dissolution of the metal.

With the exception of Peterson's explanation (297) for the gold iodide reaction, there has been general agreement that the ratio of reactants is one atomic weight of gold to two atomic weights of iodine. In all instances excess of potassium iodide has been used in both acidic and basic solutions of gold(III) chloride, Brüll and Griffi (303, 304), however, have applied successfully a procedure based on the reaction

$$2AuCl_3 + 6KI \rightleftharpoons 6KCl + Au(AuI_4) + I_2 \tag{1}$$

rather than the generally accepted reaction

$$AuCl_3 + 3KI \rightleftharpoons 3KCl + AuI + I_2 \tag{2}$$

Inexplicably, the experimental results that have been recorded for both relationships have been quite acceptable. Brüll and Griffi recommended their method for quantities of gold of the order of 1–2 mg. With about 10 mg there was some indication that reaction (2) became effective. Variation in the proportion of potassium iodide appeared to be without effect. Attempts to isolate the complex Au(AuI$_4$) failed. The authors stated that copper did not interfere (304) provided that the excess of potassium iodide was not too great. This is contrary to the findings of the present authors, whose data indicate that copper does interfere, and indeed, the reaction with copper(II) and potassium iodide under certain conditions may be complete, presumably according to the reaction

$$2Cu^{2+} + 4I^- \rightarrow 2CuI + I_2$$

It is unfortunate that the publications dealing with this important iodometric method contain confusingly contradictory information concerning the influence of acidity, iodide concentration, optimum amounts of gold, etc. Recent work (305) has indicated that Procedure 67 is relatively insensitive to variations in these three conditions. It should be noted that this procedure avoids the back-titration used by Gooch and Morley (298).

The following method is a modification of that provided by Gooch and Morley (298).

Iodine Method One

PROCEDURE 67 (298)

Reagents

Standard sodium thiosulfate solution (0.1 *N*).   Dissolve the salt in cooled, freshly boiled, distilled water containing 0.1 g of sodium carbonate per liter. Prepare 0.01 *N* and 0.001 *N* solutions by dilution as required.

Starch solution.   Add water to starch to form a paste, and dilute with the required amounts of hot water.

Potassium iodide solution.   An approximately 1% aqueous solution.

Procedure

Dissolve the sample, containing 0.4 mg or more of gold, in the minimum of aqua regia. Remove the oxides of nitrogen by several evaporations with hydrochloric acid in the presence of sodium chloride or by an air current, as described in the introduction to gravimetric methods for gold. Add to the hydrochloric acid–gold solution of pH 5 or less, enough of the 1% potassium iodide solution to redissolve the gold(I) iodide first precipitated. Titrate the solution with the 0.01 *N* sodium thiosulfate solution until the yellow color fades. Add a few milliliters of starch solution near the end point, and continue the titration until the solution is colorless.

Note.   The amount of potassium iodide added is not critical provided an excess is assured. Amounts of gold as low as 0.2 mg may be titrated if a 0.001 *N* thiosulfate solution is used and an indicator blank is determined.

Herschlag (300) recorded one of the most acceptable iodide procedures. Here the interference resulting from the nitric–hydrochloric acid dissolution of the gold was removed by treating with sodium hypochlorite or potassium chlorate, diluting, and boiling out the excess of chlorine. The solution was then neutralized with sodium carbonate, slightly acidified, and made alkaline to litmus with sodium hydrogen carbonate. Unlike the earlier treatment with ammonia, no precipitate appeared during this neutralizing process. Excess of potassium iodide was then added rapidly to avoid the oxidation of iodine by gold(III) chloride, and the liberated iodine was titrated with a standard $0.01\ N$ sodium arsenite solution. The latter was standardized against iodine, and the gold content was calculated stoichiometrically. The method was applied to pure gold solution, gold in cyanide solutions, and gold solutions containing copper, iron, silver, and nickel. Interference occurred when the colored precipitates in the basic medium affected the starch end point. The color of platinum and palladium iodide complexes also caused some interference. The recoveries on samples of 20–400 mg of gold were exceptionally accurate. The above procedure was recommended by Young (306) in a review of various titrimetric methods for gold.

Parker (307) used a method somewhat similar to Herschlag's to determine gold in plating solutions containing nickel and copper. Cyanide, however, was eliminated by treatment with nitric and hydrochloric acids, and nitrite by sodium hypochlorite. There was no interference from iron(II), iron(III), zinc, stannate, nickel, silver, copper(I), or copper(II). Sudilovskaya (308) rejected the mixed acid method for the elimination of the interference of cyanide, and recommended treatment by repeated evaporations, then fusion with metallic sodium, and treatment with concentrated sulfuric acid.

Iodine Method Two

PROCEDURE 68 (300)

Reagents

Standard arsenite solution $(0.01\ N)$.   Dissolve 0.495 g of pure arsenic(III) oxide in a 5 ml 20% sodium hydroxide solution. Neutralize the excess of alkali with dilute 1:4 sulfuric acid, to the phenolphthalein end point. Add 500 ml of distilled water containing about 25 g of sodium hydrogen carbonate. If a pink color develops, remove it with a few drops of dilute sulfuric acid. Dilute the solution to 1 liter, and standardize it against a weighed amount of pure gold. The arsenite retains its strength practically indefinitely.

Starch solution.   See Procedure 67.

Procedure

If the gold content of a solution is to be determined, measure with a pipet a sample containing about 40 mg of gold (20 ml or less). Transfer the sample to a fume hood, add

15 ml of concentrated hydrochloric acid and 5 ml of concentrated nitric acid, and heat briefly. Remove the flame, and add 25 ml of a 5% sodium hypochlorite solution. This will be accompanied by a mild effervescence of chlorine. Add 35 ml of distilled water, and boil gently for 10 min. Cool, and neutralize the acid with a saturated sodium carbonate solution or with a sodium hydroxide solution, testing by touching the stirring rod to the edge of litmus paper. Make slightly acidic with dilute hydrochloric acid. Add a saturated solution of sodium hydrogen carbonate until the gold solution reacts blue to litmus. Dissolve a few grams of potassium iodide in some distilled water, and add this solution rapidly, with stirring, to the gold solution. Add a few grams of sodium hydrogen carbonate to ensure the presence of a sufficient excess, and titrate the liberated iodine with the standard 0.01 N arsenite solution. Add a few milliliters of the starch solution when the iodine color becomes faint. The end point is indicated by the complete disappearance of the blue color. For a solid, such as a gold alloy, heat the accurately weighed sample in aqua regia until it dissolves, then continue with the above procedure.

Hydroquinone Method

Pollard (235) recorded one of the most useful titrimetric methods for micro amounts of gold. Precipitated gold was dissolved from the filter by bromine water acidified with hydrochloric acid. Aqua regia could also be used. Interfering gases were removed by impinging air on the surface of the gold solution. After dilution, the solution was buffered with potassium fluoride, o-dianisidine was added, preferably near the end point, and the gold was titrated with a hydroquinone solution. The associated base metals, copper, silver, iron, and nickel did not interfere appreciably. The method was best suited for 2 mg or less of gold; for very small quantities, the gold was initially coprecipitated with tellurium by reduction with sulfur dioxide. This precipitation was used successfully for the recovery of gold at a dilution of one part in a thousand million parts of solution. The method was found to be suitable for the determination of gold in urine. For this determination Jamieson and Watson (309) used a modification of Pollard's method. Samples of urine containing 0.085–0.522 mg of gold were used. The addition of sodium chloride to assist coagulation, and the subsequent boiling were required for acceptable recoveries.

Zvagintsev and co-workers (310, 311) used the hydroquinone titration for gold–cyanide solutions. After treatment with nitric acid and potassium chlorate, the solution was adjusted with a caustic solution to be only slightly acidic, and was then titrated with hydroquinone. The hydroquinone titration was used by Zharkova and Zhacheva (312) to determine gold in cyanide electrolytes. The solution was neutralized, and gold was separated by hydrazine hydrochloride, then redissolved, and finally titrated with hydroquinone in the presence of potassium hydrogen fluoride and o-dianisidine.

The relative value of indicators for the hydroquinone titration was discussed by Belcher and Nutten (313). Of those examined, 3-methylbenzidine gave the best color change. The end points with benzidine and 2,7-diaminofluorene were sharper than those obtained with o-dianisidine, whose change from red-violet to violet was difficult to observe. The authors also discussed the effects of acid strength and time on the color change of indicators. The relative value of indicators was also discussed by Peshkov (314), who preferred p-tolidine to either benzidine or o-dianisidine.

Milazzo (315) used the hydroquinone titration after a carrier precipitation with copper or lead by hydrogen sulfide or sodium sulfide. The combined sulfides were roasted, and the copper was selectively removed by sulfuric acid. The gold was dissolved in aqua regia, and the solution was evaporated with hydrochloric acid to remove nitrogen oxides. Potassium hydrogen fluoride was added, and the gold was titrated with hydroquinone in the presence of o-dianisidine. Iridium interfered, and under certain conditions the colors of platinum and rhodium salts interfered with the detection of the end point. The author's claim of superiority over existing methods is not justified. In the present authors' opinion, the collective precipitation with copper, and its subsequent selective dissolution, invites greater error than the coprecipitation method with tellurium proposed by Pollard (235). Concerning the question of interfering elements, the author has been too ready to predict a priori from published oxidation–reduction potentials the degree of interference (316). Anyway the description of the experimental results on interferences indicates an inadequate approach.

Various potentiometric titrations have been proposed. Ryabchikov and Knyazheva (317) used a gold wire electrode in a slightly acidic solution of gold freed from nitric acid by evaporation with hydrochloric acid and titrated the hot solution with hydroquinone or Mohr's salt in a current of carbon dioxide. Of the platinum group, only iridium interfered. Czaplinski and Trokowicz (318) used platinum and calomel electrodes with a similarly treated gold solution, and titrated at a pH of 2–5 with a $0.01 N$ hydroquinone solution. The potential change at $50°C$ at the equivalent point, after adding one drop of the titrant, was more than 100 mV. Copper and platinum did not interfere.

Bukharin (319) used the hydroquinone titration for the determination of gold in lead and copper concentrates, and slags, etc. The gold was concentrated by fusion to form a lead alloy that was then parted with nitric acid. The gold residue was dissolved in aqua regia, and the nitrogen oxides removed by the conventional evaporations of hydrochloric acid in the presence of sodium chloride. The gold chloride solution was treated with potassium hydrogen fluoride and titrated with hydroquinone in the presence of o-dianisidine. The error was found to be less than 1%.

PROCEDURE 69 (235)

Reagents

Standard hydroquinone solution.   Dissolve 0.4186 g of pure hydroquinone in 200 ml of water, add 10 ml of concentrated hydrochloric acid, and make up to 500.0 ml. The solution remains stable for at least several weeks.

o-Dianisidine indicator solution.   Dissolve 0.5 g of *o*-dianisidine in 200 ml of water, add 2 ml of concentrated hydrochloric acid, and make up to 500.0 ml.

Potassium fluoride solution.   This is a 5% aqueous solution.

Standard gold solution.   Dissolve 0.5000 g of pure gold, contained in a 500-ml flask, in 2 ml of nitric acid and 6 ml of hydrochloric acid. Remove dissolved gases by inserting a glass tube into the liquid and blowing in a strong current of purified air for 5 min. Wash the tube, remove the latter, and make up to 500.0 ml with water.

Procedure

Transfer 2 ml of the gold solution to a small beaker, add 50 ml of distilled water, 2 drops of concentrated hydrochloric acid, 0.5 ml of the 5% potassium fluoride solution, and 1 ml of the indicator solution. Titrate immediately with the hydroquinone solution to the discharge of the red color.

In those instances where the amount of gold is known approximately, add the hydroquinone to within about 0.05 ml of the end point, with the subsequent addition of 1 ml of the *o*-dianisidine solution, and after allowing the color to develop, continue with the titration. This modification is preferable for more than 2 mg of gold.

The reduction of chloroauric acid by hydroquinone proceeds according to the equation

$$2HAuCl_4 + 3C_6H_4(OH)_2 \rightarrow 2Au + 3C_6H_4O_2 + 8HCl$$

The above method can be applied directly to alloys containing less than 10 mg of copper, silver, iron, nickel, zinc, cadmium, aluminum, and tin. In the presence of lead, the aqua regia solution of gold, etc., is diluted, and a few drops of bromine water are added, the excess being removed by the passage of air through the solution. In the presence of large amounts of impurities, the gold should be isolated by precipitation.

Dithizone Extraction

A dithizone extraction titration procedure for gold was described by Titley (320). Gold ore samples were pre-extracted with sulfuric acid to remove much potential matrix interference prior to dissolution of the gold using aqua regia. Interferences are mainly mercury, palladium, platinum, silver, and tellurium. All but very small amounts of HCl and HNO_3 must be removed before extraction.

This step requires great skill to prevent loss of gold, as a precipitated metal, during the necessary fuming with sulfuric acid. The method is applicable to 0.2 to 1.5 mg of gold contained in 10 to 25 ml of solution. Results were found to be reproducible to ±1%. For details see Ref. (320) or (348).

PALLADIUM

Potassium Iodide Method

Perhaps the first volumetric method for palladium was described by Kersting (321), who titrated a palladium solution directly with potassium iodide to the disappearance of the brown iodide in the supernatant liquid. Obviously this method can be subject to considerable error. A potentiometric titration of palladium with potassium iodide was used by Pshenitsyn and Ginzburg (322, 323). When hydroquinone was used initially as a reductant there was no interference from platinum(IV), rhodium(III), and iridium(IV). The interference of platinum(IV) could also be avoided by the addition of potassium ions to form potassium hexachloroplatinate(IV).

The most satisfactory iodide procedures have been proposed by Atkinson and associates (324–326). One method (324) involved the treatment of a palladium–ruthenium alloy with hydrochloric acid containing a small amount of nitric acid. After dissolution, iron(II) sulfate solution was added to eliminate any foreign oxidizing effect on the potassium iodide, which is used as a 0.01 M solution as the titrant. The end point was determined with the assistance of a centrifuge, which encouraged coagulation and settling. The method produced an accuracy of 0.5% or better in the presence of 5% of nickel, iridium, platinum, rhodium, tungsten, molybdenum, copper, and tin.

For alloys containing gold or silver a modified technique was used. For samples containing gold the mixture was allowed to stand for 1 hr following the addition of the iron(II) sulfate. With silver, the chloride was precipitated and aliquots of the supernatant liquid were removed from the measured volume. The error from the volume of silver chloride presumably fell within the 0.5% expected from this method.

In a second paper (325), the iodide method was extended to apply to a series of binary alloys of palladium with platinum, rhodium, iridium, silver, gold, copper, tin, molybdenum, tungsten, and nickel. A slight modification was required to remove the increased amount of nitric acid used to dissolve the alloys. For the analysis of palladium–silver alloys, the hydrochloric acid concentration was adjusted (to about 4–6 N) so as to dissolve the silver chloride, in which instance the titre included both silver and palladium values, with the former determined separately by an iodide titration in an ammoniacal solution. This modification will have very limited application.

The authors (326) also recorded a general discussion of their procedures. Improvements in the technique of detecting end points were described; the possible application of a potentiometric end point was eliminated by the presence of iron(II) sulfate. Reduction to palladium metal by iron(II) occurred at elevated temperatures, low acidity, or large excesses of iron(II). Under prescribed conditions this "cloud" method was shown to have an end point corresponding to the stoichiometric value and to be free of any interference from adsorption or solubility effects.

PROCEDURE 70 (324–326)

Reagents

Standard palladium solution. There should be 1 g of metal per liter. Dissolve the metal in aqua regia, and remove the oxides of nitrogen by three successive evaporations with hydrochloric acid. Finally, make up to volume with 1 M hydrochloric acid.

Potassium iodide solution. Dissolve 1.8603 g of potassium iodide (dried by heating at 120°C for 2 hr) in 1 liter of distilled water previously treated with nitrogen to remove dissolved oxygen. The iodide solution should be similarly treated at the end of each day, after use. The solution should be kept in a dark bottle, tightly stoppered, and stored in the dark. This solution is 0.01 N, but is it best to obtain a palladium factor by standardization against a standard palladium solution. Because the solution loses strength it should be restandardized each week.

Iron(II) sulfate solution. Dissolve 20 g of iron(II) sulfate crystals (FeSO$_4$·7H$_2$O) in 100 ml of distilled water acidified with a few drops of concentrated sulfuric acid. For the iodide standardization against a palladium solution, the iron(II) sulfate solution should be 15% (w/v), so as to avoid its reducing the palladium.

Procedure

Transfer 15 mg of alloy to a 100-ml round-bottomed Pyrex flask. Add 0.20 ml of concentrated hydrochloric acid and 0.04 ml of concentrated nitric acid. Heat in a water bath at about 85°C. After dissolution of the metal, blow off the brown fumes with a current of air, add 5 ml of concentrated hydrochloric acid and 15 ml of iron(II) sulfate solution. Titrate with the 0.01 N potassium iodide solution, adding slowly to about 0.5 ml less than the expected amount while keeping the contents well stirred. Place the stopper in the flask, shake vigorously for 1 min, then transfer about 5 ml of the suspension to a centrifuge tube, and centrifuge at or above 8500 ft/min at the solution level for 15 sec. Add 1 drop of the iodide titrant to the clear solution in the tube, and estimate the intensity of the "cloud" of palladium(II) iodide. This is best done by placing the tube in the center of a beaker that has onion skin taped to a portion of its inside surface as a background. Daylight should be used, and the cloud is best seen when the drop of potassium iodide solution is allowed to fall directly onto the surface of the liquid from a short distance, rather than down the side of the tube.

Return the contents of the tube to the flask, and add more of the potassium iodide solution from the buret. After shaking the flask for 5 sec, remove a portion of the suspension, centrifuge it, and test as before. Continue the titration until the addition of the test drop fails to give a recognizable cloud.

Notes. **1.**   With experience this technique is not time consuming.

2.   This cloud method can be of special value as a "go-or-no-go" test in the hallmarking analyses of metals for jewellery.

Nickel Diethyldithiophosphate Method

Various adaptations of the palladium iodide method have been used for volumetric determinations.

Nickel diethyldithiophosphate, $[(C_2H_5O_2)PSS]_2Ni$, was used as a titrant for solutions of palladium iodide (327). The method is applicable to 1–15 mg of palladium. The reaction is quantitative and practically instantaneous in the cold. The red-brown color of the precipitate changes to a yellow at the end point. The reaction proceeds according to the equation

$$PdI_2 + [(C_2H_5O)_2PSS]_2Ni \rightarrow [(C_2H_5O)_2PSS]_2Pd + NiI_2$$

To avoid the interference of platinum, an excess of sodium hydrogen sulfite is added to form a stable platinum(II) complex. The method is suitable for the determination of palladium in alloys containing platinum, but it is not applicable to alloys containing copper.

PROCEDURE 71 (327)

Nickel Diethyldithiophosphate

Recrystallize to arrive at a melting point of 105–106°C. Dissolve 1 g of the pure salt in distilled water, and dilute to 100.0 ml. Standardize this solution either by the addition of a standard iodine solution and back-titrating the excess of iodide with a sodium thiosulfate solution or preferably with a pure palladium solution.

Procedure

Transfer 1–2 ml of the palladium chloride solution containing 1.5–4.5 mg of palladium to a small beaker, and add a potassium iodide solution to give about 0.5 g of the salt per milliliter. Add an equal volume of 2 M hydrochloric acid, and titrate the cherry-red solution with the standard nickel diethyldithiophosphate solution to the disappearance of a yellow mixture of precipitate and solution.

If platinum is present, add a 10- to 20-fold excess of sodium hydrogen sulfite after the addition of the potassium iodide, the resulting weak yellow color being due to the decomposition of the palladium iodide and the formation of a stable platinum sulfite complex. Add hydrochloric acid to the reappearance of the palladium iodide precipitate, and titrate as described above.

Differential Potentiometric Determination of Palladium in the Presence of Platinum in High-Grade Palladium–Platinum Metals

A differential potentiometric determination of palladium–platinum alloys was used successfully by Barabas and Vinaric (328). The method required the addition of potassium iodide to a hydrochloric acid solution of the alloy to remove all but a small excess of the palladium which was subsequently titrated with standard potassium iodide using platinum as the indicator electrode with the saturated calomel reference electrode. A standard sample of palladium was treated simultaneously in the same way as the unknown, so that a factor could be deduced from the weight of the pure palladium sponge; this factor was reasonably constant and could be applied to the titration value for each unknown. It was noted that if the palladium alloys with the standard are not analyzed "at the same time the potassium iodide equivalent of palladium is not exactly the same for the two runs." The acidity of the solution was adjusted to 0.2% hydrochloric acid, gold was removed by sulfur dioxide, and silver present as the complex $AgCl_2^-$ was removed by the addition of a small amount of silver nitrate. It would be interesting to know the evidence that "silver chloride acted as a sort of carrier of the soluble silver polychloride." Platinum did not interfere up to twice the concentration of palladium and copper up to 200 times that of palladium. The authors claimed an accuracy which compared favorably with that from gravimetric methods. The procedure is rather ingenious and should be useful in technical analytical work involving many determinations of palladium; however, it cannot be used indiscriminately, depending, as it does, upon the absence of the many constituents which react with potassium iodide. The statement that "there is no limitation as to the size of the sample that can be used" requires reservations.

PROCEDURE 72 (328)

Apparatus and Reagents

A Metrohm recording potentiograph Model E 336 (Metrohm, Ltd., Herisaw, Switzerland) with platinum and saturated calomel electrodes can be used along with a Schuco automatic pipet for dispensing 50-ml volumes of the potassium iodide solution with a precision of 0.0005 ml.

Standardize concentrated potassium iodide solution (18.72 g of potassium iodide per liter) and dilute potassium iodide solution (9.36 g of potassium per liter); standardize both solutions against pure palladium.

Procedure

Weigh accurately into a 100-ml beaker a sample containing from 300–350 mg of palladium. Into a separate 100-ml beaker weigh an amount of palladium sponge corresponding to the maximum amount of palladium expected in the sample (approx 350 mg). Moisten the sample with formic acid and cover the beaker. Add 15 ml of aqua regia (3 parts hydrochloric acid + 1 part nitric acid). Digest at moderate heat, then take to dryness. Take up the residue twice more with 5-ml portions of hydrochloric acid to dryness. Dissolve the residue, while still slightly wet, in water by applying some heat. If any insoluble residue is left, filter, ignite the residue, and redissolve in aqua regia as before. Combine the two solutions and adjust the volume to about 120 ml. If gold is expected in the concentrate, add 25 ml of a sulfur dioxide saturated water solution, and allow to stand 1 hr. If the presence of silver is suspected, add 2 ml of a 0.1% silver nitrate solution. Heat to coagulate and filter out gold metal and silver chloride combined. Boil to drive off sulfur dioxide, then add 50 ml of the concentrated potassium iodide solution from the automatic pipet to precipitate the bulk of palladium. Titrate the remainder potentiometrically using the dilute potassium iodide solution.

Silver and gold interfere and must be removed. Platinum will not react with the iodide as long as palladium is present.

PLATINUM

Diethyldithiocarbamate—in the Presence of Gold

The following method is intended to provide for the determination of platinum in a silver assay bead containing gold. The direct volumetric determination of platinum can, of course, be made by appropriate selections and adaptations of the described procedures.

PROCEDURE 73 (329)

Reagents

Tin(II) chloride solution.   Add 200 ml of concentrated hydrochloric acid to 400 g of tin(II) chloride and a few pieces of granulated tin. Heat on a water bath for several hours, replacing evaporated hydrochloric acid. Filter through asbestos to a clear, colorless filtrate. Cool, add 8 mg of potassium iodide, and stir thoroughly. Tin(II) chloride is more reactive with chloroauric acid in the presence of iodide, but the amount recommended should not be exceeded.

Sodium diethyldithiocarbamate solution.   Dissolve 0.591 g of the salt in water, transfer to a 250-ml standard flask, add 1 g of sodium hydroxide, and make up to volume with water. The solution remains stable for at least a week.

Procedure

Transfer the silver bead, which should weigh about 5 mg and contain no more than a total of 2 mg of platinum together with gold, to a dry test tube, and add 2 drops of concentrated nitric acid. The silver should dissolve leaving a residue of gold with most of the platinum. Add 6 ml of concentrated hydrochloric acid, and shake to dissolve the silver chloride. Heat to dissolve the gold and platinum, add 1 ml of the tin(II) chloride solution, then heat to incipient boiling. Cool, and then add 5 ml of benzene. Place a stopper in the tube, and shake thoroughly to encourage the collection of the precipitated gold by the benzene layer. Titrate with the standard sodium diethyldithiocarbamate solution from a calibrated microburet. Interrupt the titration by intermittent shaking, and thus continue to a colorless acid layer. The platinum(II)–dithiocarbamate complex being extracted completely by the benzene.

Standardize the titrant in a similar manner against a standard platinum solution of an appropriate concentration.

Notes. **1.** The amounts of nitric acid or tin(II) chloride are not critical, but a large excess of nitric acid may require an increase in the amount of the tin(II) chloride recommended.

2. In those instances where the quantitative recovery of gold is required one may use a filtering stick modified to allow the collection of the filtered platinum–silver solution. The open end of the filter stick should be attached to a capillary with a double right-angle bend, the short end of which passes through a two-holed stopper placed in a second test tube. A tube inserted in the second hole of the stopper will allow sufficient suction to filter the gold and collect the filtrate. The precipitated gold can then be redissolved, and the amount determined by a suitable method.

Diethyldithiocarbamate—in the Presence of Gold and Palladium

This modification of the above method is potentially useful, because artificial alloys such as silver assay beads frequently contain both palladium and platinum. This modification allows the collection of precipitated palladium along with gold, and by appropriate treatments both of these constituents are readily determined. The method requires the preparation of tin(II) bromide solution and the use of arsenic-free hydrobromic acid.

PROCEDURE 74 (330)

Reagents

Sodium diethyldithiocarbamate solution.   Dissolve 0.591 g of the salt in water, transfer to a 250-ml standard flask, add 1 g of sodium hydroxide, and make up to volume with water. The aqueous solution should be freshly prepared and standardized against a pure platinum solution.

Tin(II) bromide solution.   To a suitable beaker add 10 ml of hydrobromic acid (d = 1.265), 1 ml of bromine, and a few pieces of granulated tin. Transfer the solution to a tube containing granulated tin, place a stopper in the tube, and allow to stand until the solution becomes colorless. Replenish the tin(II) bromide solution, as it is used up, by the addition of tin(IV) bromide.

Hydrobromic acid.   The acid should be arsenic-free, d = 1.265, 30% hydrobromic acid. The Gutzeit test should be applied to ensure sufficient freedom from arsenic. Presumably, this is because arsenic bromide is reduced to elemental arsenic by tin(II) bromide, followed by a reaction between the arsenic and platinum(II) bromide to precipitate platinum. The hydrobromic acid may also be tested for arsenic by the addition to a solution of a few micrograms of platinum chloride of 6 ml of hydrobromic acid and 10 drops of the tin(II) bromide solution, boiling for 5 min, and titrating with the dithiocarbamate solution. The value obtained should be the same as that obtained omitting the boiling period.

Procedure

Treat the alloy as described for the chloride solutions (Procedure 73). To the nitric acid mixture add 6 ml of the hydrobromic acid, and warm to dissolve the silver bromide. Because the latter is more insoluble than silver chloride, do not allow the solution to stand for more than a few hours, after which time the silver bromide may reprecipitate. After the dissolution of the latter, heat until the contents of the tube become dark brown, and bromine and nitrosyl bromide vapors are evolved. Keep the liquid at incipient boiling for 5 min to ensure the dissolution of gold and platinum. Add a small few crystals of hydroxylammonium chloride to destroy the bromine and nitrosyl chloride. With the evolution of nitrogen the color becomes lighter; continue heating, and add a few more crystals of hydroxylammonium chloride until the solution becomes pale yellow. This treatment eliminates the formation of the tin(IV) bromide whose color interferes at the end point. Allow any of the insoluble platinum metals to settle, then add 10 drops of the tin(II) bromide solution, heat again to incipient boiling, and allow the solution to remain at water bath temperature for 5 min. Gold and palladium precipitate and can be filtered off as previously described. To the solution of the red-brown platinum complex add 5 ml of benzene, and titrate, interrupting the titration by intermittent shaking, and thus continue to a colorless acid layer. The platinum(II)–dithiocarbamate complex being extracted completely by the benzene. Standardize the titrant in a similar manner against a standard platinum solution of an appropriate concentration.

Note.   The traces of associated platinum metals usually found in the nitric acid parting solution will not interfere with the titration. It should be emphasized, however, that the results obtained by both of the above methods may be affected adversely by the presence of significant proportions of the insoluble platinum metals. The parting residue may contain platinum that resisted the action of the nitric acid solution. Where high accuracy is required, the parting insoluble should be dissolved and analyzed by appropriate methods.

Ethylenediaminetetraacetic Acid Method

Platinum in the form of hexachloroplatinate(IV) reacts slowly with EDTA in a 1:1 mole ratio. At the concentration level used (a few milligrams per 50–75 ml), favorable conditions are a solution of pH 3–4.5, and three- to eightfold molar excess of EDTA. Complete reaction required heating at 100°C for 1.5–2 hr. The reaction rate is retarded by acetate ion, but not by nitrate or sulfate. Titrimetric determination of platinum is accomplished by addition of excess standard EDTA, buffering to pH 3–4.5, heating the mixture at 100°C for 1.5 hr, buffering to pH 5.3 with acetic acid–acetate, and back-titrating with zinc acetate to a xylenol orange end point. Blank corrections are necessary to compensate for trace metal impurities in the water and/or reagents. Determinations of 0.4–3 mg of platinum per 50 ml are accurate to ±1.3% standard deviation. Both titrimetric and spectrophotometric evidence ruled out the possibility of reduction of platinum(IV) by EDTA. Titrimetric methods showed the complex to be $PtCl_4HY^{3-}$, where Y is the deprotonated EDTA.

PROCEDURE 75 (331)

Apparatus and Reagents

Small quantity analytical weighings are made on a Mettler M5 microbalance. Calibrated volumetric ware is used.

Measurements of pH are made with a Beckman Expandomatic pH meter and a saturated calomel–glass electrode system.

Absorbance measurements are made with a Cary Model 14 spectrophotometer and matched silica cells of 10.0-mm optical path.

Platinum solution.   A stock solution of hexachloroplatinate(IV) is prepared by dissolving an accurately weighed quantity (e.g., 1 g) of grade 1 platinum thermocouple wire, purity 99.99%, in hot aqua regia. After evaporation just to dryness, a small amount of concentrated hydrochloric acid is added and the solution is again taken to dryness; this treatment is repeated three times to ensure destruction of any nitroso complexes. The final residue is taken up in 10.0 ml of concentrated hydrochloric acid and diluted to 1 liter with deionized (or conductivity) water. More dilute solutions are prepared as needed by dilution of the stock solution. The platinum content of the stock solution is checked by evaporation and hydrogen reduction of aliquots of the solution. Results from triplicate aliquots show the stock solution to be 5.125×10^{-3} M in H_2PtCl_6; molarity based on weight of platinum taken is 5.126×10^{-3}. Some comparison tests, especially of absorption spectra, are made with platinum(IV) chloride and with potassium tetrachloroplatinate(II) solutions.

EDTA solution (0.0100 M).   Disodium ethylenediaminetetraacetate dihydrate (reagent grade, assay 99.9%, J.T. Baker Chemical Co.) is dried at 80°C before being weighed for preparation of the stock solution. The solution, stored in polyethylene bottles, is stable for several months.

Zinc acetate solution.   This is prepared by dissolving reagent grade zinc oxide (Allied Chemical Co. product, 99.0% minimum purity) in dilute acetic acid. The solution is standardized by comparison titration against the standard EDTA solution under the same conditions (reagent concentration, acidity, sample volume) as employed for back-titration in the platinum determination. Reproducibility of the comparison titrations is one part per thousand even with solutions as dilute as 4×10^{-4} M.

Buffers.   Use Clark and Lubs buffers of pH 2, 4, 7, and 10. The acetate buffer of pH 5.3, in the back-titration procedure is prepared by mixing 50 ml of 1.0 M acetic acid with 375 ml of 1.0 M sodium acetate.

Indicator.   Use an aqueous 0.10% solution of xylenol orange, 3,3'-bis-[*N*,*N*-di(carboxymethylaminomethyl)]-*o*-cresolsulfonphthalein.

Masking agent for aluminum.   Use Tiron (4,5-dihydroxy-*m*-benzenedisulfonic acid, disodium salt) as its aqueous 2% solution.

Conductivity water.   This is prepared by triple distillation in a Barnstead still; the conductance was 1.9×10^{-6} mho.

Procedure

To a sample solution containing 0.4–3 mg of platinum at $PtCl_6^{2-}$, add an accurately measured three- to eightfold molar excess of standard EDTA solution. Adjust to pH 3.0–4.5 with nitric acid or potassium hydroxide, as required, and heat the solution at 100°C for 1.5–2 hr. (At this stage copper(II) and/or acetate ion is to be avoided.) Cool the solution to room temperature. If the presence of aluminum(III) is suspected, add 6 drops of aqueous 2% Tiron solution. Add sodium acetate–acetic acid buffer of pH 5.3. Acetate ion sharpens the end point of the back-titration and should be used at as high a concentration as permitted by the concentration of zinc ion titrant; for $1 \times 10^{-3}\,M$ zinc, use 2.5–40 meq of acetate, but for $4 \times 10^{-4}\,M$ zinc, use 1.5–4 meq of acetate. If necessary, readjust to pH 5.3 with nitric acid or potassium hydroxide as required. Add 3 drops of aqueous 0.1% xylenol orange indicator, and titrate with standard zinc acetate of concentration $1 \times 10^{-3}\,M$ or $4 \times 10^{-4}\,M$ until the color changes from yellow to pink. Carry a blank containing EDTA and buffer, at the same volume and pH, through the heating and titration procedure, to account for traces of reacting impurities in the reagents. Because the indicator blank in the zinc titration (usually 0.05 ml or more) varies somewhat depending upon the exact composition of the solution being back-titrated, an indicator correction should be determined on an unheated solution of the same volume and composition as the sample being analyzed, but containing no platinum nor EDTA. The calculations are made as follows:

$$\text{Number of millimoles of platinum complexed} = E - Z - (E_B - Z_B)$$

where E is the number of millimoles of EDTA added to sample, Z the number of millimoles of zinc ion required for back-titration of sample (corrected for indicator blank), E_B the number of millimoles of EDTA in the heated blank, and Z_B the number of millimoles of zinc required for back titration of heated blank (corrected for indicator blank).

Precision of results.   Table 5.1 gives the results for 17 samples, containing 0.390–2.00 mg of platinum, analyzed by the procedure outlined above.

IRIDIUM

Hydroquinone Method

Hydroquinone is used frequently for volumetric determinations of iridium, with an end point being determined by potentiometric methods or by the addition

TABLE 5.1
Reliability of Platinum Determination

Pt taken (mg)	Pt found (mg)	Recovery (%)	Pt taken (mg)	Pt found (mg)	Recovery (%)
0.390	0.388	99.5	0.993	0.982	98.9
0.585	0.581	99.3	0.993	1.003	101.0
0.780	0.769	98.6	1.000	0.994	99.4
0.993	0.987	99.4	1.000	1.011	101.1
0.993	1.004	101.1	1.000	1.014	101.4
0.993	1.007	101.4	2.000	2.020	101.0
0.993	0.981	98.8	2.000	1.990	99.5
0.993	1.013	102.0	Average		99.9
0.993	0.968	97.5	Standard deviation		1.3
0.993	0.983	99.0			

of an indicator. Ryabchikov (332, 333) suggested the reaction

$$2(NH_4)_2IrCl_6 + C_6H_4(OH)_2 + 2H_2O \rightarrow 2(NH_4)_2IrCl_5 \cdot H_2O + C_6H_4O_2 + 2HCl$$

Chlorine was used to ensure the quadrivalent state, and after the removal of excess the titration was carried out potentiometrically. The hydroquinone was similarly standardized against a known weight of a pure iridium salt. There was no interference from air, rhodium, or platinum.

The indicator used by Milazzo and Paoloni (334) for the hydroquinone titration was *o*-dianisidine. These authors provided directions for the formation of a lead–iridium alloy, parting with nitric acid to obtain the impure iridium residue which was then chlorinated. The chloride solution was treated with aqua regia to ensure that the iridium was in the quadrivalent state. The alternative titrant was potassium hexacyanoferrate(II). A visual titration was also used by Pshenitsyn and Prokof'eva (335) who recommended diphenylamine as an indicator; alternative titrants were Mohr's salt or titanium(III) chloride. For the iron(II) titration, the indicator recommended was 1,10-phenanthroline. Other aminophenol titrants for iridium(IV) are *p*-methylaminophenol and *p*-aminophenol (336).

The most extensive application of hydroquinone as a titrant was reported by Pollard (337) who used the method for quantities of iridium of the order of 0.1–0.001 mg. After the isolation of iridium the solution is treated with a lithium sulfate–sulfuric acid solution and perchloric acid, with 3,3'-dichlorobenzidine as indicator, and heated according to an accurately measured time schedule to produce the mauve iridium constituent. This solution is titrated with hydroquinone, previously standardized against pure iridium, to the disappearance of the yellow color of the indicator, a drop of which is added near the completion of the titration. It is always advisable to confirm the determination by reoxidation and a second titration. The identity of the mauve iridium(IV) constituent is not known, but the color intensity is sensitive to a variety of factors, and strict attention must be paid to the exact duplication of the conditions of standardization.

PROCEDURE 76 (337)

Reagents

Titrant.   Dissolve 0.1424 g of hydroquinone in water, add 10 ml of cold 18 N sulfuric acid, and make up to 500 ml in a volumetric flask. From this solution prepare a second solution, one-tenth the strength. One milliliter of this second solution will react with 0.100 mg of iridium according to the equation

$$Ir(SO_4)_2 + C_6H_4(OH)_2 \rightarrow Ir_2(SO_4)_3 + C_6H_4O_2 + H_2SO_4$$

Acid–lithium sulfate solution.   Transfer 8 g of lithium sulfate monohydrate to a silica dish, and heat to remove water of crystallization. Add 30 ml of 98% sulfuric acid, stir, and warm to effect dissolution.

Standard iridium solution.   Add 10 mg of iridium sponge to an ignition tube 7.5 cm long and 0.5 cm i.d. Cover the iridium sample with 200 mg of pure tin and then with borax glass. Cover the mixture with dry potassium bromide powder to a depth of 1 cm. Heat the tube to a bright red, beginning at the top of the bromide and working downward until the mixture is molten. Shake the tube for 15 min to assist in mixing the molten alloy. Cool, break the tube, and clean the alloy with boiling water. Transfer the tin alloy to a boiling tube, 7 by 1¼ in., containing 30 ml of the acid–lithium sulfate solution. Place this tube within a larger tube, 8 by 1½ in., containing 30 ml of a solution of ammonium sulfate in sulfuric acid (50 g of ammonium sulfate in 70 ml of 98% sulfuric acid). Add a few small pieces of coke or similar material to the solution in the outer tube to prevent bumping. Hold the outer tube in a clamp with the rim of the inner tube supported on that of the outer. Heat the outer tube until the acid boils, and then adjust the heat so that the acid fumes condense about 3 cm above the surface of the liquid. The tin alloy should dissolve quickly, leaving a black deposit from which bubbles of sulfur dioxide arise. The black iridium deposit should dissolve in about 2 hr forming a brownish solution that becomes blue upon cooling. Dilute this blue solution, cool, and make up to 100 ml with water. The color turns to a mauve; it will fade on long standing.

Indicator.   Dissolve 0.1 g of 3,3′-dichlorobenzidine in 10 ml of (1 + 2) sulfuric acid, and dilute to 100 ml.

Procedure

Transfer the standard iridium solution to be analyzed to a clear silica test tube, 6 by ¾ in., and add 5 drops of the acid–lithium sulfate solution, 2 drops of 72% perchloric acid and 1 drop of the 3,3′-dichlorobenzidine indicator solution. Wash the wall of the tube with water, and boil to concentrate the acid. Continue heating at the top of a flame about 6 cm high, and observe carefully the color change as the indicator is oxidized, first giving a yellow color, then becoming brown, and finally fading out. After the color disappears, continue heating for 2 sec, in which time the mauve color of iridium(IV) appears. Then, immediately remove the test tube from the flame, and rotate it in a nearly horizontal position, thus allowing the hot acid to rinse the wall for about half the length of the tube. Cool the contents, add 2 ml of water, and heat to gentle boiling for 10 sec to remove chlorine and chlorine dioxide. Cool again, and titrate with the hydroquinone solution to the almost complete disappearance of the mauve color. Add 1 drop of the indicator solution, allow to stand for about 1 min, in which time a yellow color should appear. Continue the titration to the disappearance of this yellow color. Boil the contents, reoxidize as described above, and titrate again.

Notes 1.   High results are obtained if the oxidation with perchloric acid is continued for more than 2 sec.

2.   Traces of reducing impurities will react with the mauve constituent. Therefore blanks must be obtained by carrying out the complete procedure.

OSMIUM

Iron(II) Sulfate Method

A volumetric method that may be applied directly to an osmium distillate in sodium hydroxide involves the addition of sulfuric acid to make a $4-6\,N$ acidic solution. The osmium is then reduced to the quadrivalent state by passing the solution slowly through a reductor containing bismuth granules. An excess of a $0.01\,N$ ammonium metavanadate solution in $1\,N$ sulfuric acid is added, together with a few drops of a 0.2% aqueous phenylanthranilic acid solution as the indicator. The excess of metal vanadate is titrated with a $0.01\,N$ iron(II) sulfate solution. The metavanadate oxidizes osmium(IV) quantitatively to the VI-valent condition according to the equation

$$Os^{4+} + 2V^{5+} \rightarrow Os^{6+} + 2V^{4+}$$

PROCEDURE 77 (338)

Reagents

Indicator.   Use a 0.2% aqueous solution; add 0.2 g of phenylanthranilic acid to 100 ml of water, and heat.

Ammonium metavanadate solution.   Use $0.01\,N$ in $6\,N$ sulfuric acid.

Mohr's salt solution.   Use $0.01\,N$ in $6\,N$ sulfuric acid.

Reducer.   Place a thin layer of glass wool in the narrow part of a 50-ml buret, just above the tap. Add a 100-mm length of powdered bismuth whose particles are crushed to $0.3-0.6$ mm diam. Wash 2 or 3 times with $4\,N$ sulfuric acid, with a 2–3-mm layer of acid covering the bismuth.

Procedure

Transfer the alkaline osmium solution, containing from 2 to 10 mg of osmium, to a suitable vessel, and acidity with sulfuric acid to 4–6 N. Transfer quickly to the reducer, and wash 2 or 3 times with a few milliliters of 6 N sulfuric acid. The volume of the reduced solution together with the wash solution should not exceed the capacity of the reducer (about 40 ml). Regulate the volume of the solution passing through the reducer by opening or closing the tap, to provide 4–8 ml/min. Allow the osmium solution to pass until its level is 2–3 mm above the bismuth layer and then wash the reducer 3 times with 6 N sulfuric acid, using 15–20 ml each time, without exposing the bismuth layer. If the reducer has been charged correctly, the operation of reduction should not take more than 25 min.

Collect the reduced solution in a 250-ml Erlenmeyer flask, and add an excess of the 0.01 N ammonium metavanadate solution and 3 or 4 drops of the indicator. Titrate the excess of vanadate with the 0.01 N Mohr's salt solution.

Determine the relationship between the metavanadate and Mohr's salt solutions, and thus obtain the volume of metavanadate solution required for the oxidation of osmium(IV) to osmium(VI). Standardize the ammonium metavanadate gravimetrically or volumetrically.

Potentiometric Method

A useful potentiometric method for the standardization of octavalent osmium solutions has been described by Meites and Cover (339). The procedure involves a preliminary reduction of osmium(VIII) to osmium(VI) with ethanol in sodium hydroxide solution to prevent loss of the volatile OsO_4. A titration using chromous sulfate is then done under nitrogen following acidification with hydrochloric acid. The authors suggest the possible application of this procedure to osmium distillates.

PROCEDURE 78 (339)

Apparatus and Reagents

Prepare the chromous sulfate solution preferably by the method of Lingane and Pecsok (340) and standardize by the potentiometric titration of known amounts of iron(III) in 5.0 M hydrochloric acid. Use a commercially available all-glass buret and reservoir for the titrant, and store the titrant over amalgamated zinc in the reservoir under nitrogen and transfer to the buret by nitrogen pressure. The titration vessel is a 250-ml beaker and the total volume of the solution titrated always 100 ml. The beaker is stoppered with polyethylene because rubber is attacked by solutions of osmium in hydrochloric acid. Provide inlets for the introduction of nitrogen, the electrodes, and the buret tip. Use a coarse-porosity, sintered, borosilicate gas-dispersion cylinder to pass nitrogen over or through the solution. Scrub prepurified nitrogen with chromous chloride and then with water, for use in deaeration and also for protection of the titrant in the reservoir. Magnetic stirring is employed during deaeration and titration. Use a coiled platinum wire indicator electrode and a commercial saturated calomel reference electrode. Prepare stock solutions of osmium(VIII) by dissolving weighed amounts of the reagent grade tetroxide in cold water and diluting to known volume.

Procedure

The aliquots of osmium solution should contain 0.05 to 0.5 moles of osmium(VIII). Add each aliquot to a mixture of 5 ml of ethanol with 25 ml of 2.5 M sodium hydroxide in a 250-ml beaker and allow to stand at room temperature for 20 min. Add hydrochloric acid and water to give a final total of 100 ml of solution containing 0.5 M acid. Stopper the beaker, deaerate the solution, and titrate the sample.

Three inflection points are obtained, the third one is used to calculate the osmium content. The first inflection corresponds to the reduction of osmium(VI) to osmium(IV); the second is ill-defined; the third corresponds to the complete reduction of the osmium to the +3 state and occurs at −0.11 ± 0.01 V versus the saturated calomel electrode in 5.0 M hydrochloric acid. The occurrence of two inflection points during the reduction of osmium(IV) to the +3 state is probably due to the presence of two osmium(IV) species in solution which are in slow equilibrium.

CHAPTER 6

FIRE ASSAY

INTRODUCTION

The mixture of chemicals and ore used for the pot fusion provides a complex system. The chemistry of the fusion process is practically unknown. By analogy with simple systems such as a metal oxide with borax or silica one may make reasonable guesses concerning some of the reactions, but a complete explanation of the reaction for even one ore composition must await an extensive examination of these multicomponent systems. Thus the technique of a fire assay collection of the noble metals is largely an empirical process assisted to a degree by some fundamental principles.

Initially the ore to be assayed must be in an exceedingly fine state of division and thoroughly mixed with the flux constituents. These conditions are necessary to ensure the intimate contact of each ore particle with particles of the melting flux. Ideally this contact must be maintained during the early stage of the fusing process. This is necessary in order to ensure in situ a sufficiently complete

177

reaction between ore and flux and the simultaneous production of the fine globules of lead by the reduction of litharge (PbO). To bring about this fortuitous condition, the composition of the flux, the temperature, and its rate of increase must be carefully arranged. The optimum viscosity is somewhat dependent upon the proportion of borax and the character of the borates and silicates which are formed at or near incipient fusion.

In the absence of this juxtaposition of lead and noble metal one must depend upon the ability of the high density noble metal to settle to the bottom of the pot during the subsequent fusion process. This settling process is facilitated by the increase in fluidity of the mixture at the elevated temperatures, and for platinum, palladium, and gold at least, it is also facilitated by the alloying tendency of the lead finally collected in the bottom of the pot. Thus, for certain of the noble metals, a reasonably acceptable assay may be achieved even under unfavorable conditions. This opinion is supported by the fact that the classical fire assay does not accomplish the recovery of iridium through the formation of a homogeneous lead alloy. It is not unlikely that practically all of the iridium, and iridosmine, when the latter is present, are recovered by the fall of these high density metals through the low viscosity liquid in the later stages of the fusion. Anyway, there is ample evidence that in general all six of the platinum metals can be quantitatively recovered by the lead extraction.*

PREPARATION OF SAMPLES

The various techniques for the correct preparation of samples are discussed below. It is emphasized here that a high degree of fineness is essential not only from the sampling point of view but also to ensure the conditions for the initial optimum fusion. In general at least 150–200 mesh samples should be prepared. The weight of sample for an assay will, of course, depend somewhat upon the noble metal content. Modern analytical methods have developed to the degree that a careful analytical chemist can deal with a few micrograms of metals and achieve very acceptable results. On this basis small amounts of many concentrates and some ores could be used for a determination. There exists, however, the problem of securing representative samples, and this necessity is promoted by samples of the order of 20–40 g. Furthermore, one cannot justify the application of microgram methods, except as a necessity. The use of amounts of the noble metals of the order of a few milligrams to about 200 mg allow the application of methods of separation and determination which fall within the abilities of a

---

*Evaluation of various methods for analysis of platinum metals applicable to some ores (Lichadeev *et al.* (340a)).

reasonably well-trained analytical chemist. The usual sample of ore or concentrate varies between about 10 and 40 g. Where the total noble metal content of this sample is of the order of a few micrograms or less it is customary to apply some prior method of concentration such as flotation, gravity separations, etc., or prior separations by wet acid extractions.

Two methods for obtaining acceptable uniformity are now given. The procedure of Chow and Beamish (341) was used prior to wet assay, fire assay, and neutron activation methods, while that of Strasheim and Van Wamelen (342) was used prior to spectrographic determination. The ore sample received in the laboratory usually weighs less than 100 lb and is crushed and ground to 100–200 mesh.

PROCEDURES

Method of Chow and Beamish (341)

Split the ore by hand riffling and accept the two portions if their weights differ by less than 1%; otherwise recombine the portions and repeat the process. When the sample weight has been reduced to a convenient amount, place the ore on a large cellophane sheet and roll thoroughly by lifting alternate corners of the sheet, taking care to avoid losses because of too vigorous rolling. Spread the ore evenly on the cellophane and mark the surface in 1-in. squares. Obtain the sample required for an assay by taking a small portion from each square, making sure that each portion extends down to the surface of the cellophane.

Method of Strasheim and Van Wamelen (342)

Weigh 1 assay ton (approximately 29 g) of a finely ground sample passing through a 200-mesh sieve into a 150-ml glass beaker and add 50 ml of aqua regia. Slowly boil the mixture to dryness. Transfer the resultant brownish crust to a silica crucible and fire at 800°C for 2 hr while passing a current of air slowly through the furnace. Cool the sample and grind for 5 min in a Spex Mixer Mill. Divide the sample and analyze each part.

It is apparent that true homogeneity is impossible unless the noble metal is dissolved homogeneously throughout the entire ore body. In practice, real homogeneity is not obtainable by any physical procedure. The apparent homogeneity depends on the physical nature of the noble metal, the size of the ore particles, and the size of sample taken for analysis. The use of the assay ton (approximately 29 g) for the fire assay, taken from a well-mixed sample, often provides apparent homogeneity.

Note.   In order to simplify calculations it is customary to use a factor weight of sample. In Canada and the United States, the weight used is such that each milligram of noble metal recovered is equivalent to a troy ounce per avoirdupois ton of ore. Because there are 29.166 troy ounces in a 2000-pound avoirdupois ton, 1 oz of noble metal per ton is represented by 1 mg for each 29.166 g of ore.

In England and Australia the long ton of 2240 pounds is used, and the factor weight becomes 32.666 g. These factor weights are termed assay tons. It must be noted too that in some countries the metric ton, 1000 kg is used, and values are reported in grams or kilograms per ton.

FLUXES

In general the flux consituents for the fire assay for platinum metals in ores are sodium carbonate, borax, silica, litharge, and small proportions of a reducing constituent such as flour, which serves to produce the desired amount of lead by the reduction of litharge.

Ores of the platinum metals which contain high proportions of associated base metal sulfides present a special problem. Unless the assay is properly carried out the sulfides may form a matte that will be a third phase in the final slag. The matte, made up of base metal sulfides, is an excellent carrier of platinum metals. In fact, a reasonably good assay can be made in which the matte is the sole collector. With lead as the collector the presence of the matte must be avoided. This can be accomplished by the use of various fusion charges. Not infrequently, excess of litharge charges are recommended. These involve the presence in the fusion of relatively large amounts of litharge with a deficiency of acid constituents. The result is a subsilicate or basic charge. Aside from the fact that these fusions may involve losses of iridium, ruthenium, etc., there is the objectionable feature that corrosion of the pot becomes excessive, and in fact a single fusion may destroy the pot.

In so far as the platinum metals are concerned the present authors recommend either of two methods for the assay of high sulfide concentrates; the roasting procedure or the nitre* assay. The nitre assay involves the partial oxidation of the sulfides by the addition of potassium nitrate; the excess of base metals sulfides is then used to produce the lead button. In a nitre fusion the formation of a matte seems to be encouraged by an excess of acidic oxygens (e.g., SiO_2) over basic oxygens (e.g., Na_2O, CuO, NiO, and Fe_2O_3) and also by a slow fusion at a low temperature. These suggestions may find support in the relatively low formation temperatures of lead silicates and the higher temperatures required for the oxidation of sulfides by the litharge or nitrate. It is desirable to arrange the rate of heating so as to avoid reactions with the acidic constituents until oxidation of the sulfide by nitre and by litharge has been completed. In general it has been recommended that, with the nitre fusion, the pots should be placed in the furnace at 800°C, heated rapidly for about 30 min, and then removed at about 1000°C.

*Nitre is a trivial name for potassium nitrate.

For the nitre assay the authors have used the flux composition indicated in B, Table 6.1.

For nonsulfide ores add silver to obtain a bead containing the proportion of 20 silver to one of platinum metals. For sulfide ores the ratio of silver to platinum metals is also 20 to one.

Strictly the composition of the fluxes should be calculated from the determined composition of the ore or concentrate. Details for this technique are discussed by Beamish (133a) and by Bugbee (32).

FUSION TEMPERATURES

The initial temperature, the rate of heating, and the finishing temperature are important factors in a successful fire assay for the noble metals. Obviously the degree to which these variables may be controlled will depend somewhat upon the character of the heating equipment, e.g., furnaces which are heated by gas and air, by coke and air, or by electrical methods.

To the degree that heating can be controlled it is desirable to subject the pot contents to an initial sintering temperature. At about 600°C the litharge is reduced by carbon, and the resulting lead droplets are homogeneously mixed with the charge. Simultaneously, a relatively slow chemical action between the flux constituents and ore takes place. Carbon dioxide is evolved, and provides some mixing during the reaction between the base and acid constituents. This is an important phase of the assay, and may occupy some 15–20 min, varying beyond these limits with viscous slags.

TABLE 6.1
Flux Compositions

	Ores (g)			Ores (g)	
	A	B		A	B
Ore[a]	15	30	CaO	4.5	12
PbO	85.0	275	$Na_2B_4O_7$	–	–
SiO	10	15	KNO_3	–	16
Na_2CO_3	21.1	40	Flour	1–3	–

[a]In the case of blank assays for A, 15 g of silica replaced the 15 g of ore; with B, 30 g of ore was replaced by 12 g of silica, 3 g of copper sulfide, 3 g of nickel sulfide, and 12 g of iron sulfide. For ores relatively free from sulfide the composition is indicated in A.

After the sintering process the temperature is increased to a full red heat, usually about 1100–1200°C for platinum metals assays. During this period the mix continuously becomes less viscous, and the lead droplets with their noble metal content drop to the bottom of the pot to form the button.

Some assayers suggest a high finishing temperature, e.g., 1200°C. The hope is to allow the noble metals which do not alloy readily with lead to fall to the button area. This is, of course, a reasonable precaution because such metals as iridium and the alloy iridosmine do not react readily with lead. These noble metal constituents have a density comparable to that of gold and should fall readily through the fluid flux. In those instances where assays for a high iridium content are required the assayer may wisely choose to finish at about 1200°C. It has been the present authors' experience that in many instances the finishing temperature need not proceed much beyond about 1100°C. With the type of furnace used by the author (see Figs. 6.1 and 6.2 (133a)) the sintering process previously described is accomplished without any regulation of temperature. The pot charges are placed in the furnace at about 900°C. Presumably the rate of heating from the cold charge allows the simultaneous production of lead and a sufficient decomposition of the ore. Because the rate of temperature rise is normally slow the controls are immediately adjusted to allow an ultimate temperature of 1100–1200°C. Normally this period varies between about 50 and 60 min, but may be considerably extended as the heating bars become aged. The slag is quite fluid during the final period of about 20 min. In various industrial laboratories the charges are inserted at about 950°C, and are allowed to remain at this temperature for 45 min before pouring. Under these conditions the slags are usually reassayed. The problem of slag losses has been investigated for each of the six platinum metals, and the variations of these losses with the six metals and with slag compositions are discussed (133a).

LEAD BUTTONS

No data have been recorded to allow a prediction of the optimum button size for any particular ore or type of sample. It has been stated that the weight of lead button should be related to the amount of charge, increasing somewhat from certain lower unstated limits. On the other hand the data recorded for silver determinations by fire assay suggest that the collecting power of a given weight of lead bears no precise relationship to the amount of the charge.

Undoubtedly, the amounts of the noble metals in the ore sample will set limits to the lowest weight of lead required for a quantitative collection. This relationship has not been expressed numerically.

One may approach this problem from a reasonable point of view, however. It is traditional that a generally satisfactory button size varies between 25 and 35 g.

FIGURE 6.1. Globar assay furnace; connected load 15 kW (20 hp), capacity approximately 35 15-g pots, 28 20-g pots, or 24 30-g pots; Carbofrax hearth, working space approximately 14.5 in. wide by 22 in. deep by 8.5 in. to the underside of the arch. Foot operated door, manually operated vent, and chamber roof vent for cupelling. Maximum operating temperature is 2350°F.

Evidence is presented below to indicate that 10 g buttons may be ineffective for complete collection in some instances. One may guess that with high values of noble metals some advantages may accrue from amounts of lead in excess of 35 g, in that there may be some improvement in collection during the fall of the larger lead weight. Anyway, if larger amounts of lead are desired this should not be accomplished at the expense of the litharge added as a flux.

Concerning the identity of the dissolved and mechanically mixed platinum metals constituents of the lead button there are practically no data. The system of

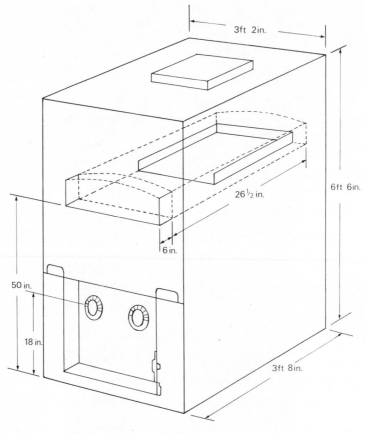

FIGURE 6.2.

lead and the six noble metals is an exceedingly complex one, and information about its character must be derived from guesses about much simpler systems. It seems reasonably certain that iridium, in some forms at least, is mechanically mixed. This guess also applies to at least some of the iridosmines. Because these constituents have a high density (Ir = 22.42) they will be found concentrated at the apex of the button. Thus the traditional method of hammering the button to form a cube cannot, in general, be recommended, because mechanical losses may well occur. Normally the button should be bright, soft, and malleable.

In those instances where slag adheres to the button it can be removed by gentle tapping with a thin metal rod.

Buttons which are obviously badly contaminated should be scorified. This contamination of buttons may arise from excessive proportions of base metals such as copper or nickel in the ore, together with an unsuitable choice of flux composition. "Dirty buttons" cannot be cupelled effectively because the base metals deposited on the surface of the cupel may carry "values," and the cupelled bead will also be contaminated. Furthermore, when wet treatments of the button are contemplated the presence of excessive amounts of base metals will complicate the wet treatment, which is already usually fraught with sufficient unavoidable difficulties.

SCORIFICATION

The scorification process may sometimes be used as a substitute for a pot assay. It is used extensively for certain types of silver and gold assays. In these instances it involves mixing the ore sample with lead, and covering with borax. The reactions involved are similar to the pot fusion, but there is also a series of reactions between air and the constituents of the charge. The oxidized lead forms part of the fusion mixture, and the residual lead acts as a collector. The scorification assay is not, in general, recommended for noble metal determination. The process is essential, however, for the reduction of button size and, as stated above, for cleaning a badly contaminated button. In this instance the button is transferred to a scorifier, and covered with a few grams of borax, the amount depending upon the degree of contamination. At the required temperature of 1050–1100°C and in the presence of air the melted lead is oxidized, and together with the borax forms a slag with the oxidized base metal contaminants. The slag moves progressively to the periphery of the scorifier, and the molten lead forms the center or *eye* of the fusion. The melt may be poured as in the pot fusion.

CUPELLATION

The classical fire assay with lead as a collector usually requires the presence of silver. Thus the lead button includes, in addition to noble metals, an amount of silver sufficient to collect the noble metals, and in proportions sufficient to allow parting procedures. The process used to produce this silver bead or *prill* is known as cupellation.

A second method of assay sometimes called the dry assay is best accomplished in the absence of silver, however, and involves a cupellation to produce a mixed residue of noble metals. In this instance the residue seldom takes the form of a spherical bead and is usually found in a noncoherent deposit. Generally the method is used for the more common noble metals, and when accurate results are not required. The subsequent method of analysis involves the process of dissolution, e.g., chlorination, and then the application of one of the separatory methods, the choice of which depending upon the composition of the residue.

The techniques and characteristics of cupellation have been studied in detail for silver and gold, particularly for the former. Detailed descriptions may be found in such standard texts as Bugbee (32), Smith (50), Fulton and Sharwood (343), and others.

Relatively little factual data have been recorded concerning the cupellation of buttons containing the platinum metals. Some of the characteristics of the cupellation for silver, however, apply also to platinum metals. As would be expected, the temperature variations and control, which are important factors for silver because of its volatility, are of relatively little importance for the platinum metals.

The process of cupellation is accomplished in vessels called *cupels*. The construction and desirable characteristics of these have been discussed (133a).

Under suitable conditions of temperature and air exposure the lead button in the cupel is oxidized to form litharge, which is liquid at its formation temperature. Differing from the noble metals constituents, the litharge wets the cupel and is absorbed, carrying with it such base metals as copper and nickel. In general suitable cupels will absorb about 98.5% of the litharge formed, the remaining 1.5% being volatilized. The process is an interesting one; under the correct conditions the lead melts quickly and then forms a gray scum of lead oxide over the surface of the melted metal. This fuses rapidly, and the melted alloy is quickly exposed, this phenomenon being known as "uncovering" or "opening." If this does not take place during the initial stages after melting, the button is said to be "frozen." Sometimes the frozen alloy may be "driven" or "cleared" again by raising the temperature, and placing a piece of charcoal adjacent to the cupel. The present authors prefer to discard frozen buttons.

Freezing is usually due to low temperatures prior to the transfer of the button. It is also sometimes caused by a temperature drop during cupellation. With the progress of the cupellation the liquid litharge flows over the surface of the melted lead and is absorbed by the cupel. It appears as a dull red or black ring on the surface of the cupel. Featherlike crystals of litharge may also appear on the sides of the cupel above the lead. At lowered temperatures these feathers accumulate and eventually cover the melted alloy, and the process results in freezing. It is generally advisable to increase the temperature toward the finish of the cupellation process. Theoretically, this is justified by the increase in noble metal proportions as the lead is removed. Whereas this precaution is an important factor in the assay for silver and to a lesser degree for gold for the removal of traces of lead, it is the present authors' experience that lead is not thus entirely removed, and in the subsequent wet treatment of the noble metal residue there is nothing to be gained by merely reducing the amount of the lead trace. Concerning gold and silver, but not always the platinum metals, the final stage of cupellation is marked by a sudden clearing of the previously driving bead. This brightening follows the removal of the final very thin layer of litharge over the surface of the bead.

The resulting play of colors is followed in a few seconds by a dullness in the bead and then a sudden brightening. With the platinum metals the final stage of cupellation is not usually clearly defined. The platinum metals, if present, will display a variety of effects on the surface of the cold silver bead.

Improper or hasty cooling of silver–noble metal beads will result in the sudden expulsion of oxygen, which is quite soluble in molten silver. The result of this action is one or more protruberances of silver from the bead. The process is called "spouting" or "spitting." With gold–silver beads the excrescences are objectionable, because the necessary coherence of the gold during the parting process is prevented. With the platinum metals violent spitting may result in mechanical loss; the various processes of parting, however, invariably produce at least a partially disintegrated residue, which introduces no particular difficulties in the wet methods of determining the platinum metals.

It may be emphasized here that the silver bead collection and its subsequent wet treatment, when properly applied, serves as an excellent method for the determination of gold, palladium, and platinum together with traces of rhodium, iridium, and ruthenium. It is quite unacceptable for osmium. For larger proportions of the more insoluble platinum metals, and where the amount of osmium is required, the direct wet treatment of the lead button is preferable. Unfortunately the wet methods at present available for lead buttons are complicated by the persistent presence of lead salts. Precipitation of lead as its sulfate from a nitric acid parting solution is by no means an effective procedure, because the evaporated filtrate yields significant amounts of the lead sulfate mixed with the palladium, traces of platinum and perhaps rhodium, etc. Furthermore, the retreatment of the lead sulfate to recover traces of platinum metals either by extraction with ammonium acetate or by scorification is cumbersome and attended by uncertain accuracy. Because the direct treatment of the lead–noble metals button offers advantages over the silver bead method, one may hope that the newer techniques will provide improved methods of separating lead from parting acids. In so far as an immediate recognition of the presence of platinum metals is concerned, however, the silver bead, as compared to the lead button, is a much better basis from which to learn not only the presence of the platinum metals but, within limits, to make reasonable guesses as to practical values.*

ASSAY OF CONCENTRATES RELATIVELY FREE FROM BASE METAL SULFIDE

PROCEDURE 79 (344)

Grind the ore or concentrate to pass a 100 mesh screen. Mix the weighed sample of ore or concentrate with the flux—see Table 6.2, columns A and E (E if improved recovery is

*Determination of gold, platinum, ruthenium and iridium during the cupellation process (Georgiev *et al.* (343a)).

TABLE 6.2
Flux Compositions

	Gold ores (g)		Cupel assay (g)
	A	B	E
Ore[a]	15	20	40[b]
PbO	85.0	275	37.5
SiO₂	(10)	15	–
Na₂CO₃	21.1	40	22.5
CaO	4.5	12	–
Na₂B₄O₇	–	–	37.5
KNO₃	–	16	–
Flour	1–3	–	2

[a]For blanks, 15 g of silica replaced 15 g of gold ore and a mixture of 12 g of silica; 3 g of copper sulfide, 3 g of nickel sulfide, and 12 g of iron sulfide replaced 30 g of nickel ore.
[b]The cupel material contained lead oxide from the cupellation.

needed). In low silver content ores add silver powder or solution (to give at least a 1 mg bead on cupellation). Transfer the mixed ore and flux to an assay pot of suitable size. Place the samples in a furnace at 1040°C and heat to 1150°C as quickly as possible (usually 45–60 min). After 30 min at 1150°C, remove the crucible and pour the melted sample into an iron mold and allow to cool. The slag can be separated from the lead button by gentle tapping with a small hammer. Cupel the buttons in magnesia cupels which had been previously heated to 980°C. This must be done quickly to prevent freezing. After "opening" for 5 min at 980°C, open the air draught and "drive" the sample for 25–30 min. Close the draught and allow 5 min for finishing and withdraw the cupels gradually to prevent spitting. Methods for the analysis of the silver bead are described in Procedures 102, 103, and 104.

PROCEDURE FOR NITRE ASSAY

PROCEDURE 80 (133a)

Add to the mixing sheet the properly sampled ore or concentrate, together with the fluxes (see Table 6.2, columns B and E (E for improved recovery)), the excess litharge for the button, the potassium nitrate, and its corresponding weight of sodium carbonate and silica. It may be preferable to screen prior to mixing.

Transfer the mixture to an assay pot of an appropriate size. When a silver bead is required, add the silver powder or solution to the pot mixture. Place the pot in the muffle at 1100°C, and fuse for about 45 min. Continue as described below in Procedure 81, pouring the fused mixture into the iron mold.

In general this nitre assay is comparable in accuracy of recovery of platinum metals to any other type of fusion.

PROCEDURE FOR ASSAYING ROASTED SULFIDE ORES

PROCEDURE 81 (133a)

Grind, mix, and sample the ore as described above. Transfer the weighed ore to a 6-in. porcelain evaporating dish, and place for a few minutes at the front of the furnace with the door open. The initial temperature should be about 600°C. Over a period of about 5 min move the dish to the furnace center, stirring intermittently. Partially close the door, and allow to roast for 2 hr at 950°C, frequently stirring for the first 30 min to avoid agglomeration of the concentrate. Remove the dish, cool, transfer the contents to a mortar and grind to a fine powder. Take care to avoid any loss of powder.

In general a bisilicate slag is a suitable medium for the collection of platinum metals in oxidizing ores. For samples with high proportions of associated base metals—copper, nickel, etc.—it may be desirable to increase the proportion of litharge. This can be done by using an increased ratio of flux to ore. For ores of the nonoxidizing or nonreducing types a flux ratio of 2 to 1 or 3 to 1 of ore is usually satisfactory. This may be increased considerably to provide a satisfactory slag. Decide on a proper ratio of flux to ore. If a silver bead is to be prepared, add to the flux an amount of silver powder or silver in solution to give a ratio of approximately 20 of silver to one of total platinum metals expected. Arrange a cellophane sheet in a suitable position, and pass the roasted ore through a 45-mesh sieve to the center of the sheet. Add a portion of the flux to the sieve to remove any traces of ore. Transfer most of the remaining flux to the mixing sheet, and mix thoroughly. Place the mixture in an assay pot of an appropriate size. Add the remaining flux to the sheet, mix, and transfer to the pot, taking care to brush the sheet free of the mixture.

Place the pots in the furnace at about 950°C, and raise the temperature at the maximum rate to 1200°C. This fusion period should be approximately 1 hr. Remove the pots, pour the mixture into conical iron molds, and allow to cool. Remove the button, taking care to retain all of the slag. Free the button of slag by gentle tapping with a small iron rod. Set the button aside. Transfer the slag to a grinding mill or to a mortar, grind to pass a No. 45 standard sieve, and place the sample on the original mixing sheet. Clean the mill, mortar, and screen with sufficient litharge to produce a second button, then transfer to the slag on the sheet. Mix well as before, and transfer the mixture to the original pot. Fuse as before, and clean the lead button as described above. If necessary a fusion of the second slag can be made to produce a third button.

Transfer the two or three lead buttons to a 3- or 4-in. scorifier previously heated to about 300°C. If there is evidence of copper or nickel in the buttons, a few grams of borax may be added to the buttons on the scorifier. Continue the scorification at about 1000°C to secure a button of about 25–30 g, and then pour into the iron mold. When the lead button is to be analyzed for osmium and ruthenium, part and treat it as described in Procedure 98 (133a).

When the noble metals are to be concentrated to form a silver bead clean the button thoroughly by gentle tapping with an iron rod, and transfer it to a bone-ash cupel which has been preheated at 900°C for at least 10 min. Continue heating the cupeling button at about 1000°C, with a plentiful supply of air. Remove the cupel over a period of a few minutes after the completion of the cupellation process (see the above description of the cupellation process). The determination of platinum metals in the silver-assay bead may be accomplished by any one of pertinent Procedures 102, 103, or 104.

Note. It should be stressed here that in general the classical methods of cleaning either a button or a bead by hammering it to form a square button or to flatten the bead are not recommended. The more insoluble platinum metals, e.g., iridium, are usually mechanically mixed with the silver or lead alloys, and losses will almost invariably occur through any interference with the shape of the alloy.

IRON–COPPER–NICKEL COLLECTION

Because of the nonquantitative collection of the more insoluble platinum metals in the classical fire assay and because in a field which involves extensive financial implications, other proven methods of isolating precious metals from ores are required, the author has in recent years attempted to develop a fire assay extraction that makes use of the naturally occurring base metals associates. In the earlier experiments (191) the iron, copper, and nickel in the roasted natural ore were reduced to form a button containing the platinum metals. The reduction was accomplished by the walls of the carbon pot placed in a high frequency furnace at 1450°C. With ores containing insufficient base metals, or with synthetic ores, the base metals were added in the form of oxides. Because the reaction time was short and difficult to control, thus producing excessively large buttons, and because the equipment was expensive and not readily available, efforts were made to devise an assay method based upon similar principles but which would allow control of button size and would require furnaces of standard types. This aim was accomplished, and a procedure was developed (133a) that was comparable in character with the classical lead collection. The base metal alloy button was prepared by reduction with mechanically mixed carbon, sodium carbonate, borax, and ore, with base metal oxides added if they were required. The standard clay crucibles were heated in a gas/air furnace to 1450°C, as determined from the top of the melt by an optical pyrometer. Button size was controlled by the amount of carbon with an efficiency equal to that obtained with the lead button. The base metal button was then parted and treated as described in the following.

In a series of papers, the above assay method has been proven for ores salted with each of the platinum metals. For either osmium or ruthenium (133a, 345) the iron–copper–nickel button is parted with 72% perchloric acid in a standard distillation apparatus. The volatile oxide is distilled from the parting medium, collected in concentrated hydrochloric acid, and precipitated by thionalide (Procedure 66 for osmium and Procedure 63 for ruthenium). For rhodium, the button is parted with hydrochloric acid and finally with a small addition of nitric acid. Base metals are removed by cation exchange and rhodium is determined gravimetrically by precipitation with mercaptobenzothiazole (Procedure 59) and spectrophotometrically with the tin(II) chloride reagent (Procedure 40).

Buttons containing iridium and platinum with palladium are prepared and parted with aqua regia. For iridium the gravimetric determinations are made with 2-mercaptobenzothiazole (Procedure 61) and spectrophotometric determinations are made with leuco-crystal violet (Procedure 42). Platinum is determined gravimetrically with hydrogen sulfide (Procedure 54) and palladium with dimethylglyoxime (Procedure 52). Spectrophotometrically, palladium is determined with p-nitrosodiphenylaniline (Procedure 28) and platinum with tin(II) chloride (Procedures 31 and 32).

In addition to the successful collection of each of the six platinum metals, recent research has indicated that the most resistant iridosmines such as the Tasmanian variety, previously used for fountain pen tips, are collected by the iron–copper–nickel button, and furthermore, that the alloy will dissolve completely in the acidic parting solutions.

The method has not yet been applied directly to sulfide ores and concentrates because the constituents of the charge would produce a matte. These concentrates are subjected to a prior roasting, and thus there is the possibility of losses of osmium and perhaps some ruthenium. Whereas the extent of these losses has not yet been investigated it is not unlikely that roasting under controlled conditions or under hydrogen will allow the collection of any volatilized osmium and ruthenium.*

The flux constituents are carbon, sodium carbonate, and borax. The ratio of acidic oxygen to basic oxygen in the final charge will, of course, depend upon the proportions of the flux constituents and the degree to which each of the base metals is carried into the button. Because no method is yet available which allows even a reasonable prediction to be made of the distribution and destiny of base metals when they are in excess over the amount required for the button collection, one cannot calculate the final slag acidity. In the present instance the flux used contains an acidic to basic oxide ratio of 0.80 to 0.56.

While different forms of carbon such as flour, etc., may be used, graphite provides the most predictable weight of metal alloy. Flour is about 25% as effective as graphite. Initially the graphite was obtained by grinding graphite rods, but aside from the dust problem the fineness of the particles made necessary some screening; commercially produced grade A-200 graphite proved consis-

---

*A successful application to sulfide ores has been reported by Van Loon (346) (see the following text).

tently satisfactory, however. The relationship between the weights of the graphite and the button must be determined experimentally as in the classical lead method. The optimum button sizes or compositions have not yet been determined. The detailed characteristics of each of the three base metals as collectors of each of the platinum metals also awaits further research. Progress is reported in Beamish and Van Loon (347). Furthermore, there is the problem of the optimum temperatures for the alloy collection. Obviously an increase in the copper content of the button will allow lower fusion temperatures, but here again the relative effectiveness of copper as a collector will have a determining influence upon optimum conditions of assaying.

The new assay procedure significantly effects the reduction of slag losses of platinum and palladium. Nickel is a selective and persistent carrier of platinum and palladium, and its presence in the slag from the classical fire assay increases the risk of some loss of platinum metals (348). With high proportions of nickel this loss may be irrecoverable by fire assay with lead as the collector (349, 350). Because nickel and copper are preferentially reduced, the new method has the advantage of a practically complete transference of the nickel to the button. The slags resulting from the application of the new fire assay to a natural ore concentrate revealed the complete removal of platinum and palladium. Fire assays made on synthetic ores salted with platinum and palladium and containing very high proportions of nickel resulted in slags containing both nickel and platinum metals. These slags may be retreated to remove completely the nickel and the remaining traces of platinum and palladium. With synthetic ores containing a high proportion of copper but no nickel or iron, the slag losses were small.

Hydrogen roasting to prevent losses of volatile osmium and ruthenium tetroxides replaces the usual air roasting of sulfides in a scheme which allows the inclusion of these elements in the iron–copper–nickel fire assay method (347). Ores and concentrates thus treated were subjected to the normal iron–copper–nickel fusion procedure and recoveries of the six platinum metals could be obtained from button solutions. The buttons were dissolved in HCl and any remaining residue solublized by chlorination. Cation exchange to remove base metals followed by distillation was used to recover osmium and ruthenium. Platinum, palladium, rhodium, and iridium remaining in filtrates and distillation pot liquids can be recovered by existing procedures.

High temperatures, beyond the range of equipment commonly available in some laboratories are necessary for iron–copper–nickel fusions. Banbury and Beamish (351) recovered platinum and palladium with a fusion temperature of 1200°C using copper collection alone. The charge contained 8 g of roasted copper sulfide ore plus the normal fluxing constituents. Quantitative recoveries of platinum and palladium in the range 50 to 10 mg were obtained. Reports have been published which indicate the loss of ruthenium during the classical fire assay procedure.* For further detailed procedures involving collection by iron–copper–nickel or by using copper as a collector see Beamish and Van Loon (347).

Some data dealing with the character of the alloys formed in the iron–copper–nickel button have been recorded by Plummer *et al.* (352). The button alloy with palladium appears to be a single phase solid solution. There is also good evidence that platinum, rhodium, and osmium form solid solutions. The black residue recovered from buttons containing approximately 20 mg of iridium or ruthenium suggest that collections of this magnitude may be at least partly a mechanical process. This remains uncertain, however; anyway, amounts of the order of 10 mg each of ruthenium and iridium in iron–copper–nickel buttons are completely dissolved by the parting acid.

PROCEDURE 82

Transfer 100 g (3.43 assay tons) of ore or concentrate to a 6-in. porcelain dish, and roast in the assay furnace at 980°C for 2 hr. Stir intermittently to encourage the oxidation. Mix the cooled calcine with 42.4 g of sodium carbonate, 27 g of borax glass, and 9.5 g of 200-mesh amorphous graphite. Crush the mix if necessary to pass a 45-mesh screen, and mix again on the mixing sheet. Transfer to a 30-g pot, and place in a gas furnace preheated to 1200°C. Turn off the gas and air, the cooling period allowing the reactions to proceed without an overflow of the charge. When the volume of the pot mixture is reduced to about one-third of the original volume, reignite the gas–air mixture, and increase the temperature over a period of about 45 min to 1250°C. Then raise the temperature over 1.5 hr to 1450°C, at which time turn off the gas, and remove the pot. Break the cooled pot, separate the button, and remove adhering slag by gentle tapping with the end of a small iron rod.

The method of parting and subsequent analysis of the button are described in Chapter 7.

THE FIRE ASSAY FOR PLATINUM METALS USING NICKEL SULFIDE AS A COLLECTOR

Robert *et al.* (353) used a flux consisting of a 2-to-1 mixture of sodium carbonate and sodium tetraborate with a collector of nickel sulfide. The presence of at least 10 g of silica is desirable for a good fusion. The silica is usually contributed by the sample, but, where it is not, it can be added as powdered silica. With a flux-to-sample ratio of 1-to-1, 20 g of nickel oxide and 10 g of

---

*Determination of ruthenium losses during assay fusion of products containing copper and nickel by a method of radioactive indicators (Lichadeev *et al.* (351a)).

sulfur, a satisfactory button is obtained at 1000°C. For chromites a suitable fusion required a 3-to-1 flux-to-sample mixture. For samples of concentrate an increase in the flux-to-sample ratio had little or no effect on the recovery of noble metals. The quantity of flux necessary for a consistently high recovery of all the noble metals lies between 90 to 105 g; below 90 the recovery is low and an excess above 105 the recoveries are erratic. The minimum weight of button is about 25 g. Buttons much larger do not increase the recovery of noble metals. Excess of sulfur produced a grayish-yellow button which sometimes disintegrated on standing.

The authors of the nickel sulfide collection have investigated the efficiency of the method relative to the classical lead collection and a wet separation involving selective extraction and recoveries with tellurium. Their data are given in Table 6.3.

The results show that the collection of noble metals with nickel sulfide is, for all the noble metals except gold, superior or equal to collection by the lead method. However, the higher results obtained for platinum by the acid-extraction procedures indicate that its recovery is incomplete by both fusion procedures. Ruthenium, osmium, and iridium are not determined by the acid-extraction procedures. A further disadvantage of the integrated nickel sulfide method is that it is time consuming compared to competitive techniques (e.g., lead assay). However claims for its superiority for some sample types justify its inclusion here.

Of the two fire-assay procedures, that using nickel sulfide as the collector has several advantages over the lead-collection method. For instance, a smaller flux-to-sample ratio and a lower fusion temperature (1000°C as against 1200°C) are used. Furthermore, the method is applicable to all six platinum-group metals and can be applied to samples high in nickel and sulfur without the pretreatment that is required in the lead method. No change in flux composition is required for different types of samples, except for chromite ores, where the quantity of flux used must be higher.

The advantages of the lead-collection procedure are that it requires less time for analysis, particularly in determinations of total platinum-group metals, and that the gold recovery is some 10 to 20% higher than with the nickel sulfide method.

Apart from the possible incomplete collection of platinum and gold, the nickel sulfide procedure offers a precise and accurate method for the concentration and isolation of the noble metals in samples of ores, concentrates, and mattes.

Its applicability to the different samples thus far encountered, together with the simplicity of the technique renders it an extremely useful procedure for the analysis of the noble metals.

The method is applicable to the determination of all six of the platinum-group metals in ores, concentrates, and mattes. The recovery of gold appears to be

TABLE 6.3
Platinum-Group Metals Recovered by Different Procedures[a]

Sample description	Method of concentration	Number of results	Average results (ppm)						
			Pt	Pd	Ru	Rh	Ir	Os	Au
Chromite	Lead collection	8	3.11 ± 0.07	1.39 ± 0.03	0.52 ± 0.05	0.53 ± 0.01			0.10 ± 0.01
	Acid extraction	18	3.70 ± 0.15	1.26 ± 0.01	1.16 ± 0.02	0.53 ± 0.01			0.12 ± 0.01
	Nickel sulfide collection	4	3.33 ± 0.09	1.42 ± 0.01		0.55 ± 0.01			0.03 ± 0.01
	Extraction by chlorine and acid	12	3.50 ± 0.10	1.45 ± 0.07		0.53 ± 0.03			—
Ore (Merensky Reef)	Lead collection	6	5.38 ± 0.10	0.89 ± 0.03	0.35 ± 0.04	0.37 ± 0.02			0.21 ± 0.01
	Acid extraction	36	6.10 ± 0.24	0.92 ± 0.02		0.37 ± 0.01			0.17 ± 0.01
	Nickel sulfide collection	12	5.41 ± 0.05	0.92 ± 0.07	0.86 ± 0.06	0.36 ± 0.02			0.16 ± 0.03
	Extraction by chlorine and acid	12	6.27 ± 0.20	1.03 ± 0.03		0.45 ± 0.02			—
Flotation concentrate	Lead collection	14	79 ± 1.5	51 ± 1.0	10 ± 0.8	13 ± 0.4			2.50 ± 0.10
	Acid extraction	12	81 ± 0.8	47 ± 1.4		10 ± 0.2			2.30 ± 0.10
	Nickel sulfide collection	12	79 ± 2.4	49 ± 1.5	14 ± 0.4	13 ± 0.2			2.20 ± 0.20
	Extraction by chlorine and acid	12	81 ± 3.2	48 ± 1.8		10 ± 0.3			—
Matte	Lead collection	4	235 ± 2	123 ± 2	23 ± 1	18 ± 0.3	4 ± 0.2	—	10 ± 0.20
	Nickel sulfide collection	4	232 ± 2	128 ± 1	34 ± 1	19 ± 1	14 ± 2	6 ± 0.5	9 ± 1

[a]All determined by atomic absorption spectrophotometry.

incomplete, and the lead-collection method is therefore preferable when gold is to be determined.

Samples containing sulfur need not be roasted before the fusion, although the sulfur content must be taken into account when the required amount of flux is calculated. If the sulfur content is unknown and the determination of osmium is not required, the sample can be roasted.

If difficulty is encountered in the obtaining of a satisfactory fusion and button (as has been experienced with samples containing unusual amounts of zinc), the sample can be leached in concentrated hydrochloric acid before the fusion. (This procedure results in a satisfactory fusion and button from Merensky Reef samples.)

For samples containing more than 10 ppm of total noble metals, platinum, palladium, rhodium, ruthenium, and iridium can be determined in one button. If the total concentration is less than 10 ppm, a separate button must be prepared for the determination of iridium. For all samples, a separate button is prepared for the determination of osmium.

The amounts of sample to be taken for the analysis of platinum, palladium, rhodium, ruthenium, and iridium are based on the estimated content of the platinum-group metals (PGM) and are shown in Table 6.4.

For the determination of osmium, where the whole button should be distilled, the correct weight of sample, based on an estimate of the osmium content, should be taken for fusion. The amounts are shown in Table 6.5.

PROCEDURE 83 (353)

The procedure up to and including the dissolution of the nickel sulfide buttons is common for all the platinum-group metals.

Transfer the required amount of sample (see Table 6.4) to a sheet of glazed paper. Add 60.0 g of fused borax, 30.0 g of soda ash (both assay quality), 32.0 g of nickel carbonate

TABLE 6.4
Weights of Sample for Determinations of All the Platinum-Group Metals except Osmium

Estimate of total PGM (ppm)	Weight of sample (g)
>1000	5.0
500–1000	10.0
250–500	15.0
100–250	25.0
50–100	35.0
<50	50.0

TABLE 6.5
Weights of Sample for Osmium Determinations

Estimate of Os (ppm)	Weight of sample (g)
<8	4
6–8	5
4–6	6
2–4	8
1–2	20
0.5–1	30
<0.5	50

(see Note 1), and 12.5 g of powdered sulfur (see Note 2). For chromite samples, add 90.0 g of fused borax and 45.0 g of soda ash.

Roll the mixture from corner to corner of the paper until it is thoroughly mixed, and transfer it to a No. 1 fireclay crucible. Wipe the surface of the glazed paper with a piece of tissue paper, and place this on top of the mixture in the crucible. With the aid of furnace tongs, transfer the crucible to a furnace at a temperature of 1000°C. Leave the mixture to fuse for 1¼ hr.

Remove the crucible and pour the contents into an iron mold. Allow the mold to cool for 30 min, and remove the button and the slag.

Weigh the button and record the weight. An ideal weight for the button is between 25 and 30 g. Break up the button by pressing it in a hydraulic press. Transfer the pieces to the small bowl of a Siebtechnik mill and grind for 2 min.

Brush the sample into a 600-ml squat beaker, add 400 ml of concentrated hydrochloric acid (C.P. grade), and cover the beaker with a watch glass. Leave the beaker on a steam bath set at a low temperature for approximately 16 hr. The operation is usually carried out overnight. Remove the beaker from the steam bath, wash down the lid and the sides of the beaker with water, and stir the contents with a glass rod (see Note 3). Allow the noble metal sulfides to settle and cool for about 1 hr.

If osmium is to be determined, proceed as described immediately below using a separate button (Procedure 84).

For the determination of platinum, palladium, rhodium, ruthenium and iridium, proceed in the following manner.

Filter the solutions under vacuum, using No. 542 11-cm filter papers in Fisher funnels. Wash out the beakers with cold 50% hydrochloric acid, and wash the papers at least five times with water to remove all traces of nickel. Discard the filtrate. With a fine jet of water, wash the residue that can be readily removed from the filter paper into the original beaker. Pour 10 ml of concentrated hydrochloric acid and 10 ml of 100 vol hydrogen peroxide onto the filter paper contained in the funnel, and cover it immediately with a watch glass.

As the noble metals dissolve, collect them in the original 600-ml squat beaker. Wash the paper three times with water (see Note 4). Retain the filter paper together with any undissolved residue. Transfer the covered beaker to a hot plate, and boil until the solution is reduced to a volume of about 20 ml. Add 2 ml of hydrogen peroxide and boil for 10 min.

Filter the solution into a 150-ml squat beaker using a No. 540 filter paper, and wash with 10% hydrochloric acid.

Ignite the filter paper, together with the above washed paper in a 15-ml porcelain crucible. Add 10 ml of aqua regia to the crucible, cover it with a watch glass, and heat strongly until the residue is dissolved. Add this solution to the original solution in the 150-ml beaker and evaporate to near dryness. Add 3 ml of hydrochloric acid and again evaporate to near dryness. Add 5 ml of 40% hydrochloric acid and warm to dissolve.

Determine the platinum, palladium, rhodium, and ruthenium by an atomic absorption procedure (see Chapter 1). For the determination of osmium proceed as described below.

Notes. 1.   When a new batch of nickel carbonate is brought into use, a number of blank determinations must be carried out to ensure that noble metals are absent.

2.   If the sulfur content of the samples is higher than 0.5%, the quantity of sulfur added as a flux will have to be decreased accordingly. If there is a large excess of the sulfur during fusion, the buttons will be too large and can disintegrate on being cooled.

3.   If the button is large, it may be necessary to add extra hydrochloric acid and continue the digestion at this stage. When the dissolution of the nickel sulfide is complete, the solution should be clear, with a layer of black residue on the bottom of the beaker.

4.   The inclusion of the filter paper, which would represent the introduction of undesirable organic matter, should be avoided.

THE DETERMINATION OF OSMIUM
IN A NICKEL SULFIDE BUTTON

PROCEDURE 84 (354)

Equipment

See Fig. 6.3.

Reagents

Hydrochloric acid (12 N).   Riedel-de Haen A.R. Grade 1.

Hydrochloric acid (6 N).   Dilute 500 ml of 12 N Riedel-de Haen hydrochloric acid (A.R. grade) to 1000 ml with water.

Thiourea solution (10% (w/v)).   Dissolve 2.5 g of thiourea in water. Dilute to 25 ml with water.

Stannous chloride solution (10% (w/v)).   Dissolve 2.5 g of $SnCl_2 \cdot 2H_2O$ (Riedel-de Haen or Merck grade) in 8 ml of hydrochloric acid. Warm gently to dissolve. Cool and dilute to 25 ml with water.

Sodium peroxide.   A.R. grade.

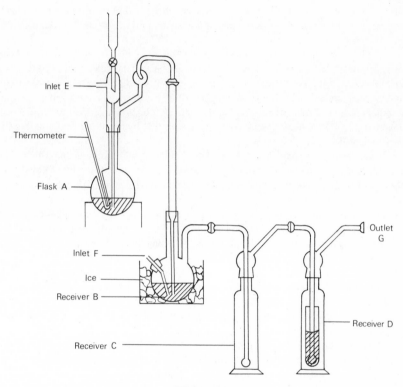

FIGURE 6.3.

Hydrogen peroxide.   50% A.R. grade.

Standard osmium solution.   Dissolve 230.81 mg of "Specpure" ammonium chloroosmate, $(NH_4)_2OsCl_6$, in 50 ml of 6 N hydrochloric acid, and heat gently until the salt has
dissolved. Cool, and dilute to 100 ml with 6 N hydrochloric acid:

$$1 \text{ ml} \equiv 1.0 \text{ mg of osmium}$$

With a pipet add 5 ml of this standard solution into a 500-ml volumetric flask, and dilute
to volume with 6 N hydrochloric acid:

$$1 \text{ ml} \equiv 10 \ \mu\text{g of osmium}$$

Procedure

The procedure consists of five steps: dissolution of the button, ignition, fusion, distillation of osmium, and color development. A reagent blank and a fire-assay blank (to allow
for any noble metals present in the nickel oxide added as collector) must be carried
through the procedure.

Place a weighed nickel sulfide button in a pellet die and break it into pieces under a pressure of 2 to 3 lb/in^2. Transfer the pieces to a swing mill (Siebtechnik), and grind for 1 min. Weigh the ground sample (see Note 1).

Place the sample in an 800-ml beaker and add 100 ml of 12 N hydrochloric acid. Warm the beaker on a water bath and when the reaction is less violent, add a further 200 ml of 12 N hydrochloric acid. To dissolve the sample, warm it gently on a water bath overnight. Cool to room temperature. If all the hydrogen sulfide has been driven off, add a few grains of sodium sulfide. Filter off the residue through a 9-cm Whatman No. 542 filter paper supported in a Fisher funnel. Wash five times with 6 N hydrochloric acid and then five times with water.

Fold the paper containing the residue and place it in a 5-ml nickel crucible which is then placed in a 10-ml porcelain crucible and covered with a Rose crucible lid. Allow a jet of burning hydrogen to pass through the crucible lid onto the filter paper for 30 min (see Note 2).

Heat the porcelain crucible with the low flame of a burner, and gradually increase the heat until full heat is obtained. Heat for 5 min at full heat. Remove the burner, and allow the crucible to cool for 10 min in the presence of burning hydrogen, which must be present in the crucible to prevent oxidation. Remove the Rose crucible lid.

Carefully crush the charred paper with a glass rod, brushing off any particles adhering to the rod. Place 1.0 g of sodium peroxide over the charred paper (do not mix). Cover the porcelain crucible with a nickel lid.

Heat the crucible gently until the reaction between the sodium peroxide and carbon ceases. The end of the reaction will be apparent from the cessation of the small explosions that occur. Remove the nickel crucible from the porcelain crucible, and add 1.0 g of sodium peroxide to the partly fused residue. Fuse again at slightly more than dull-red heat for 30 sec, swirling continuously. Cool slightly, and leach in 50 ml of water. Connect the distillation apparatus as shown in Fig. 6.3. Place 25, 35, and 10 ml of 6 N hydrochloric acid in receivers B, C, and D, respectively. Cool receiver B in ice. Transfer the leach liquor to the distillation flask. Using a vacuum water pump, draw air through the distillation flask, and sulfur dioxide and air through the receiving train for 10 to 15 min (see Note 3).

Heat the distillation flask to about 30°C, and add 12 ml of concentrated hydrochloric acid through the funnel. While adding the hydrochloric acid and later the hydrogen peroxide, adjust the flow of air so that pressure will not build up in the distillation flask and cause the contents to rise up the air inlet. Allow the mixture to stand for 1 min so that all the residue will dissolve. Increase the temperature to about 80°C, and slowly add 4 ml of 50% hydrogen peroxide while increasing the temperature to boiling point. The final 0.5 ml of hydrogen peroxide is added when the temperature is just below boiling point. Allow to distill for 50 min.

Disconnect the sulfur dioxide and turn off the vacuum water pump. Combine the contents of the three receivers in a 400-ml beaker and allow to stand overnight.

Evaporate the distillate until the volume is reduced to about 20 ml. Transfer the solution to a 125-ml Pyrex beaker and add 10 ml of 12 N hydrochloric acid. Evaporate to 1.0 ml (visual comparison) and allow to cool.

Add 1 ml of 10% thiourea solution and 0.1 ml of 10% stannous chloride solution. Transfer to a 10-ml volumetric flask and heat in boiling water for 10 to 15 min. Cool slowly to room temperature and dilute to volume with water. Centrifuge and, in a Zeiss spectrophotometer, measure the absorbance against that of water in a 5-cm cell at 480 nm.

Prepare a reagent blank in the same way as described above, and measure its absorbance against that of water as for the samples. Correct the values for the samples by subtracting the value of this blank from each.

In addition, a blank must be included from the fire-assaying step (see Note 4).

Transfer portions of standard osmium solution that contain 5, 10, 20, and 40 μg of osmium to nickel crucibles, and in the presence of 50 mg of sodium chloride evaporate to dryness on a water bath. Proceed as above with the fusion, distillation, and color development. Draw a calibration curve of optical density against micrograms of osmium.

Notes. **1.**   Weighing of the buttom before and after grinding ensures that a correction can be made for any significant portion lost during the grinding.

2.   To avoid oxidation of the osmium to the volatile osmium tetroxide, a minimum amount of oxygen must be present.

3.   Before the distillation, the 6 N hydrochloric acid in the receivers must be saturated with sulfur dioxide.

4.   Different batches of nickel oxide give different values for the blank. They range between 3.0 and 9.0 μg of osmium. To correct the sample values, the value for this blank must be subtracted from each of the sample values.

TIN AS A COLLECTOR

The scheme (Fig. 6.4) involves the collection of the precious metals in molten tin when the sample is fused at 1200–1250°C with a flux containing stannic oxide, sodium carbonate, silica, borax, and flour. The resultant tin alloy (containing the precious metals as intermetallic compounds with tin) is then treated by relatively simple wet-chemical techniques to prepare solutions in which the individual precious metals can be determined by methods of atomic absorption spectrophotometry or classical absorption spectrophotometry.

PROCEDURE 85 (355)

Apparatus

Assay furnace.   A 15-kW Globar type with suitable thermocouple and temperature controller. The furnace should be capable of accommodating 6 assay crucibles and maintaining their temperature at 1250°C.

Jelrus "Handy-Melt" portable electric furnace.   This small vertical furnace is equipped with removable graphite crucibles and is used for melting tin-base assay buttons before their granulation in water. After 4–6 months of relatively constant use, it is recommended that the bottom of the crucibles be examined for small holes.

Atomic absorption spectrophotometer.   It is recommended that this spectrophotometer have a scale expansion facility. An oxidizing air–acetylene flame is used for the determi-

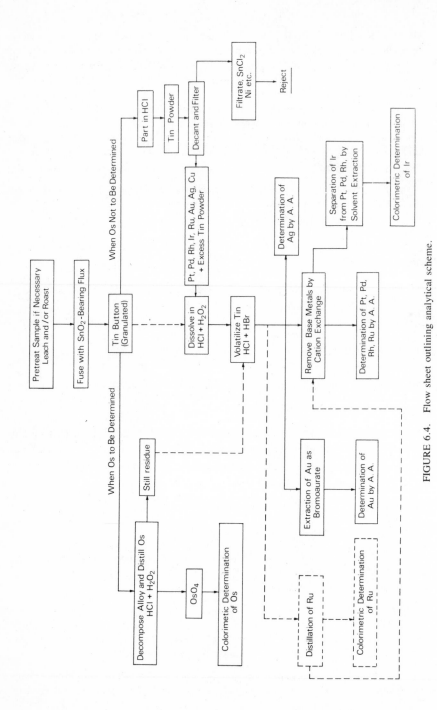

FIGURE 6.4.   Flow sheet outlining analytical scheme.

nation of gold, silver, platinum, palladium, and rhodium; a slightly fuel-rich air–acetylene flame is used for the determination of ruthenium.

Spectrophotometer. The spectrophotometer should be capable of making absorbance measurements in the range 300–800 nm.

Distillation apparatus. The design of the apparatus is shown in Fig. 6.5. Vessel A is a 500-ml round-bottomed distillation flask and receiver B is a 250-ml, two-necked, round-bottomed flask which, during the ruthenium distillation, is fitted with a 250-ml "Glas-Col" heating mantle. Receivers C and D are 125-ml, tall-form gas-washing bottles with fritted cylinders at the end of the inlet tubes. Each fritted cylinder is pierced to make a 1.5 mm hole to permit free flow of gases through the apparatus.

Note. The diameter of the air-cooled condensing bulb on vessel A should be not less than about 50 mm. A smaller bulb may not be efficient in condensing stannic chloride, which might then be troublesome in the analysis of the distillate.

Cation exchange columns. It is recommended that both large (25 × 450 mm) and small (15 × 15 mm) columns be prepared with water-washed, 20–50 mesh, Dowex-

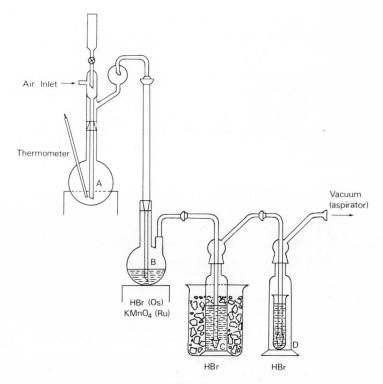

FIGURE 6.5.

50W-X8 cation exchange resin. Columns of these dimensions are suitable for the removal of approximately 4 and 0.5 g, respectively, of combined base metals (mainly copper and nickel).

To remove base metals and regenerate the resin, pass $4\,M$ hydrochloric acid through the column until the effluent solution is free from copper and nickel (test with ammonia solution). Wash the resin with water until the effluent is neutral to litmus paper.

Assay crucibles.   Catalogued as "40-gram" type.

Reagents

Details of the preparation of special reagents will be given under the following procedures in the sections describing their use.

Procedures

A. Pretreatment of Sample

Depending upon the nature of the sample, a pretreatment is frequently required to remove certain constituents or convert them into a form that will not lead to difficulties in subsequent operations.

Roasting.   This is used primarily to convert sulfides into oxides, but also to volatilize arsenic and antimony, when present, and thus prevent the formation of matte or speiss during the subsequent fusion. It is recommended that samples other than alloys and copper–nickel matte be roasted before being mixed with the flux. It is known that osmium (but not ruthenium) is lost by volatilization during the roasting operation. Fortunately, this step can be omitted when certain types of platiniferous materials are to be analyzed for osmium (e.g., copper–nickel matte).

Place the sample (200-mesh), weighing up to 1 assay ton, in a shallow fire-clay dish and roast at 750–800°C for approximately 1 hr with intermittent stirring. In cases where only a few grams of material (particularly sulfides) are to be roasted, place the sample on a thin bed of silica to prevent possible loss of the resultant calcine to the surface of the dish (include the silica used as part of the total quantity required for fluxing). When the calcine is cool, mix it with the flux and proceed with the fusion process (see the following).

Leaching of copper–nickel matte.   This can be conveniently achieved in the following manner to remove the bulk of the copper and nickel and to leave the precious metals in a concentrated form in the leach residue. Place the sample, weighing up to 2 assay tons, in a 1500-ml beaker and treat it with approximately 25 g of ammonium chloride and 100–200 ml of $12\,M$ hydrochloric acid. Heat the covered beaker and contents until the amount of insoluble matter appears not to exceed approximately 2 g. If dealing with large sample weights, it may be necessary to retreat the leach residue once or twice with fresh acid after intervening filtrations.

Dilute the combined sample solution (approximately 100 ml) with an equal volume of water, heat to approximately 70°C, and then stir the solution and dissolved residue with a polyethylene-coated, electrically driven stirrer. Add approximately 5 g of tin powder (200-mesh or finer) and stir the suspension for approximately 15 min. Filter the solution, containing most of the nickel, through a moderately fast paper. Completely wash the solids onto the paper with dilute (~5%) hydrochloric acid. Dry the washed residue and paper at 110°C for about 1 hr and then mix them with the flux for fusion (see the following).

Note. After the fusion of the matte leach residue, a small amount of a sulfide phase will adhere to the tin button but this will ultimately be decomposed during the chemical treatment of the button.

B. Crucible Fusion Process (Preparation of Buttons). Flux

The flux that has proven most satisfactory for nearly all materials so far encountered, contains the following materials in the indicated quantities.

SnO_2	35–40 g
Na_2CO_3	50 g
SiO_2	10–20 g
$Na_2B_4O_7$	10 g
Flour	40 g

The quantity of silica is adjusted according to the amount of silica in the sample.

Note. Powdered coke (100–325 mesh) is also satisfactory as a reducing agent, 6–8 g being required. In early work coke was used as reducing agent because experiments had shown that the use of relatively small quantities of flour (~10 g) resulted in the formation of inordinately small buttons. Subsequently, however, it was learned that by increasing the amount of flour to 40 g, essentially complete reduction of stannic oxide is achieved. With 120 g of the flux, 40 g of flour, and samples weighing up to 1 assay ton (29.17 g), tin buttons weighing 27–29 g are obtained. This represents 85–90% recovery of the tin.

Gold and silver content of stannic oxides and tin powder. Experience has shown that both stannic oxide and tin powder invariably contain a significant concentration of gold (e.g., 0.1–2.0 ppm). It is therefore imperative that the gold value of each batch of these materials be determined and subtracted as a blank when the determination of these metals is included in the analytical scheme.* Despite the necessity of making corrections for the blank, accurate and reproducible results are easily obtained with the procedures to be described.

Procedures for preparation of buttons. (a) *Mixing the charge.* For dry samples follow the standard assay practice by blending the dry samples with the flux on glazed paper and transfer the charge to a "40-gram" crucible.

When solutions or suspensions (e.g., from acid decomposition of meteorites) are to be mixed with the flux, place approximately one third of the standard flux in the crucible and press a square of thin, commercial, plastic wrapping film into the crucible to form an envelope, and into this envelope transfer the remainder of the flux. With a spatula, form a cavity in the bed of flux and transfer the sample slowly into the depression so as to avoid wetting the film or crucible walls. Heat the crucible and contents in a drying oven at 110°C for at least 2 hr. After the sample has dried, remove the material in the wrapping film envelope and grind it in a mortar, mix well, and place back in the film in the crucible.

---

*Silver is collected with the gold. The authors' paper deals with its treatment as part of a blank.

(b) Fusion.   Place the crucible in an assay furnace maintained at approximately 1250°C and fuse the contents for 1 hr. During the first 15 min, inspect the crucible several times to ensure that evolution of gases, or frothing, is not excessive. On the few occasions when this situation arises, swirl the molten material in the crucible several times with tongs until the melt becomes quiescent, then continue the fusion. At the completion of the fusion period, the melt should not be viscous or lumpy nor should there be extensive crust formation at the top of the melt.

Pour the melt into a conical steel mold and, when it is cool, separate the tin button from adhering slag by tapping with a small hammer. (A small file can sometimes be used to advantage in removing tenaciously adhering encrustations of slag, but loss of metal should be avoided.)

(c) Granulation of buttons.   Place the button in the graphite crucible of the Jelrus furnace, from which air is purged by nitrogen gas delivered through a suitable ceramic tube placed directly over the button. After melting the button (which may require a temperature of 600–700°C for buttons particularly rich in nickel),withdraw the nitrogen delivery tube and, using tongs to hold the crucible, quickly pour the molten tin alloy into approximately 1 gal of water contained in an enamel pail. Decant the water; dry the alloy on the hot plate (avoid remelting). Any large lumps produced can easily be reduced in size with metal shears.

Reserve the granulated alloy for analysis according to the appropriate procedures described below and outlined in Fig. 6.4.

C. Wet Chemical Procedures for Isolation, Separation, and Determination of Precious Metals

Decomposition of tin buttons (except when osmium is to be determined).   Two procedures for the decomposition of the tin button may be used, the choice being largely dependent on the nature of the sample.

The procedure used by the authors for most materials involves treatment (parting) of the tin-alloy with concentrated hydrochloric acid. A major proportion of each of the precious metals, as intermetallic compounds of tin, remains undissolved. However, suprisingly a significant fraction of rhodium, iridium, and ruthenium dissolves along with the tin matrix. The soluble species are readily recovered by precipitation with tin powder. The major advantage of the hydrochloric acid parting procedure is that much of the excess of tin, as stannous chloride, can be eliminated before the tin-volatilization procedure. If the sample contains *milligram* amounts of the platinum group metals it is advantageous to decompose the tin alloy directly with a mixture of hydrochloric acid and hydrogen peroxide. Insoluble matter which may remain is easy to recover and treat chemically. A minor disadvantage of this approach is that subsequently all the tin of the original button may have to be volatilized.

(a) Parting in hydrochloric acid.   Transfer the granulated tin alloy to a 600-ml beaker and treat it with approximately 150 ml of concentrated hydrochloric acid. Cover the beaker with a watch glass and heat until the alloy has been decomposed and vigorous evolution of bubbles from the black insoluble residue has ceased. Dilute the resulting suspension to about 400 ml with water and heat to approximately 70°C. While the suspension is being stirred with a motor-driven polyethylene-coated stirrer or magnetic stirrer, sprinkle approximately 5 g of powdered tin (200-mesh or finer) into the beaker to

precipitate any precious metals that had dissolved during the parting operation. Stir the contents of the beaker for approximately 15 min.

If iridium and/or ruthenium are to be determined, again heat the beaker to approximately 70°C and add a second 5-g portion of powdered tin; stir for an additional period of 15 min.

The insoluble matter produced in these operations should contain essentially all the copper in the sample, the excess of tin powder, and all of the precious metals. Nickel remains in solution and is ultimately discarded.

Decant the supernatant solution through a pad of paper pulp (from Whatman No. 31 paper) supported on a filter disk. Wash the solids several times by decantation with 1 M hydrochloric acid and pass the washings through the filter pad.

Add 25–30 ml of 12 M hydrochloric acid to the solids in the beaker, and then cautiously add 30% hydrogen peroxide in small portions until the solids have dissolved and excess of peroxide is present. A small amount of siliceous matter from the salt may be present but can be ignored. After the mixture has been heated to the boiling point, boil for a few minutes to ensure complete dissolution of the residue, place the beaker under the filter pad and wash the pad with a mixture of approximately 10 ml of 12 M hydrochloric acid and 2–3 ml of 30% hydrogen peroxide to dissolve any fines caught on the filter during decantation.

(b) *Decomposition in hydrochloric acid and hydrogen peroxide* (primarily for samples containing milligram amounts of the platinum group metals). Place the granulated tin alloy in an 800-ml beaker immersed in an ice bath. Add approximately 200 ml of 12 M hydrochloric acid, then while stirring cautiously add 30% hydrogen peroxide in relatively small portions until the tin alloy has decomposed and the effervescence from excess of peroxide is evident. Add an additional 50 ml of hydrochloric acid and 10 ml of hydrogen peroxide then heat the beaker and contents until the volume of the solution has been reduced to approximately 100–150 ml. Dilute the solution with approximately 100 ml of water and allow any dense, dark-colored residue to settle for about 30 min. Decant the supernatant solution through an appropriate filter paper into a 500-ml volumetric flask. Wash the beaker and residue, by decantation, with several small portions of dilute hydrochloric acid. Treat any residue left in the beaker with approximately 10 ml of 12 M hydrochloric acid and 5 ml of 30% hydrogen peroxide. Evaporate the solution to a small volume, then add 5–10 ml of aqua regia. Again evaporate the solution to 1–2 ml and, by repeated treatments with 12 M hydrochloric acid, destroy nitrogen-containing compounds and ensure that the platinum metals are present as their chloro complexes. Dilute the acid solution with a small volume of water then filter into the 500-ml flask through the paper used previously. Dry the paper and residue and then fuse to produce a second tin button. Decompose this, in turn, with hydrochloric acid and hydrogen peroxide. Filter off any siliceous matter from the resulting solution and receive the filtrate in the 500-ml flask containing the solution from the first button. Dilute to volume with water.

Because this method of decomposition is used primarily for materials rich in platinum group metals, it is often sufficient to take an aliquot of the sample solution for analysis by the procedures described below.

Volatilization of tin before the determination of gold and platinum group metals. Evaporate the above solution to a relatively small volume on the hot plate, then place the beaker in an aluminum "no-bump" solution evaporator. Heat the evaporator from below with the hot plate and from above with an infrared heat lamp to volatilize as much tin as possible.

Copious fumes of stannic halide are evolved at this point and when relatively large amounts of copper and/or nickel are present, care should be taken to avoid spattering caused by local intense heating.

When it appears that the evolution of fumes has nearly ceased, remove the beaker from the evaporator and cool for a short time. Wash down the sides of the beaker with approximately 10 ml of a mixture (7:2) of hydrochloric and hydrobromic acids and again evaporate to dryness under the heat lamp. Repeat the treatment with the mixed halogen acids (5–10 ml) and the evaporation twice or until it is evident that essentially all of the tin has been volatilized.

If gold is to be determined, simply evaporate the mixed halogen acid solution as immediately below. If gold is not to be determined, cool the beaker and residue, add 10–15 ml of 12 M hydrochloric acid, and then, while swirling the beaker in a fume cupboard, cautiously add 30% hydrogen peroxide until it is evident that an excess of peroxide is present and that the evolution of bromine vapor is not vigorous. Evaporate the solution to small volume and process according to requirements (see flow sheet in Fig. 6.4) for the removal of base metals or the extraction of gold and/or the distillation of ruthenium.

Isolation and determination of gold.   Evaporate the solution to 2–3 ml. If the solution does not contain an appreciable quantity of base-metal chlorides, place the beaker in a hot water bath, with the aid of a jet of compressed air, evaporate the solution gently to dryness to prevent baking of the gold salts on the beaker wall. When a substantial amount of base-metal chlorides is present, the solution can be evaporated gently to dryness directly on the cooler part of the hot plate. Take up the salts in 5 ml of 2 M hydrobromic acid and wash into a 60-ml separatory funnel with a further 10 ml of 2 M hydrobromic acid. Extract the gold by shaking the solution with two separate 15-ml portions of diethyl ether. Double the volume of hydrobromic acid and ether if the sample contains a relatively large quantity of copper. Combine the ether extracts and wash with three separate 5-ml portions of 2 M hydrobromic acid. Strip the gold from the ether phase by shaking it with three separate 15-ml portions of water. Treat the aqueous gold solution with approximately 2 ml of aqua regia and evaporate to a small volume. Transfer the gold solution to a volumetric flask of appropriate size, depending upon the expected gold content, and dilute to volume with water. Determine the gold content of the solution by atomic absorption (see Chapter 1).

Evaporate to dryness the aqueous phase from the ether extraction of gold and, with a small volume of hydrochloric acid–30% hydrogen peroxide mixture (2:1), destroy bromides and convert the metals into their chloro complexes. Reserve the resulting solution for the determination of platinum group metals according to Procedure 85. Determine the gold content of the final solution, containing at least 0.2 ppm of gold, by atomic absorption at a wavelength of 242.8 nm, a 0.10-m narrow-slit burner, and an oxidizing air-acetylene flame. Subtract the gold content of the blank charge from that of the sample.

Some Comments on the Analytical Scheme for the Platinum Group Metals

Most samples to which the proposed methods will be applied will require the determination of one or more of platinum, palladium, and rhodium with or without the prior determination of gold. When this is the case, the procedures described immediately below for removing base metals by cation exchange, followed by the individual determination of platinum, palladium, and rhodium by atomic absorption or colorimetric techniques, can be followed directly.

However, as will be seen subsequently, the scheme must be altered somewhat when osmium, and possibly ruthenium, is to be determined on the same assay buttons as the other precious metals. The distillation procedures for isolating osmium and ruthenium (only when ruthenium is to be determined colorimetrically) must be applied before the determination of gold, platinum, palladium, rhodium, and iridium. Although this is a perfectly feasible analytical path, it may be more convenient, if sufficient material is available, to determine osmium and ruthenium (colorimetrically) on a separate sample.

It should be noted that the comments above pertain to ruthenium only if this element is to be determined colorimetrically; if atomic absorption spectrometry is to be used for ruthenium, then this element need not be isolated by distillation and can be determined in the same solution as that containing platinum, palladium, and rhodium. The atomic absorption sensitivity for iridium is low, therefore the proposed analytical scheme involves the use of the more sensitive colorimetric determination of iridium after it has been separated from other platinum group metals by solvent extraction. If the iridium content of the sample is relatively high, however, as in precious metal concentrates, then iridium too can be determined by atomic absorption.

Removal of base metals by cation exchange (356) *before the determination of the platinum group metals.*   Evaporate the solution of metal chlorides to dryness. Take up in 5–10 ml of 12 M hydrochloric acid and again evaporate to a residue of salts. When more than approximately 0.5 g of base metals (usually copper) is present, stir the crystallizing salts with a stirring rod to aid in the removal of free hydrochloric acid. Cool the beaker and depending on the amount of base metals estimated to be present, add 0.25 or 0.50 ml of 12 M hydrochloric acid and then dissolve the salts with 50 or 100 ml of water, to give a solution with a pH of 1.0–1.5.

Following the isolation of the platinum metals a variety of methods of determination for each metal can be applied. Solvent extraction and spectrophotometric determination are particularly useful. The procedure developed by Faye (85) is especially recommended.

At this point the analyst should be aware of the possibility of hydrolysis of the anionic chloro complexes of rhodium, iridium, and especially ruthenium during the ion exchange process. In the operating time required, hydrolysis can be prevented by maintaining the chloride ion concentration at 0.1 M or higher, which is normally provided by the base metal salts present. However, when it is apparent that these are in low concentration, it is advisable to add approximately 0.5 g of sodium chloride per 50 ml of solution.

Pass the sample solution through a cation exchange column at the rate of approximately 5 ml/min. If the sample contains less than approximately 0.5 g of base metals, use a "small" column; otherwise use a "large" column as described under the apparatus heading. Collect the effluent solution in a 400-ml beaker and wash the column with sufficient dilute hydrochloric acid (pH approx 1) to quadruple the original volume of the sample solution.

To the effluent solution, add approximately 25 mg of sodium chloride and 10 ml of 12 M hydrochloric acid and reserve for the separation of platinum, palladium, rhodium, and iridium or for the isolation of ruthenium by distillation (see Fig. 6.1).

Notes **1.**   In certain cases it may become apparent, from the color developed during the subsequent evaporation of the effluent solution, that leakage of base metals has occurred. This can generally be ignored if the solution is to be analyzed by atomic absorption for one or more of platinum, palladium, and rhodium. However, it is necessary to repeat the cation exchange operation (using the small column) to remove the contaminants if colorimetric finishes are to be used.

2.   If tin were not completely volatilized in earlier steps, it would precipitate in a gelatinous or flocculent form during the ion exchange process and might contaminate both the resin bed and the effluent suspension. If the amount of precipitate is relatively small with a slightly cloudy suspension then the danger of losing platinum metals by occlusion or coprecipitation is slight, and therefore any solid material retained on the resin bed can be neglected. The small amount of tin in the effluent can be removed ultimately by repeating the volatilization procedure given above. When more than a few milligrams of tin appear to be present, strip the column with 4 M hydrochloric acid, combine the stripping solution with the original effluent and treat the whole, as described previously, for the volatilization of tin before repeating the ion exchange procedure.

THE DETERMINATION OF OSMIUM AND RUTHENIUM IN SIDERITE METEORITES (AND IN IRON–COPPER–NICKEL ASSAY BUTTONS)

The following method (133a) is applicable both to assay buttons which usually weigh about 25 g and to meteorite samples from milligram amounts to more than 100 g. The upper limit is determined by the mechanics of the distillation, e.g., the size of the sample and flask and the volume of the receiving flasks. With small samples the amount of perchloric acid is reduced to an amount appropriate for dissolution. This distillation has been examined only for determinations of milligram and microgram amounts of osmium and ruthenium. One may expect that large amounts of these metals may alter the properties of the iron alloys to the extent that perchloric acid may not completely dissolve the osmium and ruthenium. In these instances the size of the alloy sample should be reduced. In its present state of development the collection of osmium and ruthenium by the base metal alloys is recommended only for analytical purposes, and the method is best applied to samples containing from about 25 μg to 20 mg of osmium or ruthenium.

A perchloric oxidation serves to remove both osmium and ruthenium, and a proper choice of receiving liquids will facilitate the subsequent separation of the two metals.

PROCEDURE 86 (133a)

Use the distillation equipment as shown in Fig. 3.3 (page 119).
Add 100 ml of water to the trap, 30 ml of a 3% hydrogen peroxide solution to the first

receiver and 10 ml to each of the remaining two receivers. Immerse the receivers in an ice bath. Pass a stream of cold water through the condenser, and apply a gentle suction through the whole system from the exit tube of the last receiver. Add 100 ml of 72% perchloric acid to the sample of meteorite in the distillation flask, and boil gently over a low flame. If the reaction becomes too vigorous, discontinue the heating to regain control. Boil the solution for a total period of 2 hr. To ensure the complete distillation of the osmium and ruthenium, add 50 ml more of the perchloric acid to the distillation flask, and heat for another 30 min. Then add 15 ml of the perchloric acid to the trap, and boil the solution for 30 min. Transfer the receiving solution containing the osmium and ruthenium to a previously chilled second distillation flask as quickly as possible to prevent a loss of osmium by volatilization, and wash the receivers and the delivery tubes thoroughly with cold 3% sulfuric acid. Wash the tubes and receivers thoroughly with water to remove the sulfuric acid.

Add 100 ml of water to the trap and 30 ml of twice-distilled 48% hydrobromic acid to the first receiver and 10 ml of the same acid to each of the other receivers. Chill the receivers in an ice bath. Add 40 ml of a 30% hydrogen peroxide solution and 5 ml of concentrated sulfuric acid to the chilled distillation flask containing the osmium and ruthenium distillates. Boil gently for 30 min. Add 15 ml of perchloric acid to the trap, and boil for 25 min. Draw a stream of air slowly through the system (about 3–5 bubbles/sec). Transfer the contents of the receivers to a 150-ml beaker, and rinse both receivers and tubes with 10% hydrobromic acid. Evaporate the solution on a steam bath to 5 ml, transfer to a 30-ml beaker, and evaporate again to 0.5–1.0 ml. Transfer the solution to a 20-ml test tube by washing with water, and determine the osmium with thiourea (Procedure 46).

For milligram amounts of osmium, adjust the evaporated osmium distillate to the required acidity for precipitation by thionalide as directed in Procedure 66.

To recover the ruthenium in the pot liquid, add 100 ml of water to the trap, 30 ml of a 3% hydrogen peroxide solution and 1 ml of 48% hydrobromic acid to the first receiver, and 10 ml of a 3% hydrogen peroxide solution to each of the other two receivers. Cool the receivers in an ice bath. Add 10 ml of concentrated sulfuric acid to the distillation flask and then add cautiously an excess of a 10% sodium bromate solution (about 20 ml). Apply a gentle suction, and distill cautiously over a low flame for 1 hr. Then add 15 ml of the perchloric acid to the trap, and boil for 25 min. Disconnect the receivers from the water condenser, maintaining the connection between the two receivers. Add 8 ml of 48% hydrobromic acid to the first receiver and 4 ml of the acid to the second receiver. Boil the liquid for 10 min. Transfer the contents of the receivers to a 150-ml beaker, rinse the tubes and the receivers with 10% hydrobromic acid, and evaporate the combined solution to 5 ml. If a spectrophotometric method is to be used transfer the 5 ml of solution to a 30-ml beaker and evaporate again to 0.5–1.0 ml. Transfer to a 20-ml test tube, and apply the thiourea reaction (see Procedures 44 and 45). Other appropriate methods may be used.

For milligram amounts treat the evaporated ruthenium solution (0.5–10 ml) to prepare for a thionalide precipitation (Procedure 63).

The removal by perchloric acid of osmium and ruthenium leaves the remaining four platinum metals in the pot liquid. An examination of the chemistry of the pot liquid invites the application of ion exchange for the complete isolation of platinum, palladium, rhodium, and iridium. Indeed the senior author and associates have recorded what appeared to be a successful separation by cation exchange (357). Later work by the present author revealed wide variations in the recoveries, and an alternative method was pro-visionally proposed (358). The method involved first a removal of osmium and ruthenium by perchloric acid, then treating the pot liquid by a number of intermittent evaporations with hydrochloric acid and a little nitric acid added to avoid a perchloric acid explosion. The final evaporation was stopped with a volume of about 60 ml. The diluted solution was treated with sodium hydroxide solution until precipitation was completed at a pH of about 10. The precipitate was filtered and washed until free from perchloric acid as indicated by cesium chloride. The filtered and washed residue was dissolved in hydrochloric acid. The filtrate and wash water were evaporated and treated for a possible platinum content. Unexpectedly the platinum was generally collected completely along with the hydrated oxides all of which were converted to chlorides. In cases where some platinum was retained by the filtrate (at pH 10), the latter was treated by evaporation and chlorination, etc.

The hydrochloric solution of the hydrated oxides was adjusted to pH 1.2–1.5 and the separation of platinum metals from base metals accomplished by a passage through a cation exchanger (see Procedures 102 and 108). It is hoped that the above procedure may be subjected to further research.

PART II

CHAPTER 7

METHODS OF SEPARATION

THE DETERMINATION OF GOLD IN CYANIDE SOLUTIONS

Whereas a considerable variety of acceptable methods have been proposed for the determination of gold in cyanide solutions, there is a need not only for a critical comparison of these methods based on experimental data but also for researches designed to apply some of the more recently developed separational techniques. It is not improbable that investigations such as these would add profitably to large scale separations.

In this context progress is being made towards the extraction of gold from cyanide solution by ion exchangers. Whereas no generally applicable method has been proposed, Fridman and Popova (359) used 5 g of the anion exchanger AN-2F to remove 230 μg of gold from 2 liters of solution. It is significant that the presence of iron lowers the efficiency of the exchange. An unusual application of ion exchange, viz., to the determination of gold in the filter cake from an extraction plant, was also described by these authors (359). The anion exchange resin AN-2F, in the chloride form, was soaked in water for 2 hr and then mixed with a solution of the filter cake for 3 hr. The mixture was passed through a set of 35- and 100-mesh sieves. The material retained on the latter was mixed with a flux and fire assayed.

From time to time adverse criticisms have been directed toward even the classical method of determining gold in cyanide solutions (360), and there can be little doubt that some of the reactions involved in these methods are little understood. Here, then, is a potentially fertile field for fundamental and practical research.

Bugbee (32) stated that the Chiddy method with its numerous modifications is reported to be unpopular on the Rand on account of the care required to obtain satisfactory results and because it requires heating.

Alternative procedures, used in the absence of thiocyanates, involve the addition of copper(I) chloride in hydrochloric acid or the production of copper(I) ion in situ by adding consecutively a saturated solution of sodium cyanide, copper sulfate, sodium sulfite, and sulfuric acid. Potassium ferrocyanide is usually added in small amounts to provide a carrier and to aid in filtering. With these two methods the gold is first precipitated as $CuAu(CN)_2$, and the excess of cyanide is removed simultaneously as copper(I) cyanide. The precipitations are accomplished in the cold. The use of potassium ferrocyanide as a coagulant was rejected by Creed and Clayton-Boxall (361) as unnecessary and leading to losses when excess is present. These authors stated that, when the ferrocyanide was to be used, the amount should be limited to 1 ml of a 5% solution of ferrocyanide. The data provided indicated that the ferrocyanide reacted directly with the precipitated gold to produce a dissolved product.

The Copper(I) Chloride Method

King and Wolfe (362) provided the following modified method together with methods of making the copper salt solution, its effective storage, and a method of dispensing the unstable precipitant.

PROCEDURE 87 (362)

Reagents

Copper(I) chloride solution.   To a suitably large Pyrex Erlenmeyer flask fitted with a water-cooled condenser add crushed copper sulfate and concentrated hydrochloric acid in the proportion of 1 formula weight of the copper salt to 3 formula weights of hydrochloric acid. Into this mixture place a copper sheet extending to the surface of the liquid. It is preferable to use a large copper surface. Boil gently for a few hours to obtain a colorless solution. Remove the condenser, stopper the flask tightly, and allow to cool. The solution may darken as the copper(I) salt precipitates, but within 1 or 2 hr it will again become clear. Store the solution under hydrogen, and remove the necessary aliquots through an attached syphon.

An alternative method of preparing copper(I) chloride was described by Bettel (363). The procedure, which requires the use of both a storage bottle and one for current use, is the following.

Add to the storage bottle an excess of high conductivity copper, and then fill it to the neck with 1:1 hydrochloric acid saturated with copper(I) chloride. Mix this solution with an equal volume of concentrated hydrochloric acid. When the resulting solution becomes colorless transfer a suitable amount to a small bottle containing copper foil. The solution is then available for current use. As the solution is removed from the stock bottle, refill the latter with the saturated hydrochloric acid–copper(I) chloride solution.

Stock flux.

Silica	8 parts
Borax glass	4 parts
Sodium carbonate	36 parts
Litharge	40 parts
Flour	1 part

Use about 90 g of the well-mixed flux for each fusion. Add silver as required. A 15-cm paper will produce about 10 g of lead, hence only a small amount of flour is required. This flux gives a good slag at about 1000°C.

Coagulant. The $CuAu(CN)_2$ precipitated by copper(I) chloride is pale green and sometimes very finely divided. Thus it may pass through filter paper. To assist the coagulation a dilute solution of potassium ferrocyanide is added to produce the flocculent copper ferrocyanide. Excess must be avoided so as to prevent the redissolution of the gold.

A 10% solution of sodium sulfide may also be used. Five drops of this solution should suffice as a collector. There is the advantage with sodium sulfide that the formation of the copper sulfide indicates the necessary excess of copper(I), for the latter is not formed in the presence of cyanide. Too great an excess of sulfide must be avoided because the resulting excess of copper sulfide may produce high slag and cupel losses.

Filter paper pulp may also be used as a coagulant. The pulp should be added to the filter paper to form a uniform coating before filtering. In all instances the first 100 ml or so of the filtrate should be refiltered.

Procedure

To 20 assay tons of the cyanide solution add 10 ml of the clear copper(I) chloride solution. If more than 20 assay tons are used add 2 ml of the copper solution for each additional 10 assay tons. Mix thoroughly, and leave the rod in the solution. Allow to stand for 5 min, and filter through a 15-cm No. 2 Whatman paper previously treated with paper pulp. Clean the beaker with a small piece of filter paper that is then added to the solid. Allow to drain, and transfer the precipitate and paper, apex down, to a 20-g crucible containing about 20% of the flux. Add to the open cone the rest of the flux and sufficient silver for parting. Fuse, transfer the button to a cupel, and proceed according to the cupellation and parting technique described in Chapter 6, Fire Assay. Alternative Procedures 88 and 89 are given by Creed and Clayton-Boxall (361).

PROCEDURE 88 (361)

Pour 1.5–2.0 l of well-mixed slime pulp, thinned down to 1.40 density, into an enamel bucket or Winchester quart bottle. Add 15 ml of copper(I) chloride solution free of copper(II), and mix for at least 10 min.

Press off the liquor, and dry the sample in an enamel dish. Pass the dried sample through a 900-mesh screen, add silver for parting, mix well, and assay as described in Procedure 121.

Because of the inconvenience of preparation and the instability of the copper(I) chloride some assayers prefer the copper sulfate method (Procedure 89), in which the reagents are copper sulfate, sulfuric acid, and sodium sulfite.

PROCEDURE 89 (361)

To a measured volume of solution, usually containing 10 or 20 assay tons, add 5 drops of a saturated solution of potassium ferrocyanide and a slight excess of 10% copper sulfate solution. Stir or shake well. If the solution is low in cyanide it is advisable to bring it up to about 0.05% in potassium cyanide before the precipitation. Then add in succession 10 ml of a 20% sodium sulfite solution and 10 ml of 10% sulfuric acid. Stir or shake well after each addition, and allow to stand for a few minutes until sulfur dioxide is being freely evolved and the precipitate is settling readily. Twenty milliliters of the copper sulfate solution is usually sufficient, but a faint blue tinge to the filtrate is an assurance of it being in slight excess.

Filter through a 15-cm, rapid, folded filter. Wash out the beaker, then wipe the stirring rod and the beaker with a piece of moistened filter paper to ensure the collection of all the precipitate, and add to the contents of the filter. Allow the solution to drain; then dust a little mixed flux over the precipitate, and transfer the moist filter and its contents to a glazed 20-g crucible containing a little flux. Place the remainder and larger portion of a 100-g scoop of flux (see the following) on top of the filter paper, add 2 mg of silver of 99.9999% purity, if the assay is for gold alone, and fuse, cupel, and part as described in Procedure 121. The following stock flux has proved satisfactory, although the flour content may need to be varied somewhat to secure a lead button of suitable size:

Sodium carbonate	15 g
Litharge	55 g
Silica	20 g
Borax	8 g
Flour	2 g

Gold and silver are precipitated, and may be determined in the same assay. Assays may be completed in 2 hr and with the reservations mentioned in the first paragraph the results are good for both low and high grade solutions.

Notes. **1.** The potassium ferrocyanide acts as a coagulant and as a precipitant for silver. No more than the stated quantity should be added, however, because an excess leads to a loss of gold.

2. The sodium sulfite, an active reducing agent, is added to ensure complete reduction and the precipitation of all the gold.

3. The precipitate consists mainly of copper(I) cyanide; it is white, curdy, and insoluble in weak acids.

4. If the precipitate tends to pass through the filter paper, pour a little suspended paper pulp into the latter. This will ensure rapid filtration and minimize the loss of fine precipitate.

In the determination of gold in oxidized tailings or slimes which may contain ferrocyanide, Bettel (363) obtained low values with the copper(II) sulfate–sodium sulfite–sulfuric acid method. This was attributed to the production of auricyanides that resist reduction. In these instances Bettel used a combination of the copper(I) salt and lead acetate–zinc methods, the latter being applied to the filtrate from the copper(I) precipitation.

The modification requires the application of the copper(I) method, with subsequent filtering and washing. The residue is opened up, and litharge is sprinkled over the surface. The filtrate is treated with lead acetate, zinc fume, and hydrochloric acid in sufficient quantities to dissolve the zinc after the lead sponge has been formed. The latter, containing the traces of gold and copper, is filtered off, washed, drained, and added to the litharge-covered copper(I) precipitate. The whole is covered with litharge, dried, carefully heated to burn the paper, and finally assayed with a suitable flux.

The Chiddy Method

This is one of the most widely used of the older procedures, and according to Bugbee (32) it is used almost always in North America for the assay of cyanide solutions. It works equally well on strong or weak, foul or pure solutions, and almost any quantity may be taken. The method was first reported by Chiddy in 1903 in a very short communication, with claims that he originated the method and that it improved on the existing methods. The procedure has been changed very little during the past half century, although some modifications have been introduced. Wogrinz (364) rejected the method when the cyanide solutions contained ferrocyanide, but the latter, however, could be destroyed by evaporating to a residue and then fuming with sulfuric acid. Wogrinz recommended the addition of lead acetate, zinc dust, and then hydrochloric acid until the mixture was strongly acidic. Subsequently, the lead was dissolved and the gold filtered, washed, and weighed. Roodhouse (365) also slightly modified the Chiddy method by adding acetic acid to the reduced mixture, followed by boiling, decantation, and the addition of hydrochloric acid.

PROCEDURE 90 (366)

Heat a solution containing from 1 to 20 assay tons in a beaker or an evaporating dish (see Note 1). Add 10 or 20 ml of a 10% lead acetate solution containing 40 ml of acetic acid per liter. Then add 1 or 2 g of fine zinc shavings rolled lightly into a ball or an equivalent amount of zinc dust in the form of an emulsion (see Notes 2 and 3). Heat to near boiling (90–95°C) until the lead is well precipitated. This usually takes about 15 or 20 min. Then add 10–20 ml of strong hydrochloric acid to dissolve the excess of zinc. Continue heating until effervescence stops (see Note 4). It is often found that action ceases while some undissolved zinc still remains. This is entirely covered and thus protected from the acid by the spongy lead. To be sure that all the zinc is dissolved, stroke the sponge with a stirring rod, and drop a little hydrochloric acid from a pipet directly on it. As soon as the zinc is dissolved (see Note 5) decant the solution and wash the sponge two or three times with tap water. Next, moisten the fingers, and press the sponge, which should be all in one piece, into a compact mass (see Note 6). Dry by squeezing between pieces of soft filter paper or by placing on a piece of lead foil and rolling with a piece of large glass tubing. Finally, roll into a ball with lead foil, puncture to allow for escape of steam (see Note 7), add 2 mg silver for parting, place in a hot cupel, and proceed as described in Procedure 121.

Notes. **1.** The solution must be clear and free from slime or colloidal matter or the sponge may break up. If not, add a little lime water, boil for a moment to coagulate the slime, and then filter if necessary before beginning the assay.

2. Some assayers bring to boiling after adding the zinc and the lead acetate solution, then decant the solution as completely as possible, after which they add an equal volume of dilute hydrochloric acid to dissolve the excess of zinc.

3. The gold, silver, and lead will immediately begin to precipitate on the zinc. At first the solution may become cloudy, but will soon clear as more of the lead is precipitated.

4. When working with small quantities of solution it is best to add water occasionally to maintain the original volume of solution or else the hydrochloric acid will become too concentrated and cause the sponge to disintegrate.

5. Because the lead sponge begins to dissolve and break up as soon as all the zinc is dissolved, no time should be lost in decanting the solution after the zinc has disappeared.

6. In those rare instances where, because of special impurities, the sponge cannot be successfully collected by the previously described method, the solution may be filtered through rapid paper and quickly washed. The filter paper with precipitate is removed and the excess of water squeezed out. It is placed in a small scorifier with lead and borax glass together with silver for parting if gold only is sought. The filter paper is burned off, and the assay finished in the usual way.

7. Some assayers dry the assays before cupelling to avoid the danger of spitting caused by retained moisture.

The Analysis of Cyanide–Gold Plating Solutions

Whereas the problems associated with the determination of gold in mill cyanide solutions are somewhat comparable to those involved with plating solutions, with the former material the degree of contamination is subject to less control. Obviously the classical methods used for mill solutions are, in general, applicable also to plating solutions but are sometimes found to be unsuitable because of their difficulty or length of time. Thus, Kushner (366) recorded a relatively simple gravimetric method for plating solutions. The method proposed by him included the removal of free cyanide by the addition of silver nitrate and the subsequent fuming with sulfuric acid. In the presence of base metals and palladium, etc., the collected precipitate may be treated by an appropriate choice of wet methods to isolate and determine the gold content.

PROCEDURE 91 (366)

Use a 10-ml sample containing 0.5–20 g/l of gold; for lesser contents, use a suitably increased sample. Transfer the sample to a 500-ml Erlenmeyer flask, dilute with 50 ml of water, and add sufficient 0.1 N silver nitrate solution from a buret to combine completely with the free cyanide, as indicated by 5 ml of a 2% solution of potassium iodide.

Place the flask in an efficient hood, and add cautiously concentrated sulfuric acid until vigorous action ceases. Add 50 ml more of the sulfuric acid, and boil, avoiding any violent reaction. Discontinue heating when the gold becomes light brown and the acid is quite clear. Decant the supernatant liquid, treat the precipitate with an additional 50 ml of concentrated sulfuric acid, and boil to dissolve any silver sulfate. Decant the liquid, leaving as little as possible in the flask. Dilute the remaining acid with 200 ml of distilled water, and filter through a porcelain filtering crucible or a suitably prepared Gooch asbestos crucible. Wash the precipitate with hot dilute sulfuric acid and then with hot water to remove all acid. Dry, ignite to yellow gold, and weigh.

Evaporation Methods

With assay lead sheet, 0.005 in. thick, a boat suitable for the treatment of two assay tons of liquid can be made from a section 4 in. square. After the evaporation the folded boat is wrapped in a 2 × 4 in. sheet of lead, the total weight thus providing a button of 25–26 g.

In general two types of evaporation methods have been described. In one, the liquid is evaporated directly in a vessel made of an amount of lead sufficient to provide the optimum button weight for cupellation. In the second, the evaporation is made after admixture with the flux or its constituents. A fusion is then required to produce the lead button. The latter method is recommended only for rich liquids.

These methods provide the most accurate recoveries and may be used for the platinum metals as well as for gold. They are commonly used as the standards whereby one determines the effectiveness of new assay procedures for the noble metals. Whereas they are direct and simple of operation, great care must be taken to avoid losses by volatilization of the noble metals salts or, in the liquid fusion method, losses by the mechanical transference of the dried residues.

To apply the direct evaporation method, nonleaking lead vessels must be prepared; this requires a bit of practice. Furthermore, the noble metal solutions should be made neutral or basic if the acid present may attack the lead sheet. Evaporations should be carried out slowly at about 60°C, and when large volumes are to be treated this may extend overnight. More rapid evaporations will increase the probability of a loss of gold.

PROCEDURE 92a   THE DIRECT CUPELLATION METHOD

Place the boat and the measured liquid in a good draught oven at 60°C, overnight evaporations being usually sufficient for up to 2 assay tons. Cool the boat, add silver for parting, and fold the vessel carefully to protect the residue. Wrap with the 2 × 4 in. strip of lead, cupel, and part as described previously and in Procedure 121.

For the determination of platinum metals the evaporating solution should contain 20 parts of silver for each one of total platinum metals. The cuppelled bead may be treated according to one of the parting procedures. If gold only is to be determined the methods of cupellation and parting are described in Procedure 121.

PROCEDURE 92b   THE FUSION METHOD

In general a monosilicate flux (e.g., $2Na_2O \cdot SiO_2$) gives a clean button and a homogenous slag. A bisilicate flux (e.g., $Na_2O \cdot SiO_2$) may also be successful.

Use 100 g of the following mixture:

Anhydrous sodium carbonate	38 g
Borax glass	10 g
Silica	25 g
Calcium oxide	5 g
Flour*	2–3 g

Prepare a cellophane sheet 8 in. × 8 in., and place on this ½ to ¾ of the above flux. Fold the sheet and mixture into a 4-in. porcelain evaporating dish and add a solution containing 0.5 assay ton (for 1 assay ton about 150 g of the flux is recommended). If necessary add a silver salt or solution to give a gold:silver ratio of 1:10. Place the dish in a steam cabinet or oven at 60°C, maintaining a good air draught. Allow to stand overnight, then transfer the cellophane and its contents to a mortar, and grind carefully to a powder (alternatively only transfer the contents to the mortar, adding the cellophane separately to the pot). Add the ground material to an assay ton pot in which has been placed 5 g of the flux, and clean the mortar with the remaining flux. Use this portion as a cover for fluxes which have been salted. Assay for gold as described in Procedure 121.

The Electrolytic Assay of Gold–Cyanide Solutions

This method has the advantages that, compared to other methods, relatively little attention is required during the course of the determination and both small

---

*The cellophane used has a reducing value equivalent to about 1 g of flour.

and large volumes of cyanide solution may be analyzed. Gold is quantitatively deposited as a bright yellow coating on the lead cathodes which are then folded and cupelled.

PROCEDURE 93 (367)

Use as anodes $^5/_{16}$-in. arc lamp carbons held in place by clamps in the center of the beaker. Use as cathodes strips of assay lead foil 2.5 × 9 in., with the lower edge notched to encourage the mixing of the solution. Make the battery connection by cutting the upper end of the cathode into a 1-in. strip. The latter is then turned to make the terminal. Connect the two sides of the foil by folding the edges to produce a cylinder 3 in. in diameter. Add a 10-assay ton sample to a suitable beaker with the cathodes and anodes in place, and electrolyze with a 6-V accumulator source. Following complete deposition, which should require about 4 hr, remove the anodes, disconnect the cathodes and dry them on a hot plate. Fold the cathodes into a compact form with the gold protected within the fold, and cupel with silver added to obtain a gold to silver ratio of 1:10. Part with nitric acid as directed in Procedure 121.

Note. For cyanide solutions containing only traces of gold it is advisable to add a small quantity of cyanide solution to accelerate the deposition of the gold. Precautions must be taken against having the carbon anodes in contact with gold solutions in the absence of the current.

Other Methods

A number of methods have been proposed which involve the precipitation of gold from cyanide solutions in the absence of lead, followed by a filtration of the metal, which is then purified and weighed, dissolved and reprecipitated, added to lead and cupeled or added to a flux and fused.

Fraser (368) added hydrochloric acid to 522.6 ml of a cyanide solution to give an acid reaction with litmus, then 1.25 or 1.5 g of 200-mesh charcoal to reduce the gold and to provide sufficient reducing power to produce a lead button in the subsequent fusion. The suspension was warmed, stirred, and filtered through paper. The latter was burned in an assay crucible and fused with suitable fluxes. Marenkov (369) used nitric acid to isolate gold from cyanide slimes, subsequently dissolving the metal in aqua regia and reprecipitating by using hydrazine hydrochloride.

Seath and Beamish (241) used zinc to precipitate collectively the gold, tellurium, and silver in mill cyanide solutions; the mixture was then dissolved and treated with hydroquinone to remove gold selectively. Shah (370) also used zinc, the excess being removed in a sulfuric acid solution. The gold was purified by nitric acid before weighing. Rochat (371) used aluminum foil to isolate the gold, which was then purified by nitric acid to within 99.5% purity. Cadmium sulfide powder has been used to precipitate gold sulfide from an acidified solution of cyanide electrolyte (372). The precipitate was then ignited, treated with nitric acid, and finally converted to gold at 800–900°C. Wilson (373) provided simple details for the determination of gold in cyanide solutions by precipitation with ferrous sulfate and oxalic acid, after the elimination of hydrogen cyanide by hydrochloric acid.

THE MERCURY AMALGAM METHODS

The simultaneous precipitation of gold and mercury has been used as an analytical method. In particular Caldwell and Smith (374) treated a cyanide solution with a large excess of ferrous sulfate to form ferrocyanide, then added mercuric chloride solution, magnesium powder and concentrated hydrochloric acid. Presumably there was no loss of gold, even in the presence of large amounts of ferrocyanide. The probability of such losses has been emphasized by assayers (362) who advised only minimum amounts of ferrocyanide to provide a carrier in the copper(I) chloride method described above. Furthermore, Wogrinz (364) rejected the Chiddy method (375) as inapplicable in the presence of ferrocyanide. Caldwell and Smith (374), however, provided data from their method which indicated recoveries comparable in accuracy to both the copper sulfate and Chiddy methods. The cyanide solutions used for the latter methods contained 20 assay tons as compared to approximately 66–330 assay tons for the proposed method. An objection to the method is concerned with the difficulty of burning the precipitated residue prior to cupellation, because mercury evolved may well carry gold unless extreme care is taken. There are also the acknowledged slightly low recoveries with 10-1 samples of the cyanide solution. The method now described can also be used to recover silver.

PROCEDURE 94 (374)

To 2-1 samples (about 67 assay tons) of the cyanide–gold–silver solution add a solution containing ferrous sulfate about ten times the weight of the cyanide in the sample. Add 50 ml of a saturated mercuric chloride solution, 5 g of magnesium powder, and 60 ml of concentrated hydrochloric acid, adding the acid in portions to prevent bubbling over. If the volume of sample taken is much greater than the 67 assay tons, double the amount of mercuric chloride, magnesium, and acid used. Allow to stand for 6 to 8 hr or overnight. Siphon off the clear liquid, and transfer the residue from the bottle into a beaker, rinsing out any residual material with small portions of water. Let it settle for a few minutes, then filter, using a rough quantitative filter paper. As the bulk of the residue is being washed onto the filter paper, sprinkle in about 20 g of granular test lead so that the two will become intimately mixed. Allow to drain and dry.

On a 60-g bone-ash cupel spread a layer of test lead, following the general concave shape of the cupel. Remove as much of the dried residue from the filter paper as is convenient, mash the lumps, and place in the center of the cupel. With a little lead, cover the residue remaining on the filter paper, wad it up, and place on top of the cupel. Cover the residue with more test lead. The total weight of lead should not be much more than 45 g. Add silver to give a gold:silver ratio of 1:10. Introduce the cupel slowly into the muffle so that the filter paper will be burned and mercury and its salts volatilized. This last-mentioned step must be executed cautiously and requires the close attention of the analytical chemist; otherwise, a too rapid volatilization of the mercurical residue will cause spitting and serious losses, or perhaps the salting of an adjacent cupellation. When the cupel has been placed in the hottest part of the muffle, increase the temperature to nearly 1000°C, and create a reducing atmosphere by putting near the cupel bits of wood, cork, or like material. When the lead has "uncovered," as shown by its bright red appearance, cool the muffle to normal cupellation temperatures, and continue to the production of the bead. If gold only is to be determined, part, heat, and weigh as described in Procedure 121.

Even more accurate results may be obtained by the scorification of the mercury–noble metal collection residue prior to cupellation. It is recommended that the filter paper with the residue be scorified in a 6.25-cm (2.5-in.) scorifier. Place in the bottom of the scorifier a 10-g sheet of lead molded to the form of the scorifier. This prevents the absorbance of water and subsequent spattering during scorification. Add appropriate amounts of test lead and a little silica–borax glass. The scorification can be continued so as to yield an 18–30-g button in which instance smaller cupels may be used.

A mercury–gold collection was also used by Lundquist (376). He treated the cyanide solution with zinc dust and then with mercuric chloride. After warming the solution, hydroxylammonium chloride and ammonia were added, and the excess of zinc was dissolved by hydrochloric acid. The gold amalgam was treated with nitric acid to remove all traces of zinc and mercury and the gold was then annealed and weighed.

PROCEDURE 95 (376)

To the cyanide solution (29.2 ml, containing 1 assay ton) add 300 mg of zinc dust and mix well for 1 min. Add 500 mg of mercuric chloride, and heat the mixture on a hot plate. Add 500 mg of hydroxylammonium chloride and 3 ml of concentrated ammonia solution. Boil the mixture for a few minutes, cool, and acidify with 5 ml of concentrated hydrochloric acid. Boil until the solution is clear. If a flocculent precipitate appears add ammonia until the solution is alkaline to phenolphthalein and then 2 to 3 ml more. Add a further 500 mg of hydroxylammonium chloride, and boil until the solution is again clear (although mercury droplets will remain). Allow to settle, and decant the clear supernatant liquid. Add 10 ml of hydrochloride acid to the residue and boil. Dilute to 50 ml, boil for 1 min, stir well, and collect the amalgam in a large globule with a stirring rod. Decant the clear liquid, and repeat the acid treatment of the residue. Then wash the globule several times with water, transfer to a porcelain crucible, and part with nitric acid (1:4). Cool and decant the clear liquid. Add 1 ml of concentrated nitric acid, heat, dilute with water, and decant. Repeat this process until the mercury and zinc have been removed. Wash with water, decant, dry anneal, and weigh.

CATION EXCHANGE SEPARATION OF GOLD ORES

The difficult cation exchange separation of gold from the base metals was accomplished by Pitts (347). Anomalous adsorption of gold on Dowex-50W-X8 (20–30 mesh) was prevented by controlling the chloride concentration of ferric, cupric, and nickel at 0.025 M. Both microgram and milligram quantities of gold can be quantitatively separated from large excess of the base metals. For this reason the method is applicable to base metal ore materials.

THE SEPARATION OF RHODIUM FROM IRIDIUM

In most analytical schemes involving either wet and dry extractions or both techniques, rhodium and iridium are isolated simultaneously. In fire assay extractions these two metals constitute most of the so-called insolubles, and are usually contaminated by traces of base metals and some of the other platinum metals such as ruthenium. Rhodium and iridium are resistant to the usual acid dissolutions, but solutions may be obtained through chlorination in the presence of sodium chloride (377) by a bomb technique (378) (Procedure 115) or by a fusion with sodium peroxide in a silver crucible (259).

The separation of rhodium from iridium presents the most difficult aspect of platinum metals analysis. Until recently no acceptable method was available. During the past decade a good variety of methods has been recorded, all of which are capable of producing accurate results in the hands of experienced analysts. Some of the methods are intricate and time consuming; some are limited to small amounts of metals and, of course, all of them require a prior isolation of the two metals. It is a reasonably safe guess that the next decade will provide new and improved methods for this separation.

One of the earliest separatory methods for rhodium involved a fusion of the mixed platinum metals with sodium hydrogen sulfate, the rhodium being preferentially dissolved by aqueous extraction of the fusion. This method has recently been recommended by Ubaldini (379) for the separation of rhodium from platinum catalysts. This type of method, however, is subject to the common defects of selective dissolution in that with admixtures some platinum is simultaneously dissolved, and unless dissolution is complete some rhodium remains unattacked. The Gibbs method of separation using sodium sulfide (380) has recently received attention also (381), but it appears to be applicable only to small amounts of rhodium. The classical separation of iridium from rhodium as ammonium hexachloriridate is insufficiently accurate for analytical work. It has been stated that reprecipitation is not effective for purifying the precipitate (286).

A number of quantitative separations of rhodium from iridium using selective reductants have been proposed. Among these reductants are copper, antimony, titanium(III) chloride, chromium(II), and vanadium(II). In all of these instances the respective reduction potentials compared with those of iridium indicate that iridium should be reduced to the metal simultaneously with rhodium. When magnesium and zinc in 1.0 M hydrochloric acid are applied as reductants of iridium to the metal, only part of the iridium is precipitated even after boiling for 1 hr and warming on a steam bath for 24 hr. Undoubtedly some solutions of iridium contain unusually stable complexes. This view is supported by the general analytical behavior of iridium, e.g., the difficulty of arriving at complete precipitation using hydrogen sulfide. Peculiarly, the reduction of iridium seems to be encouraged by the presence of palladium. The problem seems to be a kinetic one; its interpretation would provide a distinct contribution to the chemistry of the platinum metals.

REDUCTION BY COPPER

Copper metal has been used by various authors for the general precipitation of the platinum metals. Mukhachev (382) thus determined the sum of these metals. The method involved the introduction of a copper wire into the hydrochloric acid solution of the platinum metals. The solution was boiled for 5 hr, and then the supernatant liquid was decanted. Water was added to the residue and the deposit on the copper wire was dislodged by boiling the water. The wire was removed and the residue in the water was treated with an aqueous ammonia solution, followed by hydrochloric acid to selectively remove any copper contamination. The mixture was filtered, the residue was ignited in hydrogen, and then weighed. The method is simple, but inaccurate because the final residue may contain appreciable amounts of impurities such as silica.

Aoyama and Watanabe (383) used copper powder to separate iridium from gold, platinum, and rhodium. After the reduction the iridium in the filtrate was precipitated as ammonium chloriridate. Procedures were also recorded for the complete precipitation of palladium, ruthenium, and rhodium using copper powder. In 3 N hydrochloric acid at 61°C, these metals could be quantitatively reduced. At this acidity the copper dissolved (with the evolution of hydrogen), and the rhodium metal floated as a sponge. At a similar acidity platinum could be separated from iridium. The precipitate of platinum was treated with an aqueous ammonia solution for 2 hr, and further traces of copper were removed by nitric acid. The platinum was dissolved in aqua regia and precipitated using sodium hydroxide and hydrazine. The filtrate, containing iridium and copper, was evaporated, the residue ignited in hydrogen, and the iridium purified by treatment with nitric acid. Where high accuracy is not required, copper is thus a good precipitating and separatory reagent for iridium. The selective removal of base metal impurities using mineral acids is not recommended.

Tertipis and Beamish (316) have used copper powder to separate both milli-gram and microgram amounts of rhodium from iridium. The procedure requires the precipitation of the rhodium in 1.0 M hydrochloric acid using an excess of copper powder. The rhodium is separated from admixed copper by cationic exchange after dissolution by aqua regia and then by dry chlorination. Gravimetric and colorimetric methods can be used to determine the rhodium. The iridium in the filtrate is separated from copper by cationic exchange and can also be determined either gravimetrically or colorimetrically.

PROCEDURE 96 (316)

Apparatus

Large ion exchange column.   This is a borosilicate glass tube 2.3 cm i.d., joined to a draining tube 5 mm i.d., and containing a resin bed 16–17 cm deep.

Small ion exchange column.   This is a borosilicate glass tube 1 cm i.d., joined to a draining tube 4 mm i.d., and containing a resin bed 5–6 cm deep.

Ion exchanger.   This is Dowex 50X cationic resin in the sodium form, 20–50 mesh, supported in the column by a small plug of glass wool. Regenerate the exchanger just before use with 3 M hydrochloric acid until the eluents are colorless and are free from iron and copper, as shown by spot tests with potassium thiocyanate and rubeanic acid, respec-tively. Remove the excess of acid from the exchanger by washing with water until the eluent is neutral to litmus paper.

Chlorination apparatus.   This is a Vycor tube 19 mm i.d. and 50 cm in total length, with an elongated end (outlet) to allow its connection to a rubber tube, and heated by an electric furnace, 9.53 cm long.

Procedure

For 10–20 mg of metal, make the rhodium–iridium solution, in a 150-ml beaker, 1.0 M in hydrochloric acid, and a total volume of 30 ml. For microgram amounts the total volume should be 20 ml. Add 0.5 g of freshly reduced copper (prepared by reducing copper(II) oxide powder in hydrogen at 400–420°C and grinding in an agate mortar). Cover the beaker tightly with a watch glass, and heat to the gentle evolution of bubbles (91–93°C) for 1 hr, with swirling at intervals of 1 min. Keep the volume constant by the addition of water. Add an additional 0.29 g of copper powder after the solution has been heated for 30 min, to prevent any possible redissolution of metallic rhodium and also to complete the reduction of rhodium. Filter the warm solution through a 3-ml, A2 filter crucible by suction, and decant thrice. Transfer the residue to the filter crucible using a water stream, and wash it into the crucible; 18 such washings are recommended.

Note.   When copper is added, the solution instantaneously changed from dark red to light pink, iridium probably being reduced to a low valence state.

The recovery and determination of rhodium.   Place the filter crucible containing the residue in the original beaker, and treat it with 24 ml of aqua regia on a steam bath until the reaction ceases, the beaker being well covered with a watchglass. Filter the resulting solution through the original filter crucible and wash the beaker several times using a water stream to transfer any residue (filtrate A) into the filter crucible. Dry the residue in the crucible in a steam cabinet cover it with finely ground sodium chloride, and chlorinate for 7 hr at 700°C.

Dissolve the chlorination product in 0.1 M hydrochloric acid, filter the resulting solution through a glass-wool stem, and combine the filtrate with filtrate A. Evaporate the combined solutions almost to dryness. Repeat a few times with concentrated hydrochloric acid, dilute with water, then treat again with concentrated hydrochloric acid. Evaporate to dryness to expel any nitric oxide, and to convert the metals to their chlorides. Dissolve the salts in 50 ml of water, adjust the pH of the solution to 1.3–1.5, and remove the copper by passing the solution through an appropriate cation exchange column at 1 drop/sec. Wash the large and small columns with 550 and 170 ml, respectively, of dilute hydrochloric acid (pH 1.3–1.5).

The determination of the rhodium in the effluent solution. For milligram amounts of rhodium, add 3.5 ml of concentrated hydrochloric acid to each sample, and determine the rhodium gravimetrically using 2-mercaptobenzothiazole (Procedure 59). For microgram amounts, reduce the effluent volume, transfer to a 30-ml beaker, and evaporate to dryness in the presence of 2 ml of a 2% sodium chloride solution. Destroy organic matter by digestion with concentrated nitric acid and a 30% hydrogen peroxide solution, and convert the metal to its chloride using concentrated hydrochloric acid. Determine rhodium spectrophotometrically using a solution of stannous chloride in 2 M hydrochloric acid, the absorbance being measured at 470 nm (Procedure 40). For 2 to 100 μg of rhodium, make the final volume of the solution 25 ml and use a 5-cm cell, whereas for more than 100 μg, make the volume 50 ml and use a 1-cm cell. In all instances, carry a reagent blank through the entire procedure. Use standards directly from the rhodium stock solution.

The recovery and determination of iridium. Evaporate the filtrate and washings from the rhodium precipitation to dryness in a 400-ml beaker, and dissolve the residue in 50 ml of water. Eliminate copper from the resulting solution by the cation exchange procedure described above in the recovery and determination of rhodium. Evaporate the effluent from the small column to dryness in the presence of 2 ml of a 2% sodium chloride solution, in order to remove mineral acids which interfere in the subsequent precipitation of iridium using 2-mercaptobenzothiazole.

When milligram amounts of iridium are to be determined, dissolve the residue in 30 ml of water and determine the iridium gravimetrically using 2-mercaptobenzothiazole, in the presence of one or two glass beads to prevent bumping (Procedure 61). When micrograms amounts are to be determined, dissolve the residue in water, and transfer the resulting solution to a 30-ml beaker. Destroy the organic matter by digestion with a few drops of concentrated nitric acid and a hydrogen peroxide solution, evaporate to dryness, treat the residue with concentrated hydrochloric acid, and determine the iridium colorimetrically (Procedure 42).

In both procedures carry a reagent blank through the entire procedure. Use standards directly from the iridium stock solution and, for the microgram amounts, analyze them simultaneously with the unknown samples.

Because the recorded redox potentials for iridium systems indicate that copper should precipitate iridium, it is of interest to know the amounts of iridium that can be precipitated by copper under the conditions of the previous procedure. Table 7.1 indicates the amounts which may be expected to precipitate.

THE SEPARATION OF RHODIUM AND IRIDIUM BY SOLVENT EXTRACTION

While a variety of casual references have been made to the use of liquid–liquid extractions for rhodium and iridium, only two have approached the status of analytical separations.

The Tin(II) Bromide–Isoamyl Alcohol Method

This method for the separation of rhodium from iridium requires isoamyl alcohol as a solvent for rhodium complexed with tin(II) bromide in a perchloric–hydrobromic acid medium at room temperature. Microgram quantities, $10-100$ μg of rhodium and $7.5-73.8$ μg of iridium, can be separated and subsequently determined. Rhodium and iridium are determined spectrophotometrically directly in the organic and in the aqueous phase respectively. Accurate and precise values can be obtained, and in the present authors' opinion, the method is superior to all hitherto published methods from the point of view of time required and simplicity of operation.

TABLE 7.1
Iridium Reduced by Copper (316)

Iridium		Iridium	
Added (mg)	Found (μg)	Added (mg)	Found (μg)
9.15	32	0.459	1.8[a]
	30	0.459	
	33	0.459	
4.60	8	0.092	1.4[a]
	9	0.092	
	8	0.092	
	9	0.046	0.3[a]
	8	0.046	
	9	0.046	
1.84	5		
	4		
	5		

[a]After reduction by copper, the three residues were combined, and analyzed for iridium.

PROCEDURE 97 (384)

Reagents

Tin(II) bromide–hydrobromic acid.   Dissolve 3 g of mossy tin in A.R. 48% hydrobromic acid, and evaporate the solution to dryness. Cool the residue, and dissolve it in 17.0 ml of redistilled hydrobromic acid. Dilute with water, filter into a 50-ml volumetric flask, and make up to volume with water. This solution should be freshly prepared each week.

Perchloric acid.   Dilute 19.5 ml of 70% ($d = 1.66$) perchloric acid to 50.0 ml with water.

Procedure

Transfer to a 30-ml beaker the solution of rhodium and iridium chloride salts, containing no more than 100 μg of each metal. Evaporate to dryness on a steam bath in the presence of 2 ml of a 2% sodium chloride solution.   Add 4–5 drops of redistilled hydrobromic acid, allow to stand to effect dissolution, and then transfer the solution to a 75-ml pear-shaped separatory funnel with 5 ml of water. Wash the beaker with 5 ml of the tin(II) bromide–hydrobromic acid solution and then with 5 ml of the perchloric acid (see Note 1). Allow to stand for 30 min, with frequent shaking, to produce the yellow rhodium complex. Add 25 ml of isoamyl alcohol, and shake for 2.5 min, thus leaving the iridium in the colorless, aqueous phase. Drain this aqueous layer into a 25-ml volumetric flask. Then add 3.3 ml of the tin(II) bromide–hydrobromic acid solution to the separatory funnel, shake, and drain into the 25-ml volumetric flask (see Note 2). Drain the organic phase into a 50-ml volumetric flask, and wash the separatory funnel 3 times with isoamyl alcohol. Add the washings to the 50-ml flask. Add isoamyl alcohol to the mark, and determine the rhodium immediately by measuring the absorbance of the solution at 427 nm in matched silica cells (1 cm) (see also Procedure 40). To the iridium-containing aqueous phase in the 25-ml volumetric flask add 3.3 ml of perchloric acid, shake thoroughly, and place the flask in a boiling water bath for 3 min. Remove, cool the flask in air for 5 min, and then under the tap, and make up to the mark with water. Determine the iridium spectrophotometrically by measuring the absorbance of the solution at 403 nm in matched 1-cm silica cells (see also Procedure 42). Determine standards for rhodium and iridium simultaneously.

Notes. **1.**   These washings, placed in the separatory funnel, should produce a 15 ml solution, 0.17 M in tin salt, 1.5 M in hydrobromic acid, and 1.5 M in perchloric acid.

2.   This washing removes residual traces of iridium, cleans the organic phase, and washes the stem of the separatory funnel.

THE DETERMINATION OF IRIDIUM
IN THE PRESENCE OF PLATINUM

Alloys of iridium and platinum are not readily attacked by aqua regia. The procedure, utilizing chlorination or fusion, involves a chemical separation and determination of the two constituents. Whereas this may be accomplished readily by an oxidation to form quadrivalent salts of the two metals, in which condition the iridium may be selectively precipitated as the hydrated dioxide, the method is lengthy and requires considerable experience.

One of the best known methods for the determination of iridium in the presence of platinum and other platinum metals has its origin in the classical procedures developed by Deville and Stas in the middle of the 19th century. In principle the early procedures involved the preparation of a lead alloy by the direct fusion at about 1000°C of the noble metals with lead in a graphite crucible. The alloy was then parted with nitric acid, the residue was treated with aqua regia, and the insoluble material was fused with potassium hydroxide and nitrate. From this fused mixture ruthenium was the first to be isolated—by distillation—followed by a treatment of the hot liquid to isolate iridium, iron, silica, and gold by a basic hydrolysis. Conventional methods were then used to purify the iridium. This method was examined in detail at the National Bureau of Standards and modified by Gilchrist (385). It was found that the treatment with aqua regia of the lead alloy containing iridium almost invariably produces some dissolution of iridium, whose amounts are unpredictable but sufficiently small to be ignored for most analytical purposes. For high accuracy the iridium must be separated from the aqua regia extract of the parting acid residue. The optimum ratio of nitric acid to water was not critical. Strong acid solutions produced a residue that was rather finely divided for effective filtering; a very weak acid (1:8) prolonged the parting time.

Because lead is the classical collector of noble metals in fire assay procedures, some scattered data are available which deal with the behavior of the noble metals when the alloy containing very high proportions of lead is treated with nitric acid or aqua regia.

In these assay procedures it has been determined that reasonable variations in fusion time and minor variations in lead proportions have little effect upon the recovery of noble metals. An inadequate fusion time results in a relative nonhomogeneity and very large buttons, e.g., 35–40 g, producing an increasingly finely divided parting residue. In the latter instance there is an increased susceptibility to corrosion of the insoluble by nitric acid or aqua regia. With properly prepared lead alloys the addition of aqua regia to the parting insoluble will, in general, remove platinum, palladium, gold, and such base metals as copper, nickel, etc. With fire assay buttons, the aqua regia treatment will, in general, remove little ruthenium, osmium, or rhodium. Gilchrist (385) found that rhodium was removed from the iridium by aqua regia. This difference in finding results, perhaps, from a difference in the character of the rhodium constituents and the difference in the metallurgical history of the parting residue from the lead alloys collected by fusion of natural and artificially prepared materials. Anyway the analysis of alloys of lead and noble metals, other than those containing only platinum and iridium, or palladium, platinum, and iridium or osmium, is best accomplished following a cupellation to remove the lead.

The method described below is recommended for the determination of iridium only in platinum or palladium alloys; it is not recommended when the latter two noble metals are to be determined along with iridium. The removal of lead from platinum or palladium is not a simple analytical problem, and better methods of analysis are available. In general the authors have successfully used the method described below with alloys containing 20% or less of iridium.

It can be readily adapted to the massive forms of alloys such as sheet and wire.

PROCEDURE 97a (385)

Crucibles. The graphite crucibles may be constructed from 2 × 2 in. rods cut to lengths of about 3 in., drilled to a depth of 1.5 in., and tapered toward the bottom.

Procedure

Add 0.2 to 1 g of the platinum–iridium alloy to the graphite crucible, and cover with assay lead, about 10 g of granulated or folded lead sheet being sufficient. Fuse for 1 hr at red heat, remove the crucible, and place it on a copper block to cool. Transfer the alloy, cleaned with a camelhair brush, to a 200-ml beaker. Add 75 ml of 2 N nitric acid, and heat on a steam bath to the complete parting of the lead button. Add an equal volume of water, and allow to settle. Decant through an 11-cm Whatman No. 42 filter paper, and treat the insoluble with portions of hot water each decanted through the paper. Examine the filtrate for small particles, and refilter if these are present. Transfer the paper and residue to the beaker containing the bulk of the insoluble, and add 20 ml of water, 6 ml of hydrochloric acid, and 2 ml of concentrated nitric acid. Heat the mixture on a steam bath for 2 hr, at which time the residue in the beaker should be distinctly gray. Add an equal volume of water, and filter through a 7-cm Whatman No. 42 filter paper. Wash well with hot water and then with hot 0.1 M hydrochloric acid, then again with hot water. Transfer the paper and metal to a porcelain crucible, ignite carefully in air, then in hydrogen; cool in hydrogen, and weigh as iridium metal.

Note. In the absence of iron the results tend to be a little low as determined by a caustic fusion of the alloy and a subsequent wet analysis. Iron, when present, will appear as an insoluble alloy with the iridium. For the maximum accuracy the iridium should be dissolved, and the iron determined by suitable wet methods. This may be accomplished readily by chlorination and subsequently determining the iridium.

THE DETERMINATION OF OSMIUM AND RUTHENIUM IN LEAD ALLOYS

Despite the fact that the physical and chemical properties of lead are little related to the noble metals and that lead is seldom associated with these metals in natural deposits, this metal is a suitable medium for a high temperature extraction of noble metals from a fused mixture of ore and flux. However, whereas the extraction can be made complete, the subsequent wet analysis of the lead button or regulus is complicated by partial reactions of the noble metals with the parting acid; the latter is usually nitric acid. The pertinent methods of analysis and the specific difficulties associated with the partial corrosion of each metal have been discussed (133a). It may be stated here that whereas the lead button is readily parted by nitric acid and that this acid may simultaneously volatilize octavalent osmium, the complete removal of osmium may not be accomplished. Furthermore, when ruthenium is to be determined consecutively the residual nitric acid must be destroyed. This is usually done by fuming with sulfuric acid, but obviously this is not practical in the presence of large amounts of lead. Because of these difficulties the determinations of osmium and ruthenium in lead buttons are more advantageously carried out by parting with perchloric acid, with the simultaneous distillation of both osmium and ruthenium in the presence of excess of perchloric acid. Here, as with the iron–copper–nickel alloys, the choice of receiving liquids should be appropriate to the required separation of osmium and ruthenium. For this purpose a cold hydrogen peroxide solution may be used to collect both metals, with a subsequent treatment with sulfuric acid and additional hydrogen peroxide to remove osmium selectively.

Alternatively the receiving liquid may be sulfur dioxide–hydrochloric acid in which instance an evaporation to remove sulfur dioxide is required prior to a second distillation with nitric acid, etc., to remove osmium. By this treatment the nitric acid in the distillation flask can be removed by an evaporation with hydrochloric acid. Finally, in preparation for the distillation of ruthenium, sulfuric acid is added, and evaporation to fumes is carried out to remove the hydrochloric acid.

Receiving solutions of sodium hydroxide may also be used, followed by a second distillation with nitric acid and, as stated above, treatments of the pot liquid to secure a sulfuric acid solution of ruthenium.

Other combinations of recovery liquids and oxidizing reagents will suggest themselves. The author recommends the use of a cold hydrogen peroxide solution for the collection of the combined tetroxides from a perchloric acid distillation.

PROCEDURE 98 (223, 386)

Equipment

See Fig. 7.1.

Procedure

Transfer the lead alloy (20–50 g) to the distillation flask, and add 100 ml of water to the trap, 25 ml of a 3% hydrogen peroxide solution to the first receiver and 5 ml of this

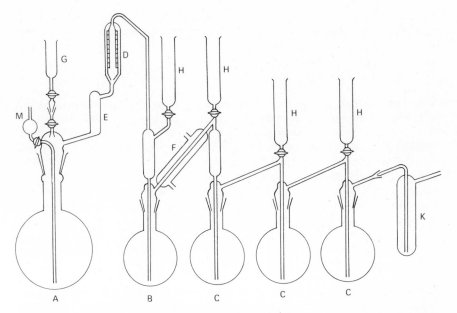

FIGURE 7.1. This apparatus is constructed from borosilicate glass in individual parts, and is mounted on a steel frame and sealed together. No rubber connections or grease are used. The distillation flask A is of 1-liter capacity; the trap B and the receivers C are of 200-ml capacity. The spray trap D and reflux device E prevent solids from being carried into the trap. The condenser F between the trap and the first receiver largely prevents the distillation of water and acid from the trap where it is heated. The oxidizing acid is introduced via the removable reservoir G. The centrifuge tubes H sealed to stopcocks at the top of the inlet tubes of the small flasks are used to introduce reagents and to flush the tubes after a distillation. Air is drawn through the apparatus at M by suction applied to a bubbler K, which contains a thiourea solution to detect any osmium tetroxide escaping from the receivers. This equipment is applicable to the distillation of both trace and macro amounts of osmium and ruthenium.

For details of the method see Procedure 98.

hydrogen peroxide solution to each of the other two receivers. Chill the receivers in an ice bath, pass water through the condenser, and apply suction to produce 2 or 3 bubbles/sec. Add 75 ml of 72% perchloric acid, and heat very gently until the lead is completely dissolved and effervescence of hydrogen has ceased. Continue the heating until the white fumes of perchloric acid have disappeared and a colorless liquid is refluxing on the still wall (see Note). Cool to about 60°C, and add 8 ml of 36% perchloric acid. Heat again to the removal of brown fumes. Repeat the addition of 8 ml of 36% perchloric acid, heating twice to ensure the complete removal of the tetroxides.

Note. With amounts of ruthenium of the order of 10 mg, the brown ruthenium tetroxide can be seen condensing in the trap and on the still wall.

A complete distillation requires 0.5–1 hr. Add 15 ml of 72% perchloric acid to the trap, and boil the solution for 30 min. Transfer the chilled receiving solution to a second chilled distillation flask as quickly as possible to prevent any loss of osmium, and wash the receivers and the delivery tubes with cold 3% sulfuric acid. Remove the sulfuric acid thoroughly from the tubes and receivers by washing with water.

Add 100 ml of water to the trap and 30 ml of twice distilled 48% hydrobromic acid to the first receiver and 10 ml of the same acid to each of the other receivers. Chill the receivers in an ice bath. Add 40 ml of a 30% hydrogen peroxide solution and 5 ml of concentrated sulfuric acid to the chilled distillation flask containing the osmium and ruthenium distillates. Boil gently for 30 min. Add 15 ml of perchloric acid to the trap, and boil for 25 min. Draw a stream of air slowly through the system (about 3–5 bubbles/sec). Transfer the contents of the receivers to a 150-ml beaker, and rinse both receivers and tubes with 10% hydrobromic acid. Evaporate the solution on a steam bath to 5 ml, transfer to a 30-ml beaker, and evaporate again to 0.5–1.0 ml. Transfer the solution to a 20-ml test tube by washing with water, and determine the osmium by thiourea (Procedure 46). For milligram amounts of osmium, adjust the evaporated osmium distillate to the required acidity for precipitation by thionalide (Procedure 66).

To recover the ruthenium in the pot liquid, add 100 ml of water to the trap, 30 ml of a 3% hydrogen peroxide solution and 1 ml of 48% hydrobromic acid to the first receiver, and 10 ml of a 3% hydrogen peroxide solution to each of the other two receivers. Cool the receivers in an ice bath. Add 10 ml of concentrated sulfuric acid to the distillation flask and then add cautiously an excess of a 10% sodium bromate solution (about 20 ml). Apply a gentle suction, and distill cautiously over a low flame for 1 hr. Then add 15 ml of the perchloric acid to the trap, and boil for 25 min. Disconnect the receivers from the water condenser, maintaining the connection between the two receivers. Add 8 ml of 48% hydrobromic acid to the first receiver and 4 ml of the acid to the second receiver. Boil the liquid for 10 min. Transfer the contents of the receivers to a 150-ml beaker, rinse the tubes and the receivers with 10% hydrobromic acid, and evaporate the combined solution to 5 ml. If a spectrophotometric method is to be used transfer the 5 ml of solution to a 30-ml beaker and evaporate again to 0.5–1.0 ml. Transfer to a 20-ml test tube, and apply the thiourea method for ruthenium (Procedure 44). Other appropriate methods may be used. For milligram amounts treat the evaporated ruthenium solution (0.5 – 10 ml) to prepare for a thionalide precipitation (Procedure 63).

THE GRAVIMETRIC DETERMINATION
OF THE SIX PLATINUM METALS

The following procedures (387) were developed by the late R. Gilchrist of the National Bureau of Standards and recorded in the Bureau journals of research over a period of several years. The method involves the initial removal of osmium by oxidation with nitric acid and distillation of its tetroxide and collection in a solution of sulfur dioxide in hydrochloric acid. The solution, residual in the flask, is treated for the removal of ruthenium by oxidation with an acid solution of sodium bromate to produce the volatile ruthenium(VIII) oxide, collected as in the case of osmium, in a solution of hydrochloric acid and sulfur dioxide. The residual solution containing palladium, platinum, rhodium, and iridium is first treated to allow the separation of platinum only. This is accomplished by the selective precipitation and separation of palladium, rhodium, and iridium as hydrated tetravalent oxides. Platinum in the tetravalent form is not precipitated and after two treatments can be isolated by filtration. It is then determined gravimetrically as insoluble platinum sulfide. The oxides of palladium, rhodium, and iridium are then dissolved by hydrochloric acid and palladium is selectively removed as the dimethylglyoximide. The filtrate from the latter is treated with titanous chloride to isolate rhodium as metal which is treated with sulfuric and nitric acids and then precipitated as sulfide. The determination of iridium can be accomplished in two ways. In one, by the elimination of titanium by cupferron and the subsequent precipitation and determination of iridium from the hydrated dioxide. The alternative method involves the dual precipitation as hydrated oxides of rhodium and iridium, and their dual determination as metals followed in a second sample by the selective removal of rhodium by titanous chloride and its determination as metal. The iridium content is then calculated from the difference between the weight of rhodium and the weight of the mixed metal.

It should be stressed that the methods previously discussed allow the adaption of many methods of determination and separation. The method has been proved for as little as 20 mg of the platinum metals but the present authors recommend the method for weights of individual platinum metals of about 100 mg upward.

PROCEDURE 99   RECOMMENDED FOR THE ANALYSIS OF THE PLATINUM GROUP (293)

I. Separation and Determination of Osmium

1. Distilling apparatus. The distilling apparatus (Fig. 7.2) consists of three main parts, namely, a 700-ml distilling flask, a set of three 300-ml absorbing flasks, and a train of inlet and delivery tubes which are sealed into one piece. The thistle tube, closed by a stopcock and placed between the first and second absorbing flasks, serves to replenish the absorbing solution with sulfurous acid if an unusual amount of nitric acid is distilled and also, at the end of the operation, to rinse the tube connecting the two flasks. The entire apparatus is constructed of Pyrex glass. The joints must be very carefully ground, and it is advantageous if they are made so that the flasks are interchangeable. It is important to note that these joints are sealed with a film of water only and not with lubricating grease, because the latter would cause reduction of some osmium tetroxide to dioxide which cannot be readily recovered. It was found necessary to grease the stopcock in the tube used for the introduction of nitric acid into the distilling flask, but this constitutes the only exception. During distillation, this delivery tube is frequently flushed with water in order to remove any osmium tetroxide which may have diffused into the column of water held in it. Three absorbing flasks are used, but the absorption of osmium tetroxide is practically complete in the first flask, and no osmium has been found to escape the second flask.

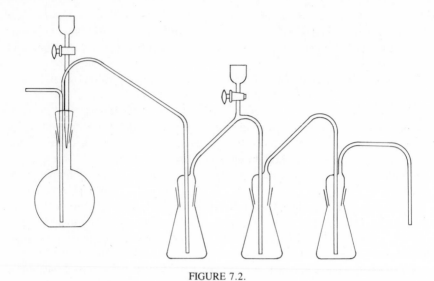

FIGURE 7.2.

2. Separation of osmium by distillation as osmium tetroxide.   Place 150 ml of diluted hydrochloric acid (1 + 1) which has been freshly saturated with sulfur dioxide,* in the first absorbing flask and 50 ml of the same reagent in each of the other two flasks. Place the solution containing the platinum metals in the distilling flask and make sure that the separate parts of the entire apparatus are properly connected. If necessary, dilute the solution in the distilling flask to about 100 ml with water, and add through the inlet tube 40 ml of diluted nitric acid (1 + 1). Flush the thistle tube and stopcock with 10 ml of water. Pass a slow current of air through the apparatus and heat the solution in the distilling flask to boiling. Continue the distillation for one hour. This length of time should be sufficient to ensure complete elimination of osmium from solutions in which it was originally present as an alkaline osmate or bromoosmate. If, however, the osmium is present as chloroosmate, the time required will be from seven to eight hours. In this case, it is preferable to distill from concentrated sulfuric acid, thereby greatly reducing the time required, or, if ruthenium is known to be absent, from concentrated sulfuric acid to which a few milliliters of nitric acid is added.

3. Recovery of osmium by precipitation as the hydrated dioxide.   Unite the portions of the absorbing solution and evaporate as far as possible on the steam bath in a clean, unetched beaker. It is important, in precipitating the platinum metals hydrolytically, that

---

*Throughout this recipe, whenever acids are mentioned, hydrochloric acid will mean the concentrated acid of specific gravity 1.18; nitric acid, the concentrated acid of specific gravity 1.42; sulfuric acid, the concentrated acid of specific gravity 1.84. Diluted acids will be designated as follows: For example, diluted hydrochloric acid (1 + 1) will mean one volume of the concentrated acid of specific gravity 1.18 diluted with one volume of water.

the beakers used do not have an etched surface. An etched beaker often becomes stained with the precipitate, and this stain cannot always be removed readily. Digest the residue with 10 ml of hydrochloric acid for 15 min, and evaporate a second time. Repeat the digestion with hydrochloric acid and the evaporation three times more. This is done to ensure complete decomposition of any sulfite compounds of osmium. Dissolve the residue from the last evaporation in 150 ml of water. Heat the resulting solution to boiling, and add to it a filtered 10% solution of sodium bicarbonate until a precipitate appears and suddenly coagulates. Add a few drops of brom phenol blue indicator solution (0.04%) to the hot solution. This indicator changes from yellow to blue at pH 4. Add the bicarbonate solution dropwise until the indicator assumes a faint bluish color. Finally, boil the solution for 5 to 6 min to ensure complete precipitation of the hydrated osmium dioxide.

4. Determination of osmium as metal. Filter the solution through a Munroe platinum crucible,* carefully pouring the supernatant liquid through first. Transfer the precipitate, and wipe the inner walls of the beaker and also the glass rod with a rubber policeman which has been thoroughly wetted so that the precipitate will not cling to it. It should be borne in mind that filter paper must not be used to wipe the beaker, although it is used when handling precipitates of any of the other five platinum metals. Wash the precipitate thoroughly with a hot 1% solution of ammonium chloride, and then cover it with solid ammonium chloride. Moisten the ammonium chloride with a few drops of the wash solution and saturate the precipitate by applying suction. If desired, a saturated solution of ammonium chloride may be used to impregnate the precipitate. Continue the suction until the bottom of the crucible is coated with solidified ammonium chloride. Wipe off this coating of salt and place the platinum cap on the bottom of the crucible.

Cover the crucible with a Rose lid, preferably of quartz. Ignite a stream of hydrogen from a Rose delivery tube, likewise of quartz, and regulate the stream so that a very small flame is produced. Then insert the tube through the opening in the lid. The hydrogen flame will probably become extinguished by this operation and must be reignited. This is done by momentarily placing a burner flame under the crucible. The hydrogen will now burn as it emerges from under the lid at the edge of the crucible. The ignited hydrogen generates the requisite amount of heat to dehydrate the osmium compound without causing deflagration. After 5 min, gradually heat the crucible with the burner flame until all of the ammonium chloride is expelled. Ignite the osmium residue strongly in hydrogen for 10 min. Remove the burner and allow the crucible to cool somewhat. Extinguish the hydrogen flame by momentarily breaking the current of hydrogen, and allow the crucible to cool to room temperature. Finally, displace the hydrogen with a current of carbon dioxide without even momentary access of air. If the hydrogen is not displaced by an inert gas, such as carbon dioxide, the reduced metal will be rapidly attacked when first exposed to the air, with significant loss of osmium. Weigh the residue as metallic osmium.

---

*The chief disadvantage in the use of a Gooch crucible with an asbestos pad is the difficulty encountered during the ignition of the dioxide to metal. Even when the Gooch crucible is protected from the burner flame by setting it in a circle of asbestos board in a larger porcelain crucible, one is not always able to extinguish the hydrogen flame readily. Another disadvantage is the inadequate protection of the reduced metal from attack by air diffusing through the holes in the bottom of the crucible.

II. Separation and Determination of Ruthenium (292, 385, 388)

1. Preparation of solution.   Evaporate to dryness on the steam bath the solution which remains in the distilling flask after the elimination of osmium. Add 5 to 10 ml of hydrochloric acid and again evaporate. Repeat the evaporation with hydrochloric acid until oxides of nitrogen are no longer evolved. Dissolve the residue from the last evaporation in about 20 to 30 ml of water. Add 10 ml of sulfuric acid and evaporate the resulting solution carefully until vapors of sulfuric acid are evolved. Transfer the solution, and any residue of platinum which may have separated, to the distilling flask that was used for the distillation of osmium tetroxide. Dilute the solution in the distilling flask to 100 ml with water.

2. Separation of ruthenium by distillation as ruthenium tetroxide.   Place 150 ml of diluted hydrochloric acid (1 + 1), freshly saturated with sulfur dioxide, in the first receiving flask and 50 ml of the same reagent in each of the other two flasks. Add 100 ml of a filtered 10% solution of sodium bromate through the thistle tube to the distilling flask. This tube should be flushed with water occasionally in order to remove any ruthenium tetroxide which dissolves in the column of liquid held in the tube. In order to maintain an excess of sulfur dioxide in the absorbing solution, a saturated solution of sulfur dioxide in water should be added from time to time through the thistle tube situated between the first and second receiving flasks. Pass a gentle current of air through the apparatus, and heat the solution in the distilling flask to boiling. Distill* in this fashion for 1½ hr. Add 25 ml of the bromate solution, and continue to distill for an additional hour.

3. Recovery of ruthenium by precipitation as a hydrated oxide.   Combine the portions of the absorbing solution and evaporate to a moist residue on the steam bath. Add 10 ml of hydrochloric acid and digest the solution, in a covered beaker, on the steam bath for ½ hr. Add 50 ml of water and heat the solution to boiling in order to dissolve completely the somewhat difficultly soluble ruthenium compound formed on evaporation. When the ruthenium compound is completely dissolved, filter the solution and wash the filter with diluted hydrochloric acid (1 + 99). Filtering ensures the elimination of a small amount of

---

*The usual way is to distill from an alkaline solution which is saturated with chlorine. If this procedure is used, one can never be sure that the elimination of ruthenium is complete, even though additional alkali is added and the distilling process repeated, owing to the precipitation, near the neutral point, of iridium hydroxide which carries with it some ruthenium. It is necessary, therefore, in order to prevent the precipitation of iridium, to distill from an acid solution. The distillation of ruthenium tetroxide may be made immediately after the elimination of osmium, without removing the contents of the distilling flask, by adding a solution of sodium bromate. From 5 to 10 ml of nitric acid should be added to increase the acid concentration of the solution. The minimum time required for complete removal of ruthenium from a boiling solution is about 3½ hr. Although the solution requires some preparation, distillation from a solution of ruthenium sulfate in diluted sulfuric acid, as given in the procedure, is preferable. The elimination of ruthenium tetroxide from such a solution is much more rapid than from a solution containing the ruthenium as chlororuthenate. A further advantage is gained by being able to conduct the entire distillation at the boiling temperature without danger of depositing ruthenium dioxide on the walls of the flask and delivery tube, since the cause of such deposition, namely, hydrochloric or hydrobromic acid, is removed during the preparation of the solution for distillation.

silica which may be present. Dilute the ruthenium solution to 200 ml, heat it to boiling, and add a filtered 10% solution of sodium bicarbonate until a precipitate begins to form. Add 3 or 4 drops of bromcresol purple indicator solution (0.04%). Continue neutralization of excess acid by adding the bicarbonate solution dropwise until the indicator changes from yellow to blue. Boil the solution for 5 to 6 min to coagulate the precipitate.

The precipitate which is formed is a hydrated oxide of ruthenium, probably of tervalent ruthenium. In general appearance, it resembles other hydrated oxides of tervalent platinum metals. It does not settle as quickly nor appear as compact as the precipitate formed on the hydrolysis of compounds of quadrivalent ruthenium. However, it is quantitatively precipitated, and no difficulty is encountered in handling it.

4. Determination of ruthenium as metal.   Filter the solution and wipe the inner wall of the beaker and also the glass rod with a small piece of ashless filter paper. Wash the filter and the precipitate thoroughly with a hot 1% solution of ammonium sulfate. Finally, wash three or four times with a cold (room temperature) 2.5% solution of ammonium sulfate.

Place the filter and precipitate in a porcelain crucible, dry them, and char the filter slowly. The dried filter will usually char completely when once it begins to smoke. This operation should be done carefully in order to prevent loss of ruthenium by deflagration. Ignite the residue strongly in air and then in hydrogen. The ignition in hydrogen is made in a manner similar to that described for the determination of osmium. Cool the resulting metal in hydrogen and leach it well with hot water. This is done to ensure the removal of traces of soluble salts. It is well to leach the residue in the crucible first and then to transfer it to a filter. Ignite the filter and metal sponge in air and in hydrogen. Cool the residue in hydrogen and weigh it as metallic ruthenium.

III. Separation and Determination of Platinum (262)

The solution which remains in the distilling flask after the elimination of ruthenium contains the four metals, platinum, palladium, rhodium, and iridium, together with sulfuric acid, sodium sulfate or acid sulfate, bromine, and undecomposed bromate. Experience shows that the platinum, rendered partly insoluble in the preparation of the solution for the distillation of ruthenium, is entirely dissolved during the distilling operation. A trace of iridium sometimes separates as the dioxide toward the end of the distilling period.

1. Preparation of the solution. (a) Treatment of the solution from the distilling flask.   In order to prepare the solution for subsequent operations, transfer the contents of the distilling flask to a liter beaker. Cautiously decompose the remaining bromate with hydrochloric acid. Unless precaution is taken in this treatment, mechanical loss may occur owing to the vigorous evolution of gas. Evaporate the solution when it has become quiescent. Make certain that any bromate remaining is decomposed, by evaporating with hydrochloric acid. Occasionally, the distilling flask is found to be slightly stained with iridium dioxide near where the level of the solution has stood. It is well, therefore, always to clean the flask with 5 to 10 ml of aqua regia, which then must be evaporated with hydrochloric acid to decompose nitroso compounds before it is added to the main solution. Finally, evaporate the solution as far as possible on the steam bath, and then dilute it to 200 ml with water.

(b) Treatment of a solution which has not contained osmium or ruthenium.   If the solution being analyzed is known not to contain either osmium or ruthenium and the parts

of the general procedure referring to these metals have not been followed, evaporate it to a moist residue on the steam bath. If the solution has contained nitric acid, add 5 ml of hydrochloric acid and again evaporate, repeating this operation to ensure the decomposition of nitroso compounds. Add 2 g of sodium chloride and 5 ml of hydrochloric acid and evaporate this time to dryness on the steam bath. Add 2 ml of hydrochloric acid and dilute the solution to 300 ml with water.

2. Separation of platinum by joint precipitation of palladium, rhodium, and iridium as hydrated dioxides.   Heat the solution containing platinum, palladium, rhodium, and iridium to boiling, and add to it 20 ml of a filtered 10% solution of sodium bromate. Carefully add a filtered 10% solution of sodium bicarbonate until the dark green solution shows evidence of the formation of a permanent precipitate. Test the acidity of the hot solution from time to time by allowing a drop of bromcresol purple indicator solution (0.01%) to run down the stirring rod into the drop which clings to it as it is lifted from the solution. Enough bicarbonate has been added when the color of the indicator changes from yellow to blue. At this stage, add 10 ml more of the bromate reagent and boil the solution for 5 min. Increase the pH of the solution slightly by carefully adding dropwise carbonate solution until a faint pink color is produced in the test drop by a drop of cresol red indicator solution (0.01%). Again add 10 ml of the bromate reagent and boil for 15 min.

On removing the beaker from the source of the heat, the mixed precipitate will settle quickly, leaving a mother liquor containing the platinum. Filter the solution by suction, using a porcelain filtering crucible* having solid walls and a porous base.

It is highly desirable to avoid the use of filter paper when repeated precipitations are to be made. The material of which the paper is composed undoubtedly reacts with acids and probably forms small quantities of organic compounds with the platinum metals which are not easily hydrolyzed. Iridium dioxide, which dissolves much less readily than either palladium or rhodium dioxide, tends to stain paper pulp. The stain cannot always be removed by washing. These difficulties are avoided if the porcelain filtering crucible is used. Furthermore, such crucibles have the advantage that concentrated hydrochloric acid can be used to dissolve the hydrated dioxides, and considerable time is saved in preparing the solution for subsequent treatment.

Pour the supernatant liquid through first, then transfer the precipitate. Rinse the beaker and wash the precipitate with a hot 1% solution of sodium chloride, the acidity of which has been adjusted to between pH 6 and 7. Place the crucible with the precipitate, and also the stirring rod, in the beaker used for the precipitation. It may be necessary to remove a small amount of the precipitate which has crept over the lip of the beaker during filtration. It is preferable to do this with moistened crystals of sodium chloride on the finger, rather than to use paper or a rubber policeman. Replace the watch glass and add from 10 to 20 ml of hydrochloric acid, pouring most of it into the crucible. Place the covered beaker on the steam bath. The rhodium and palladium compounds will dissolve quickly, the iridium

---

*This type of crucible is manufactured by the State Porcelain Works, Berlin, Germany. A crucible of convenient size is one having a height of 43 mm, a diameter at the top of 40 mm, and a capacity of 30 ml. The crucible designated by the mark "A1" filters rapidly and allows no trace of the precipitate to escape. Crucibles marked "A2," but somewhat smaller in size, were found to be perfectly satisfactory. No doubt glass crucibles of suitable porosity could also be used.

dioxide more slowly. Carefully lift the crucible with the stirring rod, wash it with water, and place it in a 250-ml beaker. Pour 5 ml of hydrochloric acid into the crucible. Cover the beaker with a watch glass and set it on the steam bath. This treatment will usually leach out the small quantity of metal chlorides in the porous bottom. This operation should be repeated with fresh acid to ensure complete removal. Combine the leachings with the main portion of the dissolved precipitate, add 2 g of sodium chloride, and evaporate to dryness on the steam bath. Add 2 ml of hydrochloric acid, dilute the solution to 300 ml with water, and repeat the precipitation of the hydrated dioxides. Two such precipitations are sufficient ordinarily to effect the complete separation of platinum from palladium, rhodium, and iridium.

3. Recovery of platinum by precipitation with hydrogen sulfide.   Add 20 ml of hydrochloric acid to each of the filtrates obtained from the hydrolytic precipitation of the dioxides of palladium, rhodium, and iridium. Carefully warm the solutions until they become quiescent. Partially concentrate the filtrates, combine them, and then evaporate to dryness. Make certain that all of the bromate is destroyed, by evaporation with hydrochloric acid. Dilute the yellow platinum solution somewhat and filter it. Wash the filter with diluted hydrochloric acid (1 + 99). Dilute the filtered solution to about 400 ml with water and have it contain 5 ml of hydrochloric acid in each 100-ml volume.

Precipitate the platinum, in a hot solution, with hydrogen sulfide as the solution cools somewhat, to ensure complete precipitation.

4. Determination of platinum as metal.   Filter the solution and wash the precipitate with diluted hydrochloric acid (1 + 99). Ignite the dried filter and precipitate in a porcelain crucible. Leach the metal residue with diluted hydrochloric acid, transfer it to a filter, and wash it thoroughly with hot water. Ignite the filter and metal again strongly in air. Weigh the residue as metallic platinum. The metal so obtained will usually contain a small but significant amount of sulfur* which cannot be eliminated by ignition in either air or hydrogen.

If the highest accuracy is desired, dissolve the metallic platinum, obtained by ignition of the sulfide, in aqua regia. Destroy nitroso compounds by evaporation with hydrochloric acid. Filter the solution into a clean, unetched beaker. Wash the filter with diluted hydrochloric acid (1 + 99). Dilute the solution to 100 ml, heat it to boiling and add to it a solution containing 3 g of sodium acetate and 1 ml of formic acid for each 0.25 g of platinum. Boil the resulting solution gently until the precipitated metallic platinum is well coagulated and the supernatant liquid is colorless. Filter the solution and wash the metallic deposit with a hot 1% solution of ammonium chloride. Place the filter and the spongy metal in a porcelain crucible and ignite them strongly in the air. Leach and wash the ignited metal as previously directed. This precaution is taken to ensure the removal of soluble salts. Finally ignite the platinum in air again. Weigh the residue as metallic platinum, which will now be free from sulfur.

*In 17 experiments, the weights of platinum (0.25-g portions) recovered by ignition of the sulfide in air exceeded those taken by amounts which ranged from 0.0 to 1.0 mg, and averaged 0.5 mg. The retention of sulfur by palladium is very marked, and the sulfide may even fuse into globules which cannot be completely decomposed by ignition. Therefore, palladium must not be determined by ignition of its sulfide. No significant amount of sulfur is retained when rhodium sulfide is ignited.

It is far better to precipitate the platinum first by hydrogen sulfide and then to reprecipitate it by formic acid in a buffered solution, than it is to attempt to precipitate the platinum directly by formic acid. The precipitation with hydrogen sulfide eliminates the relatively large amount of sodium salts which cause trouble in the formic acid reduction.

IV. Separation and Determination of Palladium (262)

1. Separation of palladium by precipitation with dimethylglyoxime.   Dissolve the precipitate of the hydrated dioxides of palladium, rhodium, and iridium in hydrochloric acid as previously directed. Filter the solution and dilute it to a volume of about 400 ml. Add a sufficient volume of a 1% solution of dimethylglyoxime in 95% ethyl alcohol to precipitate all of the palladium (2.2 g of the solid reagent is required for 1 g of palladium). An excess of the reagent amounting to 10% should be added to ensure complete precipitation. Let the solution stand for 1 hr and then filter it. The manner of filtration will depend upon the form in which the palladium is to be determined. Wash the precipitate with diluted hydrochloric acid (1 + 99) and finally with hot water. The precipitate can be washed with a considerable volume of water without a trace of it dissolving. A single precipitation of the palladium is sufficient to separate it completely from rhodium and iridium.*

2. Determination of palladium.   (a) As palladium dimethylglyoxime.   Palladium dimethylglyoxime is sufficiently stable and constant in composition to be dried and weighed. If the determination is to be made in this manner, catch the precipitate in a porcelain or glass filtering crucible, using suction. Wash the precipitate as previously directed and dry it at 110°C for 1 hr. Calculate the quantity of palladium, using the theoretical factor, 0.3167.

(b) As metal.   If the palladium is to be determined as metal, which in certain cases may be more convenient, catch the precipitate on an ashless filter. Wipe the inner walls of the beaker and also the glass rod with a small piece of ashless paper. Wrap the filter and precipitate in a second filter and place them in a porcelain crucible. Dry them, and ignite them carefully in the air. Only sufficient heat should be supplied to keep the papers smoking gently. Ignite the charred residue strongly in air, and then in hydrogen. Ignite the metallic palladium in carbon dioxide† for 2 min and cool it in carbon dioxide. Weigh the residue as metallic palladium.

V. Separation and Determination of Rhodium (262, 286)

1. Preparation of the solution.   Place the solution containing the rhodium and iridium as chlorides, together with the excess of dimethylglyoxime remaining from the precipita-

---

*If for any reason, it is desired to reprecipitate the palladium, catch the precipitate on paper. Transfer the washed filter and precipitate to an Erlenmeyer flask, closed with a short-stemmed funnel and decompose them in a mixture of sulfuric and nitric acids. Heat the solution until vapors of sulfuric acid are evolved, dilute it somewhat, and filter it. Precipitate the palladium from the diluted solution of its sulfate with dimethylglyoxime.

†Palladium possesses the property of absorbing a considerable quantity of hydrogen, so that it is difficult to obtain constant weight. If the absorbed hydrogen is eliminated by igniting the metal so obtained in an inert gas, no difficulty is encountered. Strong ignition in air is sufficient to decompose all oxides of palladium, without resort to hydrogen, but the gray metal tarnishes as it cools unless an inert gas protects. The error involved, however, amounts to only 0.1 mg with quantities of palladium weighing 100 mg.

tion of palladium, in a 500-ml Erlenmeyer flask. Place a short-stemmed funnel in the mouth of the flask. Add 10 ml of sulfuric acid and 2 to 3 ml of nitric acid, and evaporate until heavy vapors of sulfuric acid are evolved. To ensure complete destruction of organic matter, add a small quantity of nitric acid from time to time and continue to heat over a free flame, keeping the solution in constant motion. Dilute the cooled solution with 20 ml of water and again evaporate it until vapors of sulfuric acid are evolved. To ensure complete destruction of organic matter, add a small quantity of nitric acid from time to time and continue to heat over a free flame, keeping the solution in constant motion. Dilute the cooled solution with 20 ml of water and again evaporate it until vapors of sulfuric acid appear. This is done to destroy nitroso compounds which may interfere in the precipitation of rhodium by titanous chloride.

2. Separation of rhodium by precipitation with titanous chloride. Transfer the sulfate solution to a clean, unetched beaker, dilute it to 200 ml and heat it to boiling. Add dropwise a solution of titanous chloride (a 20% solution of this reagent may be purchased) until the supernatant liquid appears slightly purple. If the solution is placed over a 100-W light and stirred, observation of the end point is greatly facilitated. The metallic rhodium which is precipitated quickly coagulates into a spongy mass. If much iridium is present, the end point can be determined by the lack of formation of any further precipitate and the appearance of an orange color in the solution. Boil the solution for 2 min and filter it. Wipe the walls of the beaker and also the stirring rod with a piece of ashless filter paper. Wash the filter and precipitated metal thoroughly with cold (room temperature) diluted sulfuric acid (2.5 + 97.5).

Place the filter with its contents in a 500-ml Erlenmeyer flask, add 10 ml of sulfuric acid, char gently, add 5 ml of nitric acid, and digest the solution on a hot plate. Usually, the rhodium dissolves fairly readily. Complete the solution of the rhodium by heating the flask over a free flame, keeping the contents of the flask in constant motion. Ensure the destruction of organic matter and the elimination of nitroso compounds. If some black specks remain, dilute the solution, filter it, and return the filter to the flask. Wipe down the walls of the flask with a piece of ashless filter paper. Add 5 ml of sulfuric acid, char the paper, and destroy all organic matter with nitric acid. Heat the solution until heavy vapors of sulfuric acid are evolved. This treatment will dissolve any remaining metal and will leave only a slight deposit of colorless silica.

Precipitate the rhodium a second time in the manner described above. Redissolve the rhodium as before, dilute the sulfuric acid solution with 20 ml of water and 10 ml of hydrochloric acid, and boil the resulting solution for 15 min. This treatment is necessary to convert the rhodium into a form which will allow complete precipitation by hydrogen sulfide. During this treatment, the color of the solution will change from yellow to rose. Filter the solution and wash the filter with diluted hydrochloric acid (1 + 99). Finally dilute the solution to a volume of from 400 to 500 ml.

3. Recovery of rhodium by precipitation with hydrogen sulfide. Precipitate the rhodium as sulfide from a solution, kept just at the boiling point, by passing a rapid stream of hydrogen sulfide through it. Allow the solution to cool somewhat, with the hydrogen sulfide still passing through it.

4. Determination of rhodium as metal. Filter the solution and wash the precipitate with diluted sulfuric acid (2.5 + 97.5), and finally with diluted hydrochloric acid (1 +

99). Place the filter with the sulfide precipitate in a porcelain crucible. Ignite the dried precipitate carefully in air. Finally, ignite the oxidized residue in hydrogen, cool the resulting metal in hydrogen, and weigh it as metallic rhodium.

VI. Recovery and Determination of Iridium (262, 286)

Iridium may be determined in either of two ways. If the solution containing both rhodium and iridium can be divided conveniently into aliquot parts, the determination of iridium is greatly simplified and the precipitations of titanium by cupferron avoided. The rhodium and iridium in one portion of the solution can be recovered by hydrolytic precipitation as described in the procedure for the separation of platinum. If this is done, the mixed precipitate of rhodium and iridium dioxides is washed with a hot 1% solution of ammonium chloride, neutral to bromthymol blue (pH 7), instead of with a solution of sodium chloride. The dried filter and precipitate are impregnated with a few drops of a saturated solution of ammonium chloride, in order to prevent deflagration, and carefully ignited to a mixture of the anhydrous oxides. The oxidized residue is ignited and cooled in hydrogen, and weighed as a mixture of metallic rhodium and metallic iridium. In order to calculate the quantity of iridium, it is necessary, in addition, to determine rhodium as previously described in a separate portion of the solution.

If the iridium cannot be determined in this way, it is necessary to recover it from the filtrates resulting from the precipitation of rhodium by titanous chloride.

1. Elimination of titanium by precipitation with cupferron.   Dilute the combined filtrates from the precipitation of rhodium by titanous chloride to 800 ml. Cool the solution by placing the beaker in crushed ice. Add a chilled, filtered, freshly prepared 6% solution of cupferron (ammonium salt of nitrosophenylhydroxylamine, $C_6H_5N \cdot NO \cdot ONH_4$) in slight excess. Filter the solution and wash the titanium precipitate with chilled diluted sulfuric acid (2.5 + 97.5) containing some cupferron. The cupferron precipitate is usually slightly contaminated by iridium, but the amount does not exceed 1 mg when about 0.2 g of iridium is being handled. Return the filter and precipitate to the beaker, add 20 ml of nitric acid, and heat until the precipitate is mostly decomposed. Add 20 ml of sulfuric acid and heat the solution until vapors of sulfuric acid appear. Destroy the remaining organic matter by adding nitric acid, and heating. Dilute the resulting solution to 800 ml and repeat the precipitation of the titanium. Unite the filtrates from the cupferron precipitations and evaporate until approximately 10 ml of sulfuric acid remains. Ensure the destruction of all organic matter. Dilute the solution somewhat and filter it.

2. Recovery of iridium by precipitation as the hydrated dioxide.   Dilute the solution to 200 ml with water and neutralize most of the acid contained in it with a filtered solution of sodium bicarbonate. Heat the solution to boiling and complete its neutralization with bicarbonate to the end point of bromcresol purple, as described in the procedure for the separation of platinum. Add 20 ml of a filtered 10% solution of sodium bromate, and boil the solution for 20 to 25 min. Be sure that sufficient bromate is present to oxidize all of the iridium to the quadrivalent state. Filter the solution and wash the precipitate thoroughly with a hot 1% solution of ammonium chloride.

3. Determination of iridium as metal.   Place the filter and precipitate in a porcelain crucible. Dry them somewhat and then moisten them with a few drops of a saturated solution of ammonium chloride. Ignite the filter and precipitate carefully in the air and then in hydrogen. Leach the metallic residue with diluted hydrochloric acid, then transfer it to a filter, and wash it with hot water. Ignite the filter and metallic residue in air. Finally, ignite the resulting oxidized metal in hydrogen, cool it in hydrogen and weigh it as metallic iridium.

The procedure as herein written is designed for a high degree of accuracy relative to that obtainable by most methods for iridium. If, however, one is not interested in such accuracy, the procedure may be shortened in a number of places. For instance, some of the reprecipitations which are recommended may be omitted. Time can be saved by making only one hydrolytic precipitation of palladium, rhodium, and iridium, in separating them from platinum, and by determining platinum by simply igniting its sulfide. The second reduction with titanous chloride might be dispensed with, and also the precipitations with cupferron can be avoided by determining rhodium and iridium together in aliquot portions of the solution at this stage. It must be remembered, however, that in so doing errors of varying magnitude will be introduced, depending upon the relative proportions of the metals present.

THE CATION EXCHANGE SEPARATION OF PLATINUM
METALS FROM LARGE PROPORTIONS OF BASE METALS

The cation exchanger Ku-2 was used by Anisimov *et al*. (389) to separate copper, iron, nickel, lead, and zinc from the platinum metals. The material to be analyzed was treated with aqua regia and finally evaporated with hydrochloric acid. Sodium chloride was added, and the solution was passed through 2 g of the exchanger in a column of 15 mm diam and 20 cm height, at 50–60 drops/min.

Ion exchange techniques to provide a concentration of the noble metals sufficient for detection by spectrographic methods are discussed briefly by Brooks and Ahrens (82). The rocks bearing the traces of platinum metals were treated initially with mineral acids and subsequently by adsorption on anion exchange columns. The resin was ignited and tested spectrographically. No quantitative data were included in the report, and whereas the work was incomplete there is importance in the attempt to determine the noble metal content without recourse to fire assay methods.

In a series of publications it has been shown that platinum, palladium, rhodium, iridium, and ruthenium in the form of chloro complexes can be separated from large amounts of copper, iron, and nickel by selective adsorption of the base metals on the cation exchanger Dowex-50W-X8 (acid form). Specifically, the method has been applied to the base metal collection of the platinum metals by new fire assay procedures. These assays are now described. The base metal alloys, weighing up to about 35 g, are parted with acids, and even microgram amounts of the platinum metals may be accurately recovered from the effluent. The presence of osmium in the collecting button gives no difficulty because the parting procedure involves the use of hydrochloric acid and a final addition of nitric acid, during which process the osmium is removed as the tetroxide. Up to the present time it has been generally assumed that a resin separation of ruthenium or osmium is incomplete, and this has been attributed to such factors as the ease of reduction and particularly to the ease of hydrolysis of the metal salts. There is some evidence to indicate that ruthenium in a chloride solution at pH 1.5 may fail to pass quantitatively through the cation exchanger Dowex 50. Many attempts, under these conditions, have yielded 95–98% recovery of ruthenium. On the other hand, the recent successful use of this exchanger for the quantitative separation of ruthenium and osmium from base metals seems to be due partially to a shift in pH to about 1.0 and the use of freshly prepared solutions. Furthermore, success in achieving complete passage of ruthenium through the cation exchanger seems also to be assisted by a prior evaporation in the presence of both sodium chloride and hydrochloric acid. This guess is in agreement with other data and with opinions expressed by various authors (390).

Fragmentary reports have indicated that both anion and cation exchange columns have been used to collect ruthenium from mixed constituents. Solutions of Bikini ashes collected on board ship have been subjected to exchange reactions in columns of the cation exchange resin Dowex 50 (391). An anion exchange resin Dowex 1 was also used to separate ruthenium from tellurium, cesium, and rare earth elements (392).

The procedure recommended below for the separation of the five platinum metals from base metals is an integration of the procedures used by Coburn et al. (181) to isolate iron, copper, and nickel; Marks and Beamish (351, 393, 394) for the separation of rhodium and iridium from base metals; Plummer and Beamish (189) for the separation of platinum and palladium; and Zachariasen (351, 395) for the separation of ruthenium.

Some difficulty has been experienced in the complete separation of gold and base metals by cation resin. The cation exchanger Dowex 50 allows a quantitative passage of gold when in the form of pure gold chloride but small losses have been experienced when gold and base metals are present together. Pitts (351, 396) recently made the first quantitative cation exchange separation of gold from associated base metals. Microgram and milligram amounts of gold were recovered quantitatively from large excesses of iron, copper, nickel chlorides, which should not exceed 0.05 M. Table 7.2 indicates the percentage recovery of gold in base metal solutions.

Losses of gold due to anomalous adsorption on the resin were effectively eliminated by controlling the influent ion concentration. The chemical explanation for this recovery has not yet been recorded.

TABLE 7.2
Percentage Recovery of Gold[a] versus Base Metal Chloride Concentration

Concentration (M)	0.025	0.050	0.075	0.100	0.300
Ferric chloride	100	99.6	97.6	92.4	78.8
Cupric chloride	99.6	99.2	97.6	96.0	—
Nickelous chloride	99.2	99.6	97.6	96.4	—

[a]500 μg gold added to each sample.

The early difficulty of separating base metals from osmium has also been solved (351) and thus the cation exchange method may be used for a variety of useful analytical processes such as separation of noble metals from ores, concentrates, iron, meteorites, minerals, and other base metal products carrying large or small proportions of noble metals (351, 396).

PROCEDURE 100 (395)

Transfer 100 g (3.43 assay tons) of ore or concentrate to a 6-in. porcelain dish, and roast in the assay furnace at 980°C for 2 hr. Stir intermittently to encourage the oxidation. Mix the cooled calcine with 42.4 g of sodium carbonate, 27 g of borax glass, and 9.5 g of 200-mesh amorphous graphite. Crush the mix if necessary to pass a 45-mesh screen, and mix again on the mixing sheet. Transfer to a 30-g pot, and place in a gas furnace preheated to 1200°C. Turn off the gas and air, the cooling period allowing the reactions to proceed without an overflow of the charge. When the volume of the pot mixture is reduced to about one-third of the original volume, re-ignite the gas–air mixture, and increase the temperature over a period of about 45 min to 1250°C. Then raise the temperature over 1.5 h to 1450°C, at which time turn off the gas, and remove the pot. Break the cooled pot, separate the button, and remove adhering slag by gently tapping with the end of a small iron rod. Transfer the alloy of noble metals–iron–copper–nickel to a 600-ml beaker. The weight of alloy should not exceed about 35 g. Add 200 ml of concentrated hydrochloric acid, cover with a watch glass, and place on a steam bath out of direct contact with the steam to avoid an excessive rate of reaction. Allow to stand overnight, and then place in direct contact with the steam for 24 hr. Add carefully 5 ml of concentrated nitric acid, dropwise if necessary, to avoid spattering. After the vigorous reaction has subsided add a further 5 ml of nitric acid. Following the violent reaction, place glass hooks on the beaker edge, and evaporate to dryness on a steam bath. Cool the residue, add 8 ml of concentrated hydrochloric acid, and then small portions of water, 10 ml at a time, in order to dissolve the residue. Adjust the volume to 250 ml with water, filter into a 2-l flask, and adjust to 1.5 l with water the pH now being 1.2–1.5. Add this solution to the column of 20–50 mesh Dowex 50W-X8 cation exchanger. The tower for the solution resulting from the dissolution of a 25 g button should be 70 cm long and 4 cm diam, filled with the exchanger, whose total weight should be about 700 g. Use a passage rate of 1.5 l/hr.

Wash the exchanger with water, collect the effluent and washings in a 4-l beaker, adjust to pH 1.5, with hydrochloric acid, and evaporate on a steam bath to about 50 ml. Transfer the solution to a 150-ml beaker, and evaporate to dryness on a steam bath in the presence of 3–4 ml of a 2% sodium chloride solution. Dissolve the residue in 30 ml of water. Adjust the pH to 1.5 with hydrochloric acid, and transfer the solution to a small tower of Dowex 50W-X8, 4 cm long and 10 mm diam. Wash the column with water adjusted to pH 1.5. Collect the effluent and washings in a beaker of suitable size, and evaporate to dryness in the presence of a few grams of sodium chloride. Add concentrated nitric acid and a 30% hydrogen peroxide solution, and evaporate to dryness to remove organic matter. Repeat this oxidation three times. Add a few milliliters of concentrated hydrochloric acid, and evaporate to dryness. Repeat three times, then dilute to a suitable volume. The separation and determination of palladium, platinum, rhodium, and iridium may be accomplished by a choice of methods. For small amounts, and in the absence of ruthenium and osmium, the chromatographic method recorded in Procedure 108 is recommended. For milligram amounts the method recorded in Procedure 111 is suitable. Here one may choose a combination of methods to suit the particular purpose of the analysis.

WET SEPARATION AND DETERMINATION OF PLATINUM, PALLADIUM, RHODIUM, AND GOLD IN ORES AND CONCENTRATES

Palmer *et al.* (397, 398) used the tellurium collection method from acid extracts of low grade concentrates from sulfide ores and from chromite ores. In the latter case indiscriminate application to chromite ores could result in low values for rhodium and here the application should be checked by fire assay. In general the tellurium coprecipitation can be used for the collection of traces of platinum, palladium, rhodium, and gold from 1–6 M hydrochloric acid extracts containing large amounts of base metals. In the presence of traces of nitric acid, recoveries are low. An ore sample weighing 25 g and containing not less than 3 ppm of platinum, 0.5 ppm of palladium, 0.3 ppm of rhodium, and 0.2 ppm of gold should be taken for analysis. If the contents are lower, the sample weight should be increased proportionately. A sample of low grade concentrates weighing 10 g should contain a maximum of 150 ppm of platinum, 15 ppm of palladium, 15 ppm of rhodium, and 15 ppm of gold. If the content is higher than the above values, the sample should be diluted before determination by atomic absorption.

PROCEDURE 101 (397)

Solutions and Equipment

Concentrated hydrochloric acid (A.R. grade).

Concentrated nitric acid (A.R. grade).

Aqua regia.   Mix 3 parts of concentrated hydrochloric acid with 1 part of nitric acid.

Hydrochloric acid (1 *M*).

Tellurium solution (1 mg/ml).   On a steam bath dissolve 100 mg of tellurium in 2 ml of concentrated nitric acid and 2 ml of concentrated hydrochloric acid. Decompose the nitrates by evaporating them to dryness with 1–2 ml of concentrated hydrochloric acid. Repeat the evaporation with hydrochloric acid 5 times. Dissolve the residue in 10 ml of concentrated hydrochloric acid and dilute to 100 ml with water.

Stannous chloride solution.   Dissolve 22.5 g of stannous chloride dihydrate (A.R. grade) in 16 ml of concentrated hydrochloric acid, warming if it is necessary. Dilute to 100 ml with distilled water.

Mixed nitric and perchloric acid (1 to 1 (v/v)).

Lithium chloride solution (10 mg/ml).

Procedure

Add 25 g of sample to a flat-bottomed silica evaporating dish having a diameter of 12.5 cm. Roast the sample in a preheated muffle furnace at 600°C for one hour. Cool to room temperature. Transfer the roasted sample into a 500-ml beaker. Add 100 ml of aqua regia, stir, and cover with a watch glass, and evaporate the sample to dryness. Add carefully 50 ml of concentrated hydrochloric acid, and cover the sample with the watch glass until the reaction subsides. Remove the watch glass and evaporate to dryness. Repeat this step five times. Add 100 ml of 1 *M* hydrochloric acid, and stir. Replace the watch glass, and heat on a steam bath 4 to 5 h or overnight. Allow the insoluble material to settle.

Filter the solution by decanting it through a 7-cm Whatman GF/A glass fiber filter in a Buckner funnel having a diameter of 7.5 cm. Wash the residue in the beaker with hot 1 *M* hydrochloric acid, and stir well. Allow the residue to settle, and continue the filtration by decanting the supernatant liquid. Repeat this step until the decanted liquid is colorless. Transfer the residue to the filter, removing with moist filter paper any particles adhering to the wall of the beaker. Add the filter paper to the residue. Transfer the filtrate from the Buckner flask and the washings to a 600-ml beaker. Transfer the residue to a silica evaporation dish and set aside (see the following). Evaporate the filtrate to about 100 ml. (If the solution is evaporated to dryness, the dissolved silica precipitates out, causing difficulties during the filtration and subsequent determination of the platinum metals.) Transfer the filtrate to a 250-ml beaker for the precipitation of the noble metals with tellurium. Add 2 ml of tellurium solution to the above noble metal solution in the 250-ml beaker and bring the solution to a boil. Add stannous chloride solution from a 25-ml graduated pipet until the black precipitate appears. Then add 10 ml in excess. Continue boiling gently until the precipitate coagulates. Add a further 1 ml of tellurium solution, and continue to boil until the tellurium and tellurides coagulate. Filter the precipitate under vacuum through a 9-cm Whatman No. 542 or 544 paper in a "Fisher" filtrator funnel. Immediately wash the beaker and precipitate two or three times with hot 1 *M* hydrochloric acid until the paper is colorless, i.e., free from base metals. Do not allow the precipitate to become dry between washings. (When it stands in air, the tellurium precipitate becomes colloidal and passes through the filter paper after it has been washed. This change also occurs if filtration is too slow.) Return the precipitate on the filter paper to the original beaker. Add 10 to 15 ml of mixed nitric and perchloric acid and 1 ml of lithium chloride solution to the above beaker containing the precipitate and paper and convert the perchlorates to chlorides by evaporating to dryness with 5 ml of concentrated hydrochloric acids. Repeat this step five times. The presence of perchlorates results in a serious depression of platinum absorbance and a slight depression of palladium, rhodium, and gold absorbance during the atomic absorbance determination. Add 5 ml of aqua regia and evaporate to incipient dryness. Determine the individual noble metals by atomic absorption spectrophotometry (see Chapter 1). Prepare standards in three or more beakers adding appropriate mixtures of platinum, palladium, rhodium, and gold. Use a separate beaker to obtain a blank containing lithium chloride, etc. Combine all of the insoluble residue in a porcelain boat of appropriate size. Transfer to a vycor tube 19 mm i.d. and 50 cm in length. Construct the tube to have an elongated outlet to allow its connection to a rubber tube attached to a bubbler containing water. Mix the residue in the boat with a covering of finely divided sodium chloride and chlorinate (treat in a stream of chlorine gas) at 700°C for 7 hr at a rate of one bubble per second. Transfer the chlorinated product to a beaker of suitable size washing out the chlorinated product with 0.1 M hydrochloric acid solution. Filter and determine the platinum, palladium, rhodium, and gold as described by atomic absorption spectrography (see Chapter 1).

THE DETERMINATION OF NOBLE METALS IN SILVER ASSAY BEADS

The methods described now are recommended only for silver–noble metal alloys which contain at least 15 parts of silver to one of total platinum metals. With much smaller proportions of silver the selective dissolution of silver by nitric or sulfuric acids may be ineffective.

Parting the Alloy

Both sulfuric and nitric acids have been recommended as parting acids for the silver beads which contain platinum metals and gold. The efficient use of nitric acid as a parting acid requires some prior knowledge of the composition of the bead. If the amount of gold present exceeds that of palladium, then most if not all of the palladium can be dissolved, together with the silver, in nitric acid parting media of various strengths. The presence of gold seems to assist the dissolution of palladium and platinum in nitric acid.

Iridium, rhodium, and ruthenium may interfere with the dissolution of platinum and palladium in the nitric acid parting solution. With beads containing appreciable proportions of the more insoluble platinum metals and a ratio of 15 parts total platinum metals to one of silver, three successive cupellations and partings will not always dissolve the platinum. Beads containing only silver and palladium in a ratio of 15 to 1 require only one parting with nitric acid to dissolve the palladium completely. The presence of platinum in the bead decreases the effectiveness of the first nitric acid treatment on palladium.

The limited action of nitric acid on platinum in silver beads results in the formation of a colloidal suspension and not in a true solution. This can be made evident by allowing the nitric acid parting solution containing platinum to stand for some time; a brown solid will settle out.

In general the residue from the acid treatment must be retreated two or three times with silver to form the 15 to 1 alloy bead, and subsequently parted, if the complete removal of platinum and palladium is to be effected. Even under these conditions, rhodium and iridium may prevent the complete removal of platinum and palladium. In general gold, rhodium, and iridium are not attacked by nitric acid. This should not be interpreted to mean that microgram amounts of rhodium will not be corroded. In the absence of appreciable amounts of rhodium and iridium, aqua regia will extract gold, platinum, and palladium from the nitric acid parting residue.

Parting with nitric acid is one of the most widely used techniques for the analytical treatment of silver–nobel metal assay beads. In the present authors' opinion, however, it is relatively cumbersome and inaccurate compared to either the high temperature sulfuric acid parting or the modified sulfuric acid method. A procedure for the latter method is described below. Usually, nitric acid is used in the hope of a complete dissolution of silver, platinum, and palladium. Toward this end the recommended procedures generally involve multiple partings, with intermediate inquarting* with silver.

The Nitric Acid Parting of Silver–Noble Metal Beads

PROCEDURE 102 (399)

The silver bead should contain at least 15 parts of silver to one of total noble metals. Holding it with forceps, dip the bead into 6 N acetic acid, and then into water. Repeat until the beads's surface is free of nonmetallic impurities. Transfer the clean bead to a small beaker or to a large porcelain crucible. Add 25 ml of 1:4 nitric acid, and allow the reaction to proceed to completion, as indicated by the evolution of gas. Allow the residue to settle, and transfer the supernatant liquid to a 200-ml beaker. Add to the residue 25 ml of 1:1 nitric acid, warm slightly for 15 or 20 min, and again transfer the supernatant liquid to the 200-ml beaker. Repeat this leaching with 25 ml of 2:1 nitric acid. Evaporate the combined three parting acids on a steam bath to about 10 ml. Filter the residue, which will contain

---

*The term inquarting really refers to the process of forming an alloy of three parts of silver to one of gold. It is used throughout the following discussion, however, to indicate the process of alloying silver and noble metals irrespective of their proportions.

all of the gold and for practical purposes all of the rhodium and iridium, together with small amounts of platinum and palladium.

Inquart this residue with 15 parts of silver per one part of residue, and repeat the whole of the above parting procedure. Repeat the inquarting, cupellation, and parting a second time for maximum accuracy. Retain the filtrates.

Note.   These procedures will result in a residue of gold, rhodium, and iridium effectively free of platinum and palladium, and also give two nitric acid filtrates of silver, platinum, and palladium that together with the initial nitric acid parting solutions provide a total of three filtrates. The treatment of these filtrates to isolate silver is described now. For the maximum accuracy, each of the three resulting silver chloride precipitates should be dissolved and reprecipitated to ensure freedom from platinum and particularly from palladium. The analytical chemist may then combine all six filtrates from the silver chloride precipitations and evaporate them, or evaporate singly and mix the solutions at a suitably reduced volume. The residue of gold, rhodium, and iridium is treated as described later.

The separation of silver.   To the parting acid filtrate add sufficient dilute hydrochloric acid to precipitate the silver chloride. Allow to stand, filter the coagulated silver chloride through a 7-cm filter paper, and wash with a little hot water. Set this filtrate (No. 1) aside. Transfer the filter paper and silver chloride to a 100-ml beaker, add 10 ml of concentrated sulfuric acid, and heat carefully to char the paper. Cool slightly, and add a few drops of fuming nitric acid. Heat carefully to char, cool, and again add a few drops of the nitric acid. Repeat this procedure to obtain a clear and colorless solution. Continue to fume to a moist residue, avoiding excessive temperatures, which will interfere with the subsequent separation of platinum. Add to the moist residue of silver sulfate, etc., 100 ml of hot water, and reprecipitate the silver chloride. Filter, and add this filtrate to filtrate No. 1. Dry the silver chloride at 140°C, and weigh, if this is desired. Evaporate the combined filtrates almost to dryness, and if necessary fume to a few milliliters at as low a temperature as possible. Add 3 ml of concentrated hydrochloric acid and evaporate. Repeat this three times. Add 50 ml of water and, drop by drop, a filtered 10% sodium hydrogen carbonate solution, using bromphenol blue as the indicator. The technique is described in Procedure 65. Add 2 ml of a filtered 10% sodium bromate solution, and boil for 15 min. Add the solution of sodium hydrogen carbonate to reach pH 6, as indicated by bromcresol purple.

Boil for about 15 min and filter the hydrated oxides of palladium and the traces of lead, iron, etc., through a 7-cm filter paper. Dissolve the precipitate with 4 ml of hot 1:1 hydrochloric acid, and wash the paper free of solution. Repeat the acid treatment and washings if necessary. Repeat the precipitation of the hydrated oxides of palladium, and combine this second filtrate with the first filtrate from the hydrated oxides for the determination of platinum.

The determination of palladium.   Dissolve the brown palladium oxide in 4 ml of hot, 1:1 hydrochloric acid, wash well, and dilute the filtrate to 100 ml with water. Precipitate with dimethylglyoxime as described in Procedure 52.

The determination of platinum.   Evaporate the two filtrates from the palladium hydrated oxides on a steam bath, and add concentrated hydrochloric acid to destroy the bromate. Continue this treatment to remove bromine, and finally add 150 ml of water.

Filter the solution, and adjust the acidity to that required for a precipitation by hydrogen sulfide as described in Procedure 54.

The determination of gold. Add about 5 ml of aqua regia to the insoluble from the several parting acid treatments. Warm to dissolve the gold. Repeat the extraction once or twice, and combine the aqua regia solutions. Evaporate to remove nitrogen oxides, and prepare the gold solutions for a precipitation by hydroquinone as described in Procedure 50.

The Separation and Determination of Rhodium and Iridium

In principle the separation of rhodium from iridium may appear to be a relatively simple procedure. It is a fact that rhodium metal in various forms, e.g., finely divided, is rather readily soluble in hot concentrated sulfuric acid. However, the rhodium that has been collected by fire assay with lead and subsequently cupeled may resist a quantitative attack by either boiling concentrated sulfuric acid or by fused sodium hydrogen sulfate.

Under rather exceptional circumstances aqua regia will attack iridium. This mixed acid has no appreciable effect on cupeled iridium admixed with rhodium. Also, notwithstanding the many published recipes, strong aqua regia will not dissolve iridium. None of these methods even approaches a selective separation.

The various effective methods for the dissolution of rhodium and iridium have been described (133a). The available methods for the quantitative corrosion of this final "silver bead insoluble" are

1. a wet chlorination in a closed system,
2. an open system chlorination in the presence of sodium chloride, and
3. a fusion with sodium peroxide in a silver crucible.

The fusion method has been used successfully by the authors. The details are included in the description of the sulfuric acid parting method (see the following). In principle the procedure involves the use of a special thick-walled silver crucible, a fusion at dull red heat with sodium peroxide, an extraction by nitric acid of the blue, iridium constituent, and the treatment of the filter paper and the brown rhodium compound with sulfuric and nitric acids. The combined solutions are evaporated, and the rhodium and iridium are precipitated as the hydrated oxides at pH 6, this treatment leaving the silver in solution.

The separation of 100–200 mg of rhodium and iridium may be accomplished by titanous chloride (387). For micro amounts the separations by copper (Procedure 96), by antimony powder (Procedure 114) and by solvent extraction (Procedure 97) are recommended.

Rhodium can be determined gravimetrically using hydrogen sulfide (Procedure 58). Iridium can be determined gravimetrically using 2-mercaptobenzothiazole (Procedure 61).

The recommended colorimetric reagent for rhodium is tin(II) chloride and that for iridium is leuco-crystal violet. These methods are described in Procedures 40 and 42 respectively.

The Nitric Acid Parting Method
Using Chromatographic Separations

A nitric acid parting method followed by chromatographic separations and spectrophotometric determinations was applied by James (400) to silver assay beads containing platinum and palladium. The method requires a standard fire assay and can be used for a few micrograms of platinum or palladium. Thus it is applicable to beads obtained from an assay ton of an ore that contains 1.5552 g (1 dwt*) of platinum per ton of ore.

PROCEDURE 103 (400)

Reagents

Concentrated nitric acid.

Hydrochloric acid. This is C.P. acid passed through an Amberlite IRA-410 column to remove iron and diluted as required.

Sodium chlorate. This is an A.R., 2% aqueous solution.

Stannous chloride reagent. Dissolve 11.25 g of A.R. stannous chloride in 10 ml of 3.5 M hydrochloric acid.

Thioglycollic acid. This is a 2.5% aqueous solution.

Chromatography solvent. This is pentanol 10% (v/v), hydrochloric acid ($d = 1.18$) 30% (v/v), and hexone 60% (v/v).

Paper. These are 2.5-cm strips of Whatman No. 1 filter paper.

Apparatus. See Fig. 7.3.

Procedure

Transfer the bead to a polished block of hard steel, and flatten it by hammering it carefully with a clean-faced hammer. Place the flake into the bulb of the vessel M described in Fig. 7.3.

Add 0.5 ml of concentrated nitric acid to the bulb, and place the bulb on a heated sand bath, avoiding heating to the stems of the bulb. When the bead has been parted remove the bulb temporarily from the sand bath, and add 1 ml of concentrated hydrochloric acid to the edge of the bulb. Rotate the bulb to mix the solutions and to dissolve the platinum metals. Return the bulb to the sand bath, and evaporate the solution to <0.5 ml as determined by

*dwt = pennyweight.

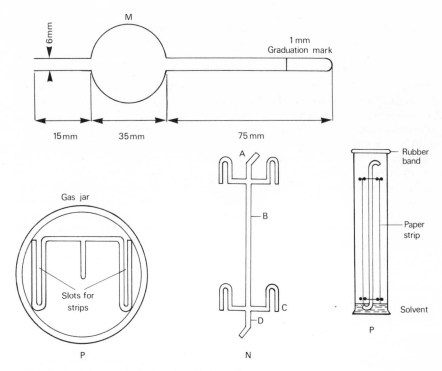

FIGURE 7.3.   A is a crooked handle for lifting rack from jar. B is a 2-mm glass rod. C are slots for paper strips. D is a tail piece to rest on jar bottom. M is a bulb for dissolution.

upending the bulb and allowing the solution to enter the marked tube. Remove the bulb, and while it is still warm add 0.5 ml of the 2% sodium chlorate solution, and mix by carefully rotating and tilting the contents. Then transfer the solution, free from residue, to the graduated stem of the bulb. Place the vessel upright in a suitable rack, and dry the bulb with an infrared lamp. Add 2–5 M hydrochloric acid to the stem to reach the 1.0 ml mark. Rotate and tilt the bulb to dissolve all soluble material.

The chromatographic separation.   For routine work it is expedient to prepare a large sheet of graph paper with lines drawn across to indicate the end of the paper strips, the spotting position and the position at which the fold to support the paper when hanging will later be made. The principal graduation marks on the graph paper should be clearly numbered, starting from zero at the spotting position. This facilitates the subsequent reading of such data as is related to solvent front and band travel. The sheet of paper is then tacked to a clean working bench and completely covered with a sheet of polyethylene film. Place in position a pair of strips of Whatman No. 1 filter paper 1 × 18 in. (2.5 × 45 cm), for each sample to be separated. Identify the strips by marking them with a pencil at the end remote from that to be spotted. Add 0.05 ml of the solution in the bulb to the strip, distributing it as evenly as possible across the full width of each strip at the spotting position, the solution penetrating to the polyethylene film, but being reabsorbed completely by the paper. Fold appropriately the ends of the paper, and dry by suspending the papers from a line by clothes pegs. In wet weather it may be necessary to dry them in a desiccator.

After drying, place each strip into its individual holder (see N in Fig. 7.3). Insert the unit into a tall cylinder, which may be a tall gas jar (see P in Fig. 7.3). Place in the bottom of the jar about 0.5 in. (1.3 cm) of the solvent mixture. Cover the jar with a lid, and seal by means of a rubber band which may be cut from a bicycle inner tube. Place the jar in a thermostatically controlled temperature compartment maintained at 25°C.

Allow the chromatograms to run overnight. Remove the holder and strips, and lightly mark the solvent front with a pencil. Suspend the strips to dry, and spray one pair with the tin(II) chloride reagent to indicate the position of the platinum band (yellow) and the palladium band (purple). Place the paired strips in position on the covered graph paper to obtain the necessary information regarding the solvent front and band travel and band width obtained. Cut off the pertinent band positions from the untreated papers and transfer to 5-ml beakers. The sprayed platinum band may also be used for a determination. To the platinum section add 2 ml of 2–5 M hydrochloric acid. To the palladium papers add 2 ml of 0.5 M hydrochloric acid. Cover the beakers, and place on a steam bath for 15 min. Filter the macerated sections by suction through No. 4 porosity filter funnels into 5-ml standard flasks. Complete the extraction as follows.

For platinum.   Rinse the beaker consecutively with 1 ml of the stannous chloride reagent and 1 ml of water, transfer each rinse through the well-packed filter cake, and make up to the mark with water.

For palladium.   Rinse the beaker consecutively with 1 ml of the thioglycollic acid solution and 1 ml of water, transfer each rinse through the funnel, and make up to the mark with water. Transfer each extracted solution to a clean spectrophotometric cell, and read its absorbance as compared to a reagent blank. Use 1-cm cells. Measure the absorbance for platinum at 403 nm and for palladium at 325 nm. Then read the metal content of each solution from a standard curve in terms of micrograms per 5 ml of solution.

Note.   If one assay ton of sample is used, the assay in dwt per ton may be obtained by multiplying the figure obtained from the standard curve by 0.4.

The Sulfuric Acid Parting of Silver–Noble Metal Beads

A treatment of the silver bead with hot concentrated sulfuric acid has the advantage of dissolving most of the silver, some of the palladium, but none of the other metals under consideration if the proper conditions are used. Two procedures are recorded in the literature for the separation and determination of palladium and silver. The more classical method involves an initial removal of silver chloride. The problem here is one of adsorption of palladium. With large proportions of the latter it is strongly adsorbed on the silver chloride precipitate. Whereas it may appear that the reprecipitation of the silver chloride following dissolution with ammonia is an effective method, the practical application of this is associated with an accumulation of salts and with the usual difficulties associated with the presence of ammonia with the platinum metals. The procedure described now incorporates a sulfuric acid parting and an initial removal of palladium from the parting acid as the hydrated oxide.

SULFURIC ACID PARTING METHOD NO. 1

Two methods of applying the sulfuric acid parting have been proposed. In one the parting acid is boiled, and retreatments are used in an attempt to dissolve both silver and palladium completely. The difficulty here is associated with the simultaneous dissolution of some rhodium, platinum, and iridium, and with the occasional failure to remove either silver or palladium completely from the main portion of rhodium, iridium, and platinum. Thus one must provide for the removal of small amounts of the four platinum metals from both parting acid and insoluble residue. In the presence of ruthenium the separations are further complicated. Osmium is seldom found in the silver bead except in very small traces, possibly alloyed perhaps with iridium, etc. In large proportions osmium will cause a violent decomposition of the silver bead. The procedure recorded below provides for a dissolution of minimum quantities of palladium, most of the silver, but generally insignificant amounts of rhodium, platinum, and iridium.

PROCEDURE 104 (259, 261)

Add to the clean bead in a 250-ml beaker, 30–40 ml of 95% sulfuric acid, and heat just sufficiently to ensure a uniform and rapid rate of parting. Heat only for 4–7 min because the complete removal of silver is not necessary. Cool, dilute to about 175 ml with hot water, filter through a quantitative paper, and wash thoroughly with hot water (see Note 1).

Evaporate the filtrate and fume to 4–5 ml, then dilute to 200 ml. Add 3 ml of a filtered 10% sodium bromate solution, and boil for 25 min. Reduce the acidity to about pH 4 with sodium hydrogen carbonate. Add 5 ml of the sodium bromate solution. Boil for 10–15 min, and now add sodium hydrogen carbonate solution to attain an acidity of pH 6 with bromcresol purple indicator added on the stirring rod. Boil to precipitate and coagulate the brown hydrated palladium dioxide. Filter, wash with freshly boiled distilled water at pH 6, and transfer the paper and precipitate to a 125-ml beaker fitted with a cover glass. Add 5 ml of concentrated sulfuric acid and a few milliliters of concentrated nitric acid, preferably fuming nitric acid, and heat to fumes of sulfur trioxide. Cool, and repeat both the addition of nitric acid and the fuming to destroy the organic matter, about 5 min being sufficient for this operation. Cool, dilute to 40 ml with water, and add a few drops of hydrochloric acid sufficient to precipitate the silver chloride. Filter and wash the precipitate. If much palladium is present in the bead, or if the parting temperature was excessive, the silver chloride may contain adsorbed palladium. If this is so, redissolve the precipitate in sulfuric and nitric acids, and reprecipitate. The filtrate from the silver chloride precipitation should now contain no more than 4 ml of sulfuric acid per 100 ml of solution. Precipitate the palladium as its dimethylglyoximate, and filter. Set this washed precipitate and paper aside to be added to the palladium dimethylglyoximate recovered from the dissolution of the parting insoluble (the details for this precipitation are described in Procedure 52). Wash the residue thoroughly from the sulfuric parting acid with ammonium acetate and water to remove lead sulfate, etc. Discard these leaching liquids. Transfer the washed residue and the paper to a 125-ml beaker fitted with a cover glass. Add 30 ml of aqua regia, and place on a steam bath for 2 hr. Dilute the solution to 20 ml, filter, and wash with hot water. Set the residue aside for the determination of rhodium and iridium (see Note 2). Add to the aqua regia extract 100 mg of sodium chloride, and evaporate to near dryness on a steam bath. Add a few milliliters of strong hydrochloric acid, and evaporate to remove nitric acid. Repeat this several times, three treatments being possible in 5 min if an excess of hydrochloric acid is avoided. Add 25 ml of water, and filter to remove the silver chloride.

Inexperience with these evaporations sometimes causes small amounts of gold to be baked out. If this occurs, burn the paper and silver chloride precipitate, treat the residue with aqua regia, evaporate the solution with three additions of hydrochloric acid, dilute with water, and then add the filtered solution to the first filtrate from the silver chloride precipitation. Acidify the total filtrate of 50–60 ml, and prepare it for the precipitation of gold by hydroquinone, as described in Procedure 50.

Notes. **1.**   The residue contains essentially all the platinum, rhodium, iridium, gold, most of the palladium if much of this metal is present, and a small amount of silver. The filtrate contains palladium and silver.

2.   The aqua regia filtrate should contain practically all of the gold and platinum and also the palladium unattacked by the sulfuric parting acid.

The determination of palladium.   The hydroquinone filtrate from the gold precipitation can usually be treated directly to precipitate palladium dimethylglyoximate. This is filtered through the paper used to collect the palladium complex from the sulfuric acid parting solution. The details of the method are described in Procedure 52.

Note. **3.**   It should be recalled here that where the amount of palladium is of the order of micrograms, the direct addition of dimethylglyoxime to the gold filtrate may produce no palladium precipitate. This eventually is dealt with after the dissolution of platinum described now.

A further difficulty with the palladium separation is sometimes caused by the coprecipitation of platinum with the palladium. The oxidation of the mixed complexes by aqua regia and the reprecipitation of palladium is ineffective when high accuracy is required. The paper and precipitate should be treated in the usual manner with sulfuric and nitric acids. If much platinum has been carried with the precipitate it sometimes appears as metal after the treatment to destroy organic matter. In this instance the residue is filtered, the paper is ignited, and the residue is treated with aqua regia. The nitric acid is then fumed with one or two additions of hydrochloric acid, and water is added. This is then filtered into the palladium solution and the latter is diluted to 175 ml. Palladium dimethylglyoximate is then reprecipitated.

The determination of platinum.   Evaporate the filtrate and wash liquid from the palladium precipitation to dryness, and fume off all the sulfuric acid. Ignite the residue to an
ash by holding the beaker over a Meker burner for 1 min. Add a few milliliters of aqua
regia to the cooled beaker, and warm to dissolve the platinum. Evaporate to dryness, and
add a few drops of concentrated hydrochloric acid. Evaporate, and repeat to remove the
nitric acid. Add about 25 ml of water, filter, and wash to a volume of 100 ml. If there is
any evidence of undissolved platinum in the above filtration, burn the paper and its
contents, treat the ash with aqua regia, remove nitric acid, and filter the aqueous solution
into the original platinum solution.

If the gold filtrate obtained from the hydroquinone precipitation showed no evidence of
palladium dimethylglyoximate, test for palladium in the platinum solution as follows.

Extract a capillary drop of the platinum solution, and remove the organic matter and
nitric acid in the usual manner. Transfer to a spot plate, and add a few drops of a saturated
ethanolic solution of 5-*p*-dimethylaminobenzylidenerhodanine. If a definite purple color is
produced the platinum solution should be treated with dimethylglyoxime to remove the
small amount of palladium whose presence was indicated.

Determine the platinum gravimetrically with hydrogen sulfide (Procedure 54).

The separation and determination of rhodium and iridium.   Place the final residue
from the aqua regia treatment in a silver crucible, and carefully burn to an ash.

Silver crucibles.   The silver crucibles available commercially frequently contain copper; they are prepared from spun silver and will allow about 12 fusions with sodium
peroxide. The amount of silver corroded during a 5 min fusion with 1.5 g of sodium
peroxide is 0.5 g or more. Thick-walled crucibles can be made by melting pure silver in a
nickel crucible of an appropriate size. The metal is then cooled slowly, and cut and drilled
to a size suitable for the fusion. Properly made crucibles may allow more than 50 fusions.
The dimensions recommended for general work are height, 2 cm; top diam, 2.4 cm;
bottom diam, 2.1 cm; and wall thickness, 0.4 cm.

Add to the black ignited residue in the silver crucible 3 g of sodium peroxide, and
maintain at a dull red heat for 10 min (if thin-walled crucibles are used, less time is
required to dissolve the residue). Allow the crucible to cool, and transfer it to a 250-ml
beaker fitted with a cover. Add water to the crucible. Following dissolution, wash the
crucible with water, and carefully transfer it to a small casserole; add sufficient 6 *N* nitric
acid to clean the crucible, and wash it well with water. Transfer the casserole liquid to the
original solution of the fusion, and add just sufficient nitric acid to dissolve the silver
oxide.

If the heavy brown residue of rhodium oxide persists, filter this, and transfer the paper
and residue to a beaker. Treat this with 6–7 ml of sulfuric acid and a few drops of fuming
nitric acid to destroy the paper, and transfer the solution to the original beaker.

If only a little rhodium is present, add 6–7 ml of concentrated sulfuric acid directly to
the original beaker, and evaporate and fume to 4 ml. Add 175 ml of hot water, and
precipitate the hydrated dioxides of rhodium and iridium as directed above for the separation of palladium from the sulfuric acid parting solution. Take care to avoid acidities much
above pH 6, otherwise hydrated silver oxide may precipitate (see Note 5). Dissolve the
oxides with hydrochloric acid as directed for the dissolution of the palladium dioxide (see
Note 4).

Notes. **4.**   Here one may encounter the usual difficulty of removing iridium from the filter paper. Careful washing with small portions of hot dilute hydrochloric acid should be satisfactory. If desired, however, the filtration of the oxide may be made with a porous filtering crucible, A2 grade.

5.   Avoid the introduction of the sulfuric acid solution to the indicator bottle by means of the capillary used to transfer the indicator to the stirring rod; otherwise the separation of silver salts will result.

The determination of rhodium.   Here the analytical chemist may choose one of several methods for the separation and determination of rhodium and iridium. In general the separation of rhodium by the selective precipitation of rhodium by copper, and the subsequent separation of the copper by cation exchange is recommended. The above two procedures for the analysis of silver assay beads are recommended only for total amounts of platinum metals of the order of 10–200 mg; however, the precipitation of rhodium by titanium(III) chloride and the indirect determination of iridium described below is also applicable. The procedure, here described and used by the authors with success, is essentially the method proposed by Gilchrist (286, 387).

Dilute the hydrochloric acid solution of rhodium and iridium to 100.0 ml. Transfer 50.0 ml of this solution to a beaker, dilute to 100 ml, and boil. Add dropwise a 20% solution of titanium(III) chloride until the supernatant liquid appears purple. Continue boiling for 2 min, and filter. Wash the filter and the precipitate with cold 1 N sulfuric acid. Transfer the paper and metal to a 50-ml beaker, and add 2.5 ml of concentrated sulfuric acid and a few drops of fuming nitric acid. Char the paper in the usual manner, and take to heavy fumes. Repeat the additions of nitric acid and fuming until a clear solution is obtained. Wash the wall of the beaker with a small amount of water, and fume again. Dilute to 100 ml, and repeat the precipitation by titanium(III) chloride. Filter, wash, and redissolve the rhodium and paper as before with 2.5 ml of concentrated sulfuric and fuming nitric acids. Carefully wash the wall of the beaker with a few milliliters of water, and fume again. Add 20 ml of water, 2 ml of concentrated hydrochloric acid, boil for 15 min, and filter. If the residue in the paper suggests the presence of rhodium, treat again with sulfuric and nitric acids, etc. (see Note 6). Determine the rhodium gravimetrically with hydrogen sulfide (Procedure 58).

Note. **6.**   When very small amounts of rhodium are present, not an unusual condition with ores, etc., the pink rhodium color may not appear.

The determination of iridium.   The following is an indirect method, and is of course not recommended for less than 2 or 3 mg of iridium if high accuracy is required.

Add to the second 50.0-ml portion of the rhodium–iridium solution 5 ml of a 10% sodium bromate solution. Boil for 20 min (see Note 7). Add a filtered 10% solution of sodium hydrogen carbonate to bring the pH of the solution to 4, as indicated externally on the stirring rod by bromphenol blue. Add a few milliliters of the sodium bromate solution, and boil. Add the sodium hydrogen carbonate solution to obtain a pH of 6, using a 0.04% solution of bromcresol purple as the indicator. Boil to coagulate the hydrated dioxides of rhodium and iridium, and filter through paper, or preferably through a porous porcelain crucible. Wash with a hot 1% ammonium chloride solution. If a porcelain filter is used, add a saturated solution of ammonium chloride to the filtered and washed oxides. If paper is used, transfer the paper and the oxides to a porcelain crucible, and moisten it with the saturated ammonium chloride solution. Ignite very carefully, an indication of a suitable heating rate being the production of a thin line of ammonium chloride vapor during the initial heating process (see Note 8). Finally, ignite in air, then in hydrogen. Leach the mixed metals with 2 M hydrochloric acid, transfer to a filter, and wash with hot water. Dry, ignite in air, then in hydrogen, cool in hydrogen, and weigh. The weight of iridium is the difference between the weight of the combined metals and that of rhodium (see Note 9).

Notes. **7.**   If the original liquid was blue it will now turn amber. The identity of the blue constituent is not known.

8.   The ammonium chloride is used to prevent deflagration of the hydrated dioxides.

9.   As would be expected, high errors may result when the proportion of rhodium is large. The method is of acceptable value when the proportions of metals are roughly comparable.

THE SULFURIC ACID PARTING METHOD NO. 2

A new procedure for the determination of platinum, palladium and gold in silver beads was proposed by Barefoot and Beamish (401). The method requires standard methods of determination for each constituent but the order of separation of silver, gold, and palladium is reversed. After the parting method of the bead using sulfuric acid (Parting Method No. 1), palladium is removed from the acid parting solution as its dimethylglyoximate, followed by silver if the determination of the latter is required. The parting residue is dissolved in aqua regia, and then palladium is removed as its dimethylglyoximate, followed by gold by using hydroquinone, then platinum by using zinc. The aqua regia insoluble, containing rhodium and iridium, may be chlorinated according to Procedure 115, and rhodium and iridium separated by Procedure 96 or 97. For good accuracy it is desirable to produce a blank silver bead by simultaneous fusion with lead. The full procedure should be applied to the blank bead, and the blanks subtracted at the appropriate places. This blank helps to account for

1.   the ash content of filter papers when these are burned,

2.   the presence of silica resulting from the corrosion of laboratory glassware by the chemicals used in the analysis, and

3.   the presence of small quantities of impurities in the chemicals.

The time required to complete an analysis can be reduced by arranging a series of weighings which result in the determination of one constituent by differences.

PROCEDURE 105

Clean the silver bead by brushing it with a fine brush, then dipping it in acetic acid. Transfer the bead to a 50-ml beaker, add 2 ml of concentrated sulfuric acid, and heat carefully on a hot plate until the rapid evolution of bubbles ceases and the solution begins to turn orange, thus indicating the presence of palladium. Avoid continued heating that may result in some dissolution of platinum. Remove the beaker from the hot plate, and cool to room temperature. Add 25 ml of cold water, and filter through a 7-cm Whatman No. 42 filter paper. Transfer the residue to the filter, and wash with 50 ml of hot water. Combine the wash water and the filtrate in a 150-ml beaker. Set this solution A aside for the determination of palladium. Wash the parting residue 5 times with a hot 20% ammonium acetate solution, and then 3 times with hot water. Discard these washings.

Palladium. Dilute filtrate A to 100 ml with chloride-free distilled water. Add 3 ml of a 1% aqueous solution of sodium dimethylglyoximate, cool immediately to about 10°C, and let stand for 30 min. Transfer immediately the palladium dimethylglyoximate to a 7-cm Whatman No. 42 filter paper, and wash with 75 ml of hot water. Dry with filter under a lamp, transfer to a porcelain microcrucible, and treat with 6 drops of mixed acids, prepared by mixing 3 volumes of concentrated sulfuric acid, 3 volumes of nitric acid, and 2 volumes of water, this treatment allowing the subsequent ignition of the precipitate without any loss of palladium. Heat the crucible slowly in a muffle furnace to char the contents. Then raise the temperature to 800°C, and ignite for 30 min. Cool, then heat in hydrogen, cool in nitrogen, and weigh as palladium metal. Subtract a blank at this stage.

Notes. **1.** The palladium may contain microgram amounts of silver which are accounted for by the blank. This contamination is the result of chloride ion in the distilled water.

2. If desired the silver in the parting acid may be determined after the removal of the palladium. The above reversed procedure is preferred to the usual initial precipitation of silver followed by that of palladium dimethylglyoximate, because silver chloride is particularly susceptible to contamination by palladium.

3. The amount of palladium that escapes precipitation by the above procedure should not exceed a few micrograms.

The treatment of the parting insoluble. Transfer the filter and parting residue, partially dried, to a tared porcelain microcrucible. Heat slowly in a muffle furnace at about 500°C for 1.5 hr. Cool, heat in hydrogen for 5 min, cool in nitrogen, and weigh (see Note 4). Transfer this residue to a 50-ml beaker, add 5 ml of aqua regia, cover the beaker, and place it on a steam bath for 1 hr. Add 5 drops of aqua regia to the crucible, place the crucible in a covered beaker, and heat on a steam bath for 30 min. After this digestion period wash the contents of the crucible into the main aqua regia extract, and dilute to 15 ml. Filter through a 7-cm Whatman No. 42 filter paper, and retain the filtrate in a 50-ml beaker. Transfer the solids to the filter, and wash with 10–15 ml of hot water. Combine the washings and filtrate. Reserve the filter and residue B which is used as described below.

To the filtrate add 0.1 g of sodium chloride, and evaporate on a steam bath to a moist residue, avoiding any drying or baking. Treat the moist salts four times with 3 or 4 drops of concentrated hydrochloric acid, with careful evaporations between the additions. Dissolve the salts in 10 ml of 0.1 M hydrochloric acid and 10 ml of water, and heat on a steam bath to coagulate the small quantity of silver chloride. Filter this through the above filter containing residue B, and wash with 50 ml of water. Place the filter and residue B in the original crucible, and heat slowly to 600°C. Cool the crucible, heat in hydrogen, cool in nitrogen, and weigh as before. Subtract the blank. The resulting weight represents the quantity of silver and aqua regia insolubles, including rhodium and iridium, in the original parting residue. Subtract this weight from the total weight of the sulfuric acid parting residue to obtain the combined weight of palladium, gold, and platinum not dissolved in the sulfuric acid parting solution. Add the weight of palladium in the parting acid to obtain the combined weight of platinum, gold, and palladium present in the bead (see Note 5).

Notes. **4.** After the subtraction of the blank this weight is made up of platinum, gold, rhodium, iridium, undissolved palladium, silver, and a small quantity of other insolubles.

5. This technique has proved to provide accurate results as indicated by salted samples.

The separation of palladium from gold and platinum. Add slowly and with stirring to the filtrate from the above silver chloride filtration 10 ml of 0.1 M hydrochloric acid, 1 drop of nitric acid, and 2 ml of a 1% aqueous solution of sodium dimethylglyoximate. Chill for 30 min at 10°C, then filter immediately through a 7-cm Whatman No. 42 filter paper. Retain the filtrate in a 250-ml beaker. Wash the palladium dimethylglyoximate with 80–90 ml of hot water, adding the washings to the filtrate. Determine the palladium as previously described. This palladium will show only insignificant contamination by gold and none by platinum.

The separation of gold and platinum. To the palladium filtrate add immediately 1 ml of concentrated hydrochloric acid per 100 ml of filtrate. The solution should be 0.1 M in hydrochloric acid. Heat to boiling and add dropwise 3 ml of a freshly prepared 1% aqueous solution of hydroquinone, stirring continuously. Boil for 15 min, place on a steam bath for 30 min to coagulate the gold precipitate, cool, and filter immediately through a double filter composed of a 5.5-cm Whatman No. 44 filter paper superimposed on a 7-cm No. 42 paper. Transfer the gold to the filter with the aid of small sections of filter paper. Wash the precipitate with 40 ml of hot water, setting the filtrate and washings for the determination of platinum. Partially dry the filter and residue, place in a tared microcrucible, and ignite in a muffle furnace. Cool, weigh, and subtract the blank. This gold should reveal no significant contamination by platinum.

The determination of platinum. Evaporate the filtrate from the gold precipitation to 80 ml, and adjust the pH to 1.3 by means of a 1 N sodium hydroxide solution. Determine platinum as described in Procedure 116.

Note. **6.** As stated above, it is possible to obtain the weight of platinum by difference; however, if there is any doubt as to the accuracy of the results for gold and palladium, the determination of platinum should be carried out. Table 7.3 indicates the accuracy which can be obtained by the above method. Experiments 12–15 record the results of the analysis of a solution of palladium, gold, and platinum prepared by mixing known volumes of standard solutions.

The determination of rhodium and iridium. The published procedure does not include directions for the determination of rhodium and iridium. This analysis, however, is readily made by chlorination and the subsequent separations, as described in Procedure 115.

TABLE 7.3
The Analysis of Silver Beads for Palladium, Gold, and Platinum

Expt. No.	Metals taken[b]			Metals found					
	Palladium (mg)	Gold (mg)	Platinum (mg)	Palladium (mg)			Gold (mg)	Platinum (mg)	
				Parting acid	Residue	Total		By analysis	By difference
1	2.98	—	—	2.98		2.98			
2	2.71	—	—	2.70		2.70			
3	2.33	—	—	2.27		2.27			
4	—	2.13	3.86			Total wt of Au + Pt: 6.58			
5	—	0.85	1.23			Total wt of Au + Pt: 2.28			
6[a]	—	2.89	4.25				2.84		4.21
7[a]	—	1.41	1.24				1.41		1.17
8[a]	1.30	1.65	2.97	0.79	Total wt Pt, Au, Pd in residue		0.74	5.05	Error = −0.08
9[a]	1.07	1.48	3.01	0.65	Total wt Pt, Au, Pd in residue		1.10	4.89	Error = −0.02
10[a]	1.42	0.75	2.37	0.28	1.14	1.42	1.01	2.42	2.43
11[a]	0.74	1.07	1.64	0.22	0.54	0.76	0.97	1.64	1.62
12	0.71	0.98	1.99			0.75	1.00	2.04	
13	0.71	0.98	1.99			0.74	0.99	2.09	
14	0.71	0.98	1.99			0.72			
15	0.71	0.98	1.99			0.75			

[a] 0.1 mg Rh + Ir present in the silver bead.
[b] Experiments 1–11, weight of silver taken = 0.1 g.

Attention should be paid to the identity of the constituents which contaminate the aqua regia insoluble. Associated base metals such as copper, iron, and nickel are readily removed by cation exchange (Procedure 100), and the appropriate treatment of the chlorinated residue will remove silica.

THE SOLVENT EXTRACTION OF GOLD, PALLADIUM, PLATINUM, AND SILVER BY DITHIZONE

For the consecutive determination of palladium, gold, platinum, and silver in silver assay beads, Young (187) used a combination of extractive and volumetric techniques. The method involves a sulfuric acid or aqua regia parting of the bead and a selective extraction of palladium dimethylglyoximate by chloroform from the acid parting solution. The palladium is then determined by a titration with dithizone in carbon tetrachloride. The aqua regia solution of the parting acid residue is prepared for titration of the gold with the dithizone–carbon tetrachloride titrant, and subsequently divalent platinum is similarly determined.

The method is rapid and provides reasonable accuracy. It is applicable to beads containing microgram amounts of gold, palladium, and platinum. The method should not be used indiscriminately, however. Whereas it has been stated that the remaining platinum metals do not react with dithizone, their presence in significant amounts in the bead may introduce appreciable errors, because of the resistance toward dissolution by aqua regia of the platinum and palladium incorporated in the final insoluble. The most accurate methods for assay bead or lead button analysis require the complete dissolution of the bead constituents prior to any separation.

PROCEDURE 106 (187)

Reagents

Dithizone solutions. *(a) Concentrated stock solution.* Stir thoroughly 45 mg of dithizone (diphenylthiocarbazone) powder with 200 ml of carbon tetrachloride, and filter into a 250-ml separatory funnel. Cover with 30–40 ml of sulfurous acid solution.

(b) Dilute standard solution. Dilute the stock dithizone solution 10 times with carbon tetrachloride, and standardize against weighed quantities of gold, palladium, and platinum by the procedures described below. The solutions may conveniently contain 0.01–0.02 mg of gold, palladium, or platinum per milliliter, and the factors of the dilute standard dithizone solution will be approximately 1 ml $\equiv$ 0.01 mg of gold or platinum and 0.05 mg of palladium. The standard metal solutions can be made by dissolving the metals in aqua

regia, evaporating to dryness, taking up in hydrochloric acid, and diluting to appropriate volumes so that the hydrochloric acid concentration is about 1%.

Stannous chloride.   Dissolve 80 g of $SnCl_2 \cdot 2H_2O$ in 180 ml of warm hydrochloric acid, and dilute with 300 ml of water. Place a stick of tin in the bottle to keep the solution in the reduced state.

Procedure

Weigh the silver bead if the weight of silver is required, in which instance it is determined by the difference between the combined weight of gold, platinum, and palladium and the weight of the bead. Transfer the bead to a 50-ml beaker, add 5 ml of 1:1 sulfuric acid, and fume strongly on a hot plate. In favorable instances the palladium and silver will dissolve, and the gold and platinum will remain as a residue. In those instances where parting is incomplete, apply the alternative procedure described below. Cool, carefully dilute with water, and separate the soluble silver and palladium from the remaining metals by decanting the solution into a 100-ml separatory funnel. Wash the insoluble residue with water by decantation. If the gold and insoluble platinum metals are so finely divided that it is impossible to separate and wash by decantation, filter through a small Whatman No. 40 paper containing pulp; wash, dry, and ignite the paper and residue in a small porcelain crucible. To the solution containing the silver and palladium, in about 50 ml in the separatory funnel, add 2 ml of a 1% aqueous solution of sodium dimethylglyoximate. Stand for 10 min, shaking the separatory funnel occasionally to hasten the reaction. Extract the palladium dimethylglyoximate by shaking twice with 4–5 ml portions of chloroform and drawing off the lower layer into a 50-ml beaker. Evaporate the chloroform by placing the beaker on the edge of a hot plate, add 3–4 ml of hydrochloric acid and 2–3 ml of nitric acid, and evaporate to dryness.

Dissolve the palladium in 5 ml of hydrochloric acid with gentle heating, cool, and transfer the solution to a 50-ml separatory funnel. Rinse the beaker with 15 ml of water so that the final concentration of hydrochloric acid is 25%. Add 0.2 ml of stannous chloride solution, and run in the standard dithizone solution in small portions from a 10-ml buret, shaking vigorously between additions, and withdrawing the lower layer from the separatory funnel. When all the palladium has been extracted the carbon tetrachloride layer in the funnel changes from the dull olive-green of palladium dithizonate to the unmistakable brilliant emerald green of dithizone.

Notes. 1.   The volume of dithizone solution required to extract the palladium is the measure of the quantity of the latter present. If the palladium titration shows a color change from olive-green to *yellow* near the end point it is probable that the reducing action of the stannous chloride has been exhausted and the yellow oxidation product of dithizone is appearing. The addition of a drop or two of stannous chloride will restore the true green color of unchanged dithizone or the gray-green of palladium dithizonate.

2.   The dithizone is standardized against known quantities of palladium by the procedure used for the sample. The usual precautions regarding cleanliness of glassware and reagent blanks, which apply to all dithizone procedures, must be observed.

3. By a simple modification of the above procedure, palladium may be determined not only in the presence of silver but also in the presence of gold and platinum. This is of importance, for instance, where the method used to decompose the sample, or to part the assay bead, leaves gold, platinum, and most of the palladium unattacked. Dissolve the mixed metals in dilute aqua regia, and to the resulting solution containing silver, gold, platinum, palladium, etc., and about 0.2 ml of hydrochloric acid and 0.1 ml of nitric acid in a volume of 10 ml, add 2 ml of dimethylglyoxime solution, and proceed as previously described. The presence of the small quantity of nitric acid prevents the precipitation of gold and the coprecipitation of platinum with the palladium dimethylglyoximate. After the withdrawal of the two chloroform extractions, retain the upper layer in the separatory funnel for the determination of the gold and platinum as described below.

Gold. Dissolve in aqua regia the residue of gold and platinum obtained in parting the assay bead, evaporate the resulting solution to dryness, and take up the chlorides in the minimum quantity of hydrochloric acid. Dilute this solution, or the solution of gold and platinum from which palladium has been removed by extraction with dimethylglyoxime and chloroform as previously described, to 10–15 ml in a 50-ml separatory funnel, add 0.2 ml of hydrochloric acid and 0.1 ml of a 10% sodium bromide solution (see Note 4), and titrate with a standard dithizone solution (see Note 6) as in the palladium analysis, adding 0.2-ml portions at a time, shaking vigorously (see Note 5) until the lower layer shows the yellow color of gold dithizonate, and then withdrawing the layer as for palladium. Continue the additions, shaking and drawing off until the dithizone layer remains green. This indicates that all the gold has been extracted.

Notes. **4.** The addition of sodium bromide prevents the interference of silver, which if present, would also give a yellow dithizonate.
5. Excessively long shaking may cause a slight fading of the dithizone color at the end point in the gold titration; reducing agents must be absent in order to keep any platinum present in the platinic state.
6. The dithizone is standardized against a known gold solution under the same conditions.

Platinum. After removing gold, as described above, and drawing off any excess of dithizone from the separatory funnel, add sufficient hydrochloric acid to bring the acidity to 25% that of concentrated acid, and then add 0.2–0.3 ml of stannous chloride solution, shake, and allow to stand for several minutes. Now titrate the platinum with standard dithizone solution as just described, until the carbon tetrachloride layer no longer acquires the light yellow color of platinous dithizonate. Determine the titration value of the dithizone solution against a known platinum solution under the same conditions.

Note. **7.** With materials like assay beads, silver is usually determined by the difference from the initial weight of the bead. The other platinum metals, viz., osmium, ruthenium, rhodium, and iridium, do not react with dithizone under the conditions described above. They will therefore be counted as silver in any difference calculation unless the gold and platinum metal residue are weighed after the parting of the bead (see discussion of "insoluble" that follows).

THE WET DETERMINATION OF THE SIX PLATINUM METALS
IN HIGH CONCENTRATIONS

The method recorded below is the first complete procedure for the direct quantitative determination of the six metals in concentrated natural deposits. Payne (402) also applied the procedure to synthetic samples containing amounts of the metals from a few milligrams to about 100 mg. The recovered values indicate very acceptable accuracy.

The method involves the fusion of the sample with sodium peroxide, acidification with hydrochloric acid, and oxidation to remove osmium and ruthenium simultaneously. The pot liquid is treated hydrolytically in the presence of nitrite to precipitate the base metals. The mixture is made up to a known volume. Half this volume is filtered and treated for the remaining four platinum metals. This 50% aliquot is passed through a cation exchanger, and the effluent is added to a cellulose column to separate each of the four platinum metals. For the column and technique see Procedure 107 (402).

For a few milligrams of metal the original procedure directs the use of spectrophotometric finishes.

The recommended gravimetric methods are a hydrolytic precipitation for osmium, rhodium, and iridium, a magnesium reduction for platinum, and a dimethylglyoximate precipitation for palladium. For ruthenium the evaporated distillate is treated directly by ignition in hydrogen.

The absorptiometric reagents recommended are thiourea for osmium and ruthenium, and tin(II) chloride for platinum. Palladium and rhodium are determined as their chloro complexes.

The determination of iridium was accomplished by measuring the difference in absorbance between the oxidized and reduced forms of the chloro acid. The method involved the use of two aliquots of the iridium solution. Fifty milliliters of 1-to-1 hydrochloric acid were added to each, the solutions were boiled for 5 min, cooled and each transferred to a 100-ml calibrated flask.

One solution was diluted to about 90 ml with 1-to-1 hydrochloric acid, and 4 ml of freshly prepared chlorine water were added. The flask was placed in a boiling water bath for 30 min, cooled, and adjusted to volume with the diluted hydrochloric acid.

The second solution was treated with 2 drops of a 5% hydroquinone solution, and diluted to the mark with the 1-to-1 hydrochloric acid. This solution was used as the reference solution. The absorbance of the first or oxidized solution was measured at 4900 Å in 4-cm cells.

It should be noted that if the method described below is applied to materials with high proportions of iridium, the latter may contaminate the rhodium. It is then necessary to determine the iridium in the rhodium extract. This may be accomplished by the method described below.

Absorptionmetric methods were used by Payne (402) for amounts of a few mg to about 10 mg. Over this range gravimetric methods remain applicable and these are recommended by the present authors. The most suitable gravimetric method for each metal is indicated in the appropriate part of the procedure described now. Although the procedure is not a suitable one for microgram amounts of the six platinum metals, a reference is included to indicate a suitable spectrophotometric method for each metal. In general the procedure follows the original, but appropriate variations are integrated and comments are interjected where they may clarify difficulties experienced by the analytical chemist.

It should be noted that the procedure has been found satisfactory for concentrates containing high proportions of platinum metals and presumably of low sulfur content. One may expect some difficulty with samples containing osmium and appreciable sulfur. In any case this problem requires further investigation.

PROCEDURE 107 (402)

Apparatus

Ruthenium distillation train. This consists of a 500-ml distillation flask with a dropping funnel, four 250-ml receivers, and a Drechsel bottle. Connections are made via Quickfit & Quartz spherical joints. The complete train is shown in Fig. 7.4.

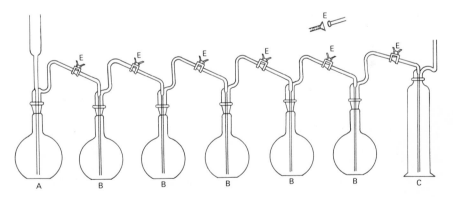

FIGURE 7.4. A is a 500-ml flask. B are 250-ml flasks. C is a Drechsel bottle. E are spherical ground glass joints.

Although five of these flasks are illustrated, this number may be reduced to meet the specific requirements. Procedure 107 directs the use of three receivers and the Drechsel bottle C. It should be noted that although (1 + 1) hydrochloric acid is recommended, more efficient collectors can be used that would, of course, result in the simultaneous collection of osmium and ruthenium and require a modification of the author's procedure.

Ion exchange column.   The column is made by joining a length of Pyrex glass tubing ($14 \times 1\frac{1}{4}$ in., 35×3.1 cm) to the cut off top and bottom portions of a standard Quickfit & Quartz CR/32/20 chromatography column (see B in Fig. 7.4). It is filled with Zeo-Karb 225, and prepared by acid washing, etc., in the usual manner.

Cellulose column.   This is a standard Quickfit & Quartz CR/32/40 column equipped with a reservoir and tap adapter. The preparation of the column is described in the following.

Reagents

Hydrochloric acid.   This is a $(1 + 1)$ solution.

Ethanolic sodium hydroxide solution.   This is a 10% solution of sodium hydroxide in 10% industrial ethanol.

Hydroquinone.   This is pure, as a 5% (w/v) solution.

Dimethylglyoxime solution.   This is as its salt, 5% (w/v) solution in water.

Acid solvent.   This is isobutyl methyl ketone containing 2% (v/v) of hydrochloric acid.

Reducing solvent.   This is acid solvent plus 0.05% of hydroquinone.

Oxidizing solvent.   This is freshly prepared (see the following) cellulose powder and is Whatman standard grade.

Procedure

The dissolution of the sample.   Weigh the crushed sample (containing not more than 0.25 g of platinum metals) into a nickel crucible, and moisten with a few drops of water. Introduce about 5 ml of hydrofluoric acid, and evaporate the mixture to dryness. Heat the crucible gently, and raise the temperature slowly to a bright red heat (under cover of hydrogen if osmium is to be determined), taking care to avoid loss by spitting.

When cool, add 10 g of sodium peroxide, and mix the sample intimately with the aid of a glass rod. Heat the crucible slowly until the contents sinter, maintain in this state for 30 min, and then raise the temperature to bright red heat, swirling the melt to ensure complete attack. Avoid prolonged heating at this temperature, to minimize attack on the crucible, 1 or 2 min usually being sufficient.

Set the crucible aside to cool, then place it in a 400-ml beaker, and just cover the crucible with water. When the violent reaction has subsided, rinse the cover and sides of the beaker, remove the crucible with a pair of forceps, wash it inside and out, clean the surface with a rubber-tipped glass rod, and wash again.

Transfer the contents of the beaker, with washings, to the ruthenium distillation flask, and connect the flask to the train. Place 150 ml of $(1 + 1)$ hydrochloric acid in each receiver and 150 ml of the ethanolic sodium hydroxide solution in the Drechsel bottle. Meanwhile, place 20 ml of the hydrochloric acid in the nickel crucible, and heat to dissolve any particles adhering to the surface. With a suction pump, draw a slow current of air through the distillation train, and transfer the acid to the distillation flask. Rinse the crucible with a further 15 ml of acid, and add the rinsings to the contents of the flask.

The distillation of osmium and ruthenium.   Heat the flask slowly, and boil the contents gently for about 10 min, maintaining a steady flow of air through the train. Add 10 ml of a 20% (w/v) sodium chlorate solution, a few drops at a time, and then 35 ml of a 10% (w/v) sodium bromate solution in 5-ml portions at intervals of 2 min. Boil the solution continuously while the additions are made and then for a further 30 min to complete the volatilization of osmium and ruthenium. Some 15 min before the end of the distillation, heat the contents of the first receiver to boiling to reduce most of the ruthenium to the tervalent state, any ruthenium still unreduced, together with osmium and oxides of chlorine, etc., passing into the second receiver. Heat the contents of this in turn, and so on until the osmium and other volatile products are finally absorbed in the ethanolic sodium hydroxide solution. Most of the ruthenium will be in the first receiver, a little in the second, and only a trace in the third. The distillation is them complete.

Combine the distillates containing the ruthenium in a 1-liter beaker, evaporate to a small volume, transfer to a 150-ml beaker, and again evaporate to a small volume, but not to dryness. Set the solution aside for the subsequent determination of ruthenium as described below.

Transfer the solution containing the osmium to a beaker, and set it aside for the determination as described in the following.

The treatment of the pot liquid from the ruthenium distillation.   Transfer the residual liquor from the distillation to a 600-ml beaker; rinse the dropping funnel and flask with 50 ml of hydrochloric acid, and cautiously add the rinsings to the main solution. Rinse the funnel and flask with hot water, cover the beaker, and boil the solution vigorously to reduce the volume to about 100 ml, taking care to remove the beaker before crystallization of the salts causes severe bumping. Transfer the beaker to a low heat hot plate, and continue the evaporation until the contents are almost dry. On no account allow the salts to bake.

The nitrate separation.   Dissolve the salts in about 150 ml of water, boil for 5 min, and then dilute to about 400 ml with hot water. To the boiling solution add sodium nitrite a little at a time until the pH changes to 7 (measured with test paper), and then, after another small addition of nitrite, boil for a further 5 min. When cool, transfer the solution and precipitate to a 500-ml calibrated flask, and make up to the mark with distilled water.

Mix the contents of the flask well, and filter through a Whatman No. 40 filter paper, rejecting the first few milliliters of filtrate. Collect 250 ml of the filtrate in a dry calibrated flask, and set aside; then continue the filtration, and allow the filter paper and precipitate to drain. Wash the precipitate once with cold water, and discard the remaining filtrate.

Transfer the filter paper and precipitate to a 400-ml beaker, add 2 ml of the hydrochloric acid and 20 ml of water, and bring to the boil. Boil for 5 min, dilute to about 100 ml, and repeat the nitrite treatment as before. When cool, transfer the solution, precipitate, and filter paper to a 200-ml calibrated flask, and filter as before, but this time collect 100 ml of filtrate. Discard the remaining filtrate and precipitate, any gold in the original sample being in this precipitate.

Transfer the two filtrates (which now represent one half of the original sample, see Note 1) to a 600-ml beaker, and rinse the flask with distilled water. Add 4 ml of hydrochloric acid, and heat the solution to boiling. Boil for not more than 1 min to remove excess of nitrous fumes, then cool as rapidly as possible, and adjust the pH to about 3 with sodium hydroxide (see Note 2).

Notes. **1.** It will be recognized that this method of providing an aliquot is in error by that amount of platinum metals contained in the volume of solution displaced by the precipitate of base metal hydroxides. This error can be calculated from the densities of the solution and the solid, and its significance will depend upon the proportions of the two mechanically mixed constituents. For very small amounts of precipitate, and where good accuracy is not required, this error can be ignored.

2. Excess of nitrite results in the disruption of the column owing to the liberation of free nitrous acid. On the other hand if insufficient nitrite is present some palladium may be lost in the cationic form.

The removal of base metals by ion exchange.   Regenerate the resin (Zeo-Karb 225) with the (1 + 1) hydrochloric acid, and then elute with water until the effluent is at pH 7. Place the reservoir in position, and fill with the solution of the nitrites of the platinum metals. Pass the solution through the column at about 1 drop /sec, refilling the reservoir as necessary. When all the solution has been transferred, rinse the beaker 2 or 3 times, and add the washings in the same manner. Immediately before the last few milliliters of solution become absorbed, rinse the reservoir with about 20 ml of water, and allow this to pass through. Repeat with several successive 20-ml portions, fill the reservoir with water, and continue the elution until the effluent is neutral.

The treatment of the ion exchange effluent.   Evaporate the effluent (now free from sodium, nickel, etc.) to a small volume, and then transfer to a 250-ml beaker; add 2 ml of a 10% (w/v) lithium chloride solution, and continue the evaporation to dryness. Moisten the residue with a few drops of water, add 10 ml of 60% perchloric acid, and evaporate until copious fumes are evolved. Continue heating until all the free perchloric acid has been expelled, and then set aside to cool. Rinse the wall of the beaker with the minimum amount of water, and evaporate again until the evolution of fumes ceases completely (see Note 1).

Treat the cooled residue with 5 ml of (1 + 1) hydrochloric acid, and evaporate to dryness. Repeat this treatment at least 6 times (up to 10 times for materials very rich in iridium). Dissolve the residue in 20 ml of concentrated hydrochloric acid, and bring to the boil in a covered beaker. Continue boiling until the volume has been reduced to about 15 ml, remove the cover, and evaporate gently to 5–10 ml (see Notes 2 and 3).

Notes. **1.** Perchloric acid is required for the necessary destruction of the nitroso complexes of the platinum metals. To remove the perchloric acid and thus allow the formation of chloro complexes an evaporation is necessary. Lithium chloride allows the evaporation under conditions that prevent the reduction of the chloro complexes of the platinum metals.

2. Evaporations to smaller volumes, e.g., 2 ml as stated in the published procedure, may result in the extraction of rhodium and its transfer to the column, especially when the rhodium value is as high as 100 mg. In a private communication the author recommends a 5 N acid concentration at this stage.

3. It is most important not to allow the contents of the beaker to solidify at this stage, otherwise the final treatment with 20 ml of hydrochloric acid will have to be repeated. If, however, the residual liquor tends to crystallize on cooling, add hydrochloric acid one drop at a time until the solid redissolves.

The chromatographic separation. The cellulose column should be freshly prepared as follows: Prepare a quantity of acid solvent by adding 20 ml of 1:1 hydrochloric acid to 1 liter of isobutyl methyl ketone. Take 200 ml of this acid solvent, and add Whatman standard grade cellulose powder until a thin slurry is formed. Pour the mixture into the column in the usual manner, adding sufficient material to form a bed of cellulose about 35 cm deep. Allow to drain, leaving about 1 cm of solvent over the cellulose, and then set aside until required.

Add to the cold solution of platinum metals a small amount of solid hydroquinone and then 10 ml of reducing solvent. Stir vigorously until the iridium has been reduced, adding more solid hydroquinone if necessary.

Begin the column separation by opening the tap fully. As soon as the solvent has drained, introduce the first extract by careful decantation, avoiding the transfer of any of the aqueous phase. Repeat the extraction with a further 10 ml of reducing solvent, and, as soon as the first extract has been taken up, transfer in the same way. Continue the extractions in this manner, with 10 ml of reducing solvent at a time, and avoiding the transfer of the aqueous phase. Do not allow the column to drain at any time to such an extent that air is introduced between the column wall and the cellulose.

As the elution proceeds the platinum band will move ahead, with the palladium band following somewhat more slowly. Collect the platinum fraction, however, as soon as the extractions are begun, for an almost invisible platinum band moves ahead of the main one, and may otherwise be lost.

As soon as it is evident that all of the palladium has been extracted and transferred to the column, make a final extract from the beaker with the acid solvent not containing hydroquinone, then place the reservoir in position, fill with reducing solvent, and continue the elution. Meanwhile set aside the beaker containing the iridium and rhodium.

When the main platinum band has reached the lower end of the column, a gap of some inches should have appeared between it and the following palladium band. Continue to collect the platinum fraction until just before the palladium starts to come through.

At this point, change the receiver, and collect the palladium fraction in the same way, replenishing the reservoir with reducing solvent as necessary. Immediately the last of the palladium comes through, remove the reservoir, and elute with two or three successive portions of the acid solvent (nonreducing) to remove the hydroquinone and render the ensuing oxidation treatment fully effective. The column is then ready for the second stage, and meanwhile elution is stopped.

Extract the iridium by elution with an oxidizing solvent. Freshly prepare this solvent as follows. Take 100 ml of the acid solvent, add 2 ml of hydrochloric acid and then 4 g of sodium chlorate. Stir thoroughly until the solution becomes cloudy, add 10 g of cellulose powder, and macerate. Decant the clear liquid into another beaker, and dilute 50 ml of this to 250 ml with the acid solvent. Add 10 ml of the above oxidizing solvent to the original beaker which contains an aqueous rhodium solution, and shake. Add to the column.

Repeat the extraction with 10 ml of oxidizing solvent exactly as was done above for platinum and palladium. When all of the iridium has been extracted, and the dark brown band is well clear of the residual pink rhodium band (see Note 1), place the reservoir in position, and continue eluting with the acid solvent containing a little of the oxidizing solvent. When it is certain that all of the iridium has been collected, dilute the residual rhodium-containing phase with a little water, remove the reservoir, and transfer the solution to the column. Rinse the beaker 2 or 3 times. When these washings have almost been absorbed, fill the reservoir with water, and continue eluting. When the pink band has almost reached the lower end of the cellulose, collect the rhodium fraction in a separate beaker.

Notes. **1.**   Despite care in avoiding any transfer of the aqueous phase, a small amount of rhodium usually finds its way on to the column.

2.   In a private communication Payne states that the use of an acid solvent containing 1 vol of tri-*n*-butyl phosphate with 2 vol of isobutyl methyl ketone containing 2% of hydrochloric acid allows the use of a cellulose column about 10 cm long. Platinum and palladium, however, cannot then be separated; this is conveniently accomplished by conventional methods. The technique of reducing with hydroquinone followed by oxidizing with the appropriate solvent enables platinum and palladium to be eluted together, then the iridium, leaving rhodium on the column as before. Also the platinum metals can be back-extracted from the solvent with water, provided that petroleum spirit (bp 60–80°C) is added to the solvent in the proportions of 1 vol of petroleum to 2 vol of solvent. The aqueous extract is then cleaned by extracting twice with carbon tetrachloride.

The treatment of individual fractions.   To the platinum and palladium fractions add 1 g of lithium carbonate per 100 ml of ketone. To the iridium and rhodium fractions add 5 ml of a 10% (w/v) lithium chloride solution. Evaporate the respective fractions by gentle boiling on an electric hot plate in a well ventilated fume cupboard. Continue the evaporation to dryness, but avoid heating the residue beyond the point necessary just to remove the ketone.

When cool, add 10 ml of water to each and then 50 ml of nitric acid. Cover the beakers, and heat gently. When the first vigorous reaction has subsided, bring the solutions to the boil, and continue boiling until brown fumes are no longer evolved, adding more nitric acid if required. Then add 25 ml of 60% perchloric acid, boil until most of the nitric acid has been expelled, remove the covers, and evaporate the solutions until fumes are evolved. Continue fuming until the volume has been reduced to 5 ml for iridium or until the salts have begun to crystallize for the other metals. Set aside the iridium fraction for subsequent determination. Add 50 ml of (1 + 1) hydrochloric acid to the others; heat the solutions to boiling to reconvert the metals to the chloro complexes and treat the rhodium and the platinum and palladium fractions as described in the following.

Gravimetric methods.   Gravimetric methods are recommended for 2 mg or more of each metal. Whereas the method described above is not specifically recommended for microgram amounts of metals, a suggested list of spectrophotometric methods for each metal is included. These may be useful for quantities of metals of the order of 1 mg or less.

Ruthenium. Treat the evaporated ruthenium–hydrochloric acid distillates in the 150-ml beaker with thionalide (Procedure 63).

Osmium. Acidify the sodium hydroxide–ethanol solution of osmium with hydrochloric acid, and evaporate to a small volume. Transfer to a 150-ml beaker, and precipitate osmium by using thionalide as directed in Procedure 66.

Rhodium. Transfer the rhodium solution to a 150-ml beaker, and precipitate with hydrogen sulfide acid, as described in Procedure 58. Alternatively the hydrolytic precipitation now described for iridium may be applied.

Iridium. Dilute the perchlorate solution containing the iridium to about 200 ml with hot water, and add a sodium hydrogen carbonate solution until the pH changes to about 4. Add 10 ml of a 10% sodium bromate solution, and boil gently for about 30 min to coagulate the precipitate, adding more sodium bromate solution if necessary to keep the pH between 6.5 and 7.0. Allow the precipitate to settle, and then filter through a Whatman No. 40 filter paper, washing well with hot water. Then wash with a 5% ammonium chloride solution to prevent deflagration during the subsequent ignition. Dry the paper and precipitate, ignite carefully, and treat the iridium dioxide with hydrofluoric and nitric acids in the usual manner. Collect the precipitate once again, ignite, reduce under hydrogen, and weigh as usual.

Platinum. Treat the chloro complexes to precipitate platinum by using zinc, as described in Procedure 116. An alternative procedure uses hydrogen sulfide (Procedure 54).

Palladium. Prepare the solution of the chloro complex for a hydrolytic precipitation, as described for rhodium in Procedure 60. Then dissolve with hydrochloric acid and precipitate with dimethylglyoxime, as described in Procedure 52.

Spectrophotometric methods. Ruthenium. Dilute the original solution to 100 ml in a calibrated flask, and extract an aliquot equivalent to about 1 mg of ruthenium. Transfer the aliquot to a 150-ml beaker, add 5 ml of (1 + 1) sulfuric acid, and evaporate until copious fumes are evolved. After fuming for 2 or 3 min, cool the solution, add 60 ml of (1 + 1) hydrochloric acid and boil for 5 min.

Cool the solution, and transfer to a 100-ml calibrated flask, rinsing the beaker with water to bring the volume to about 85 ml. Add 10 ml of a 10% thiourea solution, mix well, and heat the flask in a water bath at 50 ± 2°C for 30 min. Cool, make up to the mark with water, and measure the absorbance at 6750 Å in a 4-cm cell. Calculate the amount of ruthenium by comparison with a standard.

The standard is prepared as follows. Add 1.00 ml of a standard ruthenium solution (1% in hydrochloric acid and 1% in ruthenium chloride standardized by thionalide) to a 150-ml beaker, then add 5 ml of (1 + 1) sulfuric acid. Evaporate to fumes for 2 or 3 min, cool, and continue as previously stated.

Note. In a private communication the senior author states that the thiourea method has been replaced by sodium thiocyanate (Procedure 120).

Osmium. Boil the alkaline osmate solution from the distillation to expel most of the ethanol, cool, and make up to a suitable volume from which an aliquot containing 1 mg of osmium may be taken. Transfer the aliquot to a 100-ml calibrated flask, and neutralize with hydrochloric acid, keeping the solution as cool as possible. Add 10 ml of concentrated hydrochloric acid, then 10 ml of the 10% thiourea solution, and dilute to the mark.

Mix well, immerse the flask in a water bath at 75 ± 2°C for 30 min, cool, and measure the adsorbance of the solution at 4800 Å in a 4-cm cell. Calculate the amount of osmium present by comparison with the standard. To prepare the standard add 1 ml of the standard osmium solution (a sodium hydroxide distillate standardized gravimetrically, and containing 1 mg of osmium per milliliter) to a 100-ml calibrated flask, add 60 ml of (1 + 1) hydrochloric acid and then 10 ml of the 10% thiourea solution, and dilute to the mark with water.

Add the small amount of iridium found in the rhodium solution (now discussed) to this result.

Rhodium.   The rhodium solution may contain a trace of iridium if the original material contained a high percentage of this metal; it is necessary to determine it to correct the main iridium figure. Transfer a small portion of the previously oxidized and diluted solution (see the treatment of individual fractions given earlier in this procedure) to a dry beaker, and add a few crystals of hydroquinone. With this reduced solution in the reference or water cell, measure the absorbance of the oxidized solution at 4900 Å in a 4-cm cell. Calculate the amount of iridium present by reference to the iridium standard, and correct the main iridium figure as necessary. Then filter the reduced solution through a dry Whatman No. 540 filter paper into a dry beaker, and measure the absorbance of this filtrate against water at 5150 Å in a 4-cm cell. Calculate the amount of rhodium present by comparison with the standard. Prepare the standard from rhodium chloride to contain about 1 mg of metal per milliliter, and standardize by hydrogen sulfide (Procedure 58).

Iridium.   Adjust the perchlorate solution to a suitable volume from which an aliquot containing 1 mg of iridium may be taken. Transfer 2 aliquots to 150-ml beakers, add 50 ml of (1 + 1) hydrochloric acid to each, and boil for 5 min. When cool, transfer each solution to a 100-ml calibrated flask, make the first solution up to about 90 ml with (1 + 1) hydrochloric acid, and add 4 ml of freshly prepared chlorine water. Place the flask in a boiling water bath for 30 min, then cool, and adjust the volume with (1 + 1) hydrochloric acid. Meanwhile, treat the second solution with 2 drops of a 5% hydroquinone solution, and dilute to the mark with (1 + 1) hydrochloric acid. With the reduced solution in the reference or water cell, measure the absorbance of the oxidized solution at 4900 Å in 4-cm cells. Calculate the amount of iridium by comparing the absorbance with that of the standard. Prepare the standard by precipitating hydrolytically from an iridium–chlorine solution (see the preceding) sufficient oxide to produce a hydrochloric acid solution containing about 1 mg of iridium per milliliter of solution.

Palladium.   Measure an aliquot representing about 10 mg of palladium, and make up to 100 ml in a calibrated flask with (1 + 1) hydrochloric acid. Mix well, filter a portion through a dry Whatman No. 540 filter paper into a dry beaker, and measure the absorbance of the filtrate at 4700 Å in 4-cm cells. Calculate the amount of palladium present by comparison with the standard.

Prepare the standard from a palladium chloride solution standardized gravimetrically by dimethylglyoxime (Procedure 52). Transfer 10 ml of this solution, which should contain about 1 mg of palladium per milliliter, to a 250-ml beaker. Add 5 ml of nitric acid and then 5 ml of perchloric acid, and evaporate to about 2 ml of acid. Cool, add 50 ml of (1 + 1) hydrochloric acid, and boil for 10 min. Cool, dilute to 100.0 ml with the (1 + 1) hydrochloric acid. Measure the absorbance at the same time as that of the unknown.

Platinum. Transfer an aliquot representing about 0.35 mg of platinum to a 150-ml beaker, add 5 ml of perchloric acid, and evaporate until fumes are evolved. While fumes are being evolved, add about 10 drops of hydrobromic acid, and continue heating until most of the excess of perchloric acid has been removed; then cool, add 50 ml of (1 + 1) hydrochloric acid, and boil to reconvert the platinum to the chloro complex. Cool again, transfer the solution to a 100-ml calibrated flask, and dilute to about 80 ml with (1 + 1) hydrochloric acid. Add 10 ml of a 20% tin(II) chloride solution, and dilute to the mark. Mix well, set aside for 30 min, and then measure the absorbance at 4030 Å in a 4-cm cell; compare with a standard. Prepare the standard from platinum sponge and aqua regia. Remove the nitric acid as usual, and dilute to provide 0.1 g of platinum per liter.

CHROMATOGRAPHIC METHODS

The Chromatographic Separation of Rhodium, Iridium, Palladium, and Platinum

The most complete chromatographic methods yet recorded for rhodium, iridium, palladium, and platinum were developed by Kember and Wells (403) and co-workers. These methods and their modifications will find increasing applications where microgram amounts of the noble metals are to be determined, and particularly when traces of base metals are associated with noble metals. Despite this potential usefulness, however, one cannot use the methods indiscriminately. Methods of dissolution and the proportions of the metals present can be significant.

Some conception of the limitations imposed by the separational methods were discussed by Kember and Wells. They recorded the following data.

Platinum. Qualitative tests with the hexone solvent show that when more than 1.5 mg of platinum are present on the strip severe trailing of the metal occurs, particularly along the extreme edges. With the *n*-butanol solvents less platinum can be tolerated, owing to the closeness of the platinum and palladium bands. Excellent separations are obtained with the hexone solvent on a mixture of platinum (1.4 mg of platinum on the strip) containing about 0.1% each of rhodium and palladium.

Palladium. The width of the palladium band when the hexone solvent is used increases if the quantity of palladium exceeds 200 μg, but unlike the other metals, both the leading and trailing edges of the band remain very sharp. A mixture containing palladium (3 mg of palladium on the strip) and about 0.1% each of platinum and rhodium can be analyzed successfully. With the *n*-butanol solvents, the same limitations apply as for platinum.

Rhodium. With all solvents the width of the rhodium band increases with an increase in loading by diffusion towards the palladium band. The upper limits are

3 mg for the hexone solvent and 5 mg for the *n*-butanol solvents. Mixtures of rhodium containing about 0.1% of platinum, palladium, and iridium can be successfully analyzed.

Iridium. The upper limits for iridium are lower for both solvent mixtures. With the *n*-butanol solvent the limit is 0.5 mg, and above this backward trailing causes interferences.

Base metals. Analyses have been carried out in the presence of large amount of commonly occurring base metals which remain comparatively immobile in one or other of the solvents. Tests have shown that up to 5 mg of some of these metals can be tolerated before distortion of the palladium band occurs. Sodium is only slightly soluble in the solvents and is regarded as an immobile metal.

In the presence of large amounts of metals that move in the center of the strip, distortion of the platinum metal bands occurs. This can be avoided by a change of solvent.

Metals that move in or near the solvent front impose a similar practical limitation. As their quantity increases the bands occupy more fully the space between the wet and dry solvent fronts, increasing this area above its normal limits. This reduces the space available for the platinum metal bands, so that not more than 5 mg of base metals can be present on the strip.

The procedure now described relies on the absence of osmium and ruthenium. The chromatograms are unacceptable when the hexone solvent is used and iridium is to be determined in the presence of more than 20 μg of ruthenium. With the *n*-butanol solvent, however, iridium and ruthenium are separated. Concerning rhodium, not more than 2 μg of ruthenium can be tolerated with either the hexone or butanol solvents. Osmium will interfere with platinum, both through its position on the strip and in the subsequent determination. It should be noted here that these interferences are usually disregarded with the claim that both osmium and ruthenium can be removed as their volatile oxides, but this removal from chloride solutions in which form the noble metals are separated chromatographically is not generally recommended. The distillation of ruthenium(VIII) oxide is usually made from sulfuric or perchloric acid solution, and chromatographic separations in these media have not been recorded. Recent researches have indicated, however, that the distillation of at least microgram amounts of both osmium and ruthenium can be made successfully from a hydrochloric acid medium.

PROCEDURE 108 (403)

Solvents

The solvent used may be one of the following.

The first is 60 parts by volume of isobutyl methyl ketone (hexone), 10 parts of *n*-pentanol and 30 parts of hydrochloric acid, $d = 1.18$. It yields well-defined bands for rhodium, palladium, platinum, and gold in this order of decreasing R_f values. Iridium in low concentrations is reduced to the trivalent state, and in this condition it remains with rhodium as a partially separated band. The solvent is therefore recommended for the separations of either rhodium or iridium from the remaining noble metals.

The second is n-butanol saturated with 3 M hydrochloric acid and containing hydrogen peroxide. It is used for the separation of rhodium from iridium previously isolated by the hexone solvent. The peroxide effectively moves the iridium away from the rhodium because the iridium is selectively oxidized to the more mobile iridium(IV).

The third is n-butanol saturated with 3 M hydrochloric acid. It is useful for separations involving base metals because the base metals are thus made relatively immobile.

Paper

Whatman No. 1 filter paper in 45-cm long, 3-cm wide strips is satisfactory. No. 4 paper is faster, No. 20 is slower, and Nos. 3 MM and 15 are thicker but offer no advantages.

Apparatus

Use a glass jar, 50 cm high and 7.5 cm i.d. The jar is fitted with a stopper from which is suspended a 30-cm solvent container and two strip supports. Downward diffusion is generally used, but upward diffusion is equally satisfactory.

Procedure

To the test chloride solution, free from cyanide, potassium and ammonia, in 1–5 M hydrochloric acid, add a 1% (w/v) solution of sodium chlorate, and make up to a suitable known volume (see Note 1). Transfer accurately 0.05 ml aliquots of the solution to the required number of strips, drawing the tip of the pipet across the strip along a line 7 cm from one end of each strip of filter paper (see Note 2). Set the strips aside to dry for 1 hr in air or for 30 min in a vacuum desiccator, desiccator drying being necessary if the atmosphere is humid or if deliquescent salts are present in high proportions in the mixture. Fold each strip 4 cm from the sample end, and place it in the chromatographic jar containing the required solvent for development by downward diffusion. Place the jar in a protected position in which the temperature can be maintained between 22 and 26°C, and leave overnight (14–18 hr) (see Note 3). Remove the strips from the jars, and set them aside to dry, preferably hanging in clips to avoid contamination (see Note 4). Spray one strip with a stannous chloride solution (11.25 g of A.R. material in 100 ml of 3.5 M hydrochloric acid). Warm gently to indicate the position of the platinum metal bands, viz., rhodium, yellow-orange; palladium, brown; platinum, yellow; gold, yellow. Lay the other strip or strips on a glass plate and divide them into sections at points between the bands and 1 cm beyond the furthest detected position, using the control strip as a guide. The strip can be divided by placing a small piece of glass across it and lifting the free end with a pair of tweezers, breaking it cleanly along the line. Transfer each section to a 10-ml microbeaker, and carry out determinations on these sections as previously described.

Notes. 1.　The chlorate serves to oxidize platinum(II) to platinum(IV). This avoids double bands.

2. The number of strips depends on the analysis required, e.g., if both iridium and rhodium are to be determined. In general two strips are prepared for each analysis, one as a control.

3. Much lower temperatures encourage diffuse bands and double platinum bands. Higher temperatures offer no advantages. By using strips of No. 4 paper the time of standing can be reduced to 6 or 7 hr. This gives slightly wider bands and is not advisable when the quantity of any component of the mixture approaches the upper limit.

4. The strips are very fragile at this stage if the hexone solvent has been used, and they become more brittle as the solvent evaporates.

The aqueous extraction apparatus (Fig. 7.5) is used to remove the metals from each portion of the strip. The same general instructions apply to all metals, the only variation being the type and order of extracting solutions; these are given in the following section. The volume of the calibrated flask A should be chosen to provide a concentration which will give a reasonable light transmission in the final colorimetric determination, 5- or 10-ml flasks normally being used. The extract for the determination of iridium is collected in a tall tube (made from a specimen tube 7.5 × 2.5 cm in diameter) provided with a lip.

Add the first extraction solution to the paper, cover the beaker with a watch glass, and warm on a steam bath for 10 min unless otherwise stated. Break up the paper to a coarse pulp with a small glass rod, which is then left in the beaker. Do not allow prolonged heating, otherwise a fine pulp is formed that is difficult to filter. Allow the pulp to settle, and filter the mixture through the funnel of the extraction apparatus directly into a calibrated flask A, applying gentle suction. Then transfer the pulp to the funnel, pressing it into a wad with the glass rod. Release the vacuum D, and add the second extraction solution to the funnel, transferring it via the beaker. Leave to soak for 1 min, then filter into the flask. Repeat this stage with the other extraction solutions. Remove the flask, make up to the mark with water, and measure the transmittance using the spectrophotometer. For iridium analyses remove the tube, and wash the filtrate into the titration vessel with the minimum amount of water.

Extraction solutions. The volumes given are for a 5-ml flask; alter the volumes in proportion for other sizes of flask.

Rhodium (upper section).

(1) 1 ml of stannous chloride solution + 1 ml of 5 M hydrochloric acid. Heat for 30 min.

(2) 1 ml of 5 M hydrochloric acid.

(3) 1 ml of water.

Palladium (center section), by thioglycollic acid method.

(1) 1 ml of M hydrochloric acid + 1 ml of water.

(2) 1 ml of thioglycollic acid solution.

(3) 1 ml of water.

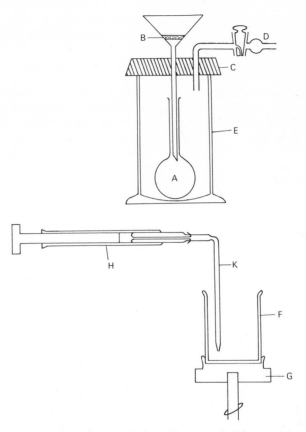

FIGURE 7.5. A is a calibrated flask (5–10 ml), see Procedure 49. B is a no. 4 sintered disk. C is a rubber stopper. D is a line to vacuum. E is a chromatographic jar, 50 cm tall, 7.5 cm i.d., fitted for downward diffusion. F is a titration vessel. G is a chuck turned from polyethylene, and slightly eccentrically mounted. H is a micrometer syringe. K is a microburet. See Procedure 108.

Palladium (center section), by p-nitrosodimethylaniline method.

(1) 1 ml of sodium acetate solution + 1 ml of water.

(2) 0.1 ml of *p*-nitrosodimethylaniline = 0.9 ml of water.

(3) 1 ml of water.

Platinum (lower section).

(1) 1 ml of 5 *M* hydrochloric acid + 1 ml of water.

(2) 1 ml of stannous chloride solution.

(3) 1 ml of water.

Iridium (upper or lower section according to solvent used).

(1)   Add 0.2 ml of liquid bromine to the paper, leave to soak for 5 min with the beaker covered, then add 1.8 ml of water.

(2) and (3)   1 ml of 1 *M* hydrochloric acid.

Determinations.   Make all final measurements on a Unicam spectrophotometer or a similar instrument, using 1-cm cells (silica for palladium by thioglycollic acid and glass for the others), and making comparisons against reagent blanks. Prepare standard curves, using the same reagents as were used in the aqueous extraction method over the ranges given in the following. In preparing the standard curve for palladium by the *p*-nitrosodimethylaniline method, add 0.5 ml of 0.1 *M* hydrochloric acid to each 5 ml of solution to give the required pH. During the actual determinations the residual acid in the paper is sufficient. The values of wavelength and range involved are as follows.

	Wavelength (nm)	Range in 5 ml of solution (μg)
Rhodium	480	0–100
Palladium (thioglycollic acid method)	325	0–50*
Palladium (*p*-nitrosodimethylaniline method)	525	0–1.5
Platinum	403	0–100

With the procedure described above and with the hexone solvent, iridium and rhodium appear as two closely associated bands. No method of selectively separating these two bands has been developed. To determine both rhodium and iridium two chromatograms are necessary. By the volumetric method described below for iridium there is no appreciable interference from rhodium. Thus the combined bands from the hexone separation may be used for the iridium determination. A second chromatogram obtained as described above, but with the solvent No. 3 (*n*-butanol/peroxide) will move the iridium to the trivalent state, so that the rhodium band is isolated.

To the bromine water extract of iridium add 0.4 ml of a lithium sulfate solution,† and heat to fumes of sulfur trioxide, preferably under infrared lamps, and in the titration vessel, although some charring will still probably occur, because of traces of solvent and decomposed cellulose present. Add 0.2 ml of perchloric acid, and heat to fumes, repeating this until a clear solution is obtained. Add 0.2 ml of perchloric acid, 1 drop of 3,3′-dichloroben-

---

*The present authors record their inability to find in the literature any quantitative spectrophotometric procedural data using thioglycollic acid for palladium. It is recommended that, for the range of 0–50 μg of palladium, *p*-nitrosodimethylaniline be applied with appropriate measured diluents. Presumably Kember and Wells (403) used the report on thioacids as spot test reagents for palladium by Konig and Crowel (403a). However, see Procedure 103.

†Acid–lithium sulfate solution. Transfer 8 g of lithium sulfate monohydrate to a silica dish, and heat to remove water by crystallization. Add 30 ml of 98% sulfuric acid, stir, and warm to effect dissolution.

zidine solution as indicator, and heat to 300–320°C (see Note). Leave for 10 sec at this stage, then remove and cool rapidly. Add 2 ml of water, and cool again. Transfer the vessel to the titration apparatus, and titrate with the hydroquinone solution (diluted 10 times from stock solution, i.e., 0.01 ml of hydroquinone solution $\equiv$ 1 μg of iridium, see Procedure 76). Standardize the diluted hydroquinone solution by a series of titrations against known quantities of iridium, first to the discharge of the purple color. Then add one drop of indicator to produce a yellow color, and add further hydroquinone solution until this color is just discharged.

The titration may be repeated after evaporating the solution again to fumes of sulfur trioxide and repeating all stages.

Note. The temperature is fairly critical. If a lower temperature is used, a longer time of heating is necessary; this must be determined by experiment. The solution first turns yellow then brown. This color is discharged and the final solution is colorless or purple, according to the amount of iridium present.

The Chromatographic Separation of Gold, Platinum, Palladium, and Silver

PROCEDURE 109 (404)

One drop of the noble metal solution should contain approximately 0.1 mg of each metal. Weigh the alloy, and transfer it to a 10-ml volumetric flask. Add a few milliliters of aqua regia, and following dissolution add an equal volume of water and make up to volume. Measure an appropriate volume of the solution with a calibrated pipet, and transfer to a small porcelain crucible; evaporate carefully under a lamp to about 1 drop. Prepare a cylinder of Whatman No. 1 paper, 28 cm high and 26 cm wide. Transfer the drop of the evaporated solution to a position 2.5 cm from the bottom of the cylinder, previously marked with a pencil. Wash the crucible with water added by means of a dropper drawn to a capillary at one end and containing a small rubber bulb at the opposite end. With care the washing can be accomplished with a few drops of water. If necessary the drops on the paper can be evaporated intermittently by careful exposure to a lamp prior to the addition of further drops. Clean a 5-gal (imperial) crock and place within it a glass dish containing sufficient *n*-butanol saturated with 1 *M* hydrochloric acid. Add the *n*-butanol solution to the bottom of the crock to ensure an atmosphere saturated with the gaseous mixture. Place the cylinder within the glass dish in a position to avoid any direct contact of the stain with the butanol solution. Allow to stand for 20–24 hr (see Notes 1 and 2). Following the 24-hr period, expose the cylinder to ammonia, then to hydrogen sulfide (see Note 3). To determine each constituent, cut out the areas of deposition, transfer each section to a 50-ml beaker, and keep the latter overnight at about 400°C. Add a few drops of aqua regia to dissolve the ash, evaporate in the presence of hydrochloric acid to remove nitric acid, and prepare for the desired spectrophotometric method. For silver, treat the ignited ash with nitric acid, the subsequent treatment depending on the method of determination.

Notes. **1.** On ascent two liquid fronts are observed, one being the front of the dehydrated butanol and the second being the front of the aqueous phase, about 5 cm below the upper front. The aqueous phase has a dark front, which does not fade on drying, and is convenient to use as a reference for R_f values. Extracted substances in the upperbutanol layer will thus have an R_f value greater than unity.

2. Silver is deposited at the place where the drop is applied ($R_f = 0$). Copper moves slowly ($R_f = 0.1$). Palladium has an R_f of 0.6, platinum an R_f of 0.72–0.80, and gold an R_f of 1.05–1.13.

3. Silver and copper produce black spots which fade in about 1 hr. The spots of gold and platinum are yellow-brown and that of palladium orange-brown. A very small amount of gold is reduced; it leaves a trail of purple colloidal gold. This small amount is usually insignificant. The spots of platinum, palladium, and gold deepen on storage and can be kept as permanent records of the analyses.

A Modified Chromatographic Determination of Platinum, Palladium, Rhodium, and Iridium

This method was designed to provide improved procedures for the determination of platinum and palladium, and particularly improved methods of determining rhodium and iridium in order to avoid the necessity of two chromatograms.

To produce the chromatogram, sodium chlorate was replaced by sodium chloride with the result that iridium was quantitatively fixed in a band adjacent to the immobile rhodium band. It is not unlikely that this substitution is the cause of the appearance of a few micrograms of platinum with the palladium band. This contamination is of no significance because the two bands are combined and may be analyzed for platinum and palladium by a variety of methods. The additive absorbance method is described in Procedure 37, and the p-nitrosodimethylaniline –chloroform extraction method is described now (Procedure 110). In the present authors' opinion this approach provides for rapid and accurate determinations.

PROCEDURE 110 (403)

After the removal of the chromatographic strips from the jar (Procedure 108) leave them suspended in air for at least 15 min. Spray the control strip with the tin(II) chloride solution, and gently heat the strip with a lamp maintained at an appropriate distance. Cut the sample strip with scissors to isolate rhodium together with iridium and platinum together with palladium. Place the platinum–palladium band in a 30-ml beaker, and ignite overnight in an oven at 400–450°C. Treat the ash with 8 ml of aqua regia, and with a cover glass on place on a steam bath and evaporate to near dryness. Repeat this treatment. Treat the residue three times with 5 ml of concentrated hydrochloric acid, with intermittent evaporations to dryness, then evaporate three times with 12 drops of concentrated hydrochloric acid. Cool the evaporated residue, and rinse the watch glass and the wall of the beaker with a few milliliters of water whose acidity has been adjusted to pH 3 with hydrochloric acid. Transfer the solution to a 75-ml, pear-shaped separatory funnel, and wash with 8 ml of water at pH 3.

Add 0.5 ml of the p-nitrosodimethylaniline solution to the 10 ml of solution (see Procedure 29), followed by 5 ml of 95% ethanol. Shake for 5 min, remove the red palladium complex by an extraction with 10 ml of chloroform, and return the extract to the original beaker. Add to the separatory funnel 0.5 ml of chloroform, extract as before, and add the extract to the original beaker. For the determination of palladium cover the beaker with a Speedvap cover glass and evaporate to a wet residue. Remove the beaker, and allow the residue to dry at room temperature. Cover with a watch glass, and destroy the organic matter with fuming nitric acid and a 30% hydrogen peroxide solution in the usual way. Convert to the chloride form, and determine palladium by p-nitrosodimethylaniline (Procedure 29).

For the determination of platinum, transfer the aqueous phase from the palladium separation to a 50-ml beaker, covered with a Speedvap cover glass. Wash the separatory funnel thoroughly with water, and add the washings to the platinum solution. Evaporate to dryness on a steam bath, and continue as with the chloroform–palladium extract, including the destruction of organic matter. Determine platinum by the tin(II) chloride method (Procedure 26).

The separation and determination of rhodium and iridium. After the removal of the platinum–palladium section of the chromatogram, place the remaining strip with the rhodium–iridium band over a Kleenex sheet of similar size, and fold carefully to protect the metal-containing bands. Place the folded paper in an A2 porosity porcelain filtering crucible, and set the latter inside a larger porcelain crucible. Place the crucibles in a furnace at 400–500°C as described above. Place the filtering crucible containing the ash in a Vycor tube, and heat for 30–45 min to about 600°C in a current of hydrogen. Remove the crucible, add about 10 mg of sodium chloride, transfer the crucible to a Vycor tube, and pass chlorine over the residue at about 650–700°C for 4 hr.

Raise the level of the tube to allow the crucible to slip into a 250-ml beaker. Rinse into this beaker the inner tube surface with 1 M hydrochloric acid. Evaporate on a steam bath to about 20 ml, and remove the crucible, washing thoroughly with 1 M hydrochloric acid. Filter the solution through the washed crucible, and wash with water to a volume of 75 ml. Evaporate the filtrate on a steam bath to a small volume, transfer to a 30-ml beaker, and evaporate to dryness. Add a few drops of 47–49% redistilled hydrobromic acid, and treat by solvent extraction with isoamyl alcohol as described in Procedure 97.

The Chromatographic Separation of Rhodium, Iridium, Palladium, and Platinum on the Macro Scale

This method allows the separation of each of the four metals in quantities of the order of 75 mg. The solvent mixtures are acid–hexone and acid–oxidant–hexone. The columns contain cellulose powder, the volume of which varies with the separation required. Both oxidizing and reducing conditions are applied; under oxidizing conditions platinum and iridium move together. A separation requires a second column and reducing conditions. The reduction is accomplished by tin(II) chloride, which allows the collection of iridium as a first fraction, and subsequently platinum is eluted with the acid solvent. Palladium is collected as a second fraction. Rhodium is almost immobile with both oxidizing and reducing solvents, but is readily eluted with water. The method of separating the four metals is about comparable in speed and accuracy to the better chemical separations.

The authors of the method now described recommended a prior nitrite separation of small proportions of such base metals as copper, nickel, tin, etc. (Procedure 117). In the presence of large amounts of associated base metals the present authors recommend a prior cation exchange separation (Procedure 100).

PROCEDURE 111 (405)

Equipment

Cellulose.   This is fine grade Whatman ashless cellulose powder.

Columns.   The adsorption tubes are 1.8 ± 0.1 cm i.d. Columns 15 cm long are suitable for the separation of two-component systems. Thirty-centimeter columns are used to separate three- or four-component systems. The columns are treated by passing 5 ml of acid solvent per centimeter of cellulose column height. When oxidizing conditions are required, a minimum of 50 ml of oxidizing solvents are used to treat the cellulose. The glass tubes may be widened at the upper end to form a funnel. The lower end is narrowed and closed by a short length of polyvinyl chloride tubing fitted with a tap. The inside surface of the glass extraction tube is treated with dichlorodimethylsilane to encourage water repelling properties (406).

Solvents.   The acid solvent is hexone containing 3% (v/v) of concentrated hydrochloric acid. The oxidizing solvent (generator solvent) is 100 ml of hexone containing 4% (v/v) of concentrated hydrochloric acid added to a mixture of 4 g of sodium chlorate and 12 g of cellulose powder to absorb the water formed by the reaction, the solvent then being decanted from the cellulose and mixed with an equal volume of hexone containing 4% (v/v) of concentrated hydrochloric acid.

Procedure

Transfer the chloride solution of the four metals, preferably in their fully oxidized states, to a 400-ml beaker. Add zinc chloride in an amount at least equal to the combined weight of platinum and iridium. If nitric acid is present evaporate to dryness on a steam bath, add a few milliliters of strong hydrochloric acid, and again evaporate. Repeat this three times. Add 5 ml of concentrated hydrochloric acid. In the absence of nitric acid, evaporate to dryness on a steam bath in the presence of zinc chloride, and add about 5 ml of concentrated hydrochloric acid.

In either of the above instances saturate the solution with chlorine.

Evaporate again, avoiding any overheating of the residue, remove excess of acid and condensate from the wall of the beaker at as low a temperature as possible. Dissolve the residue in 1.0 ml of concentrated hydrochloric acid saturated with chlorine, and add 10 ml of generator solvent and then 10 ml of hexone. Replace each withdrawal of solvent by the addition to the generator of the same volume of hexone containing 4% (v/v) of concentrated hydrochloric acid, and stir the mixture (see Notes 1, 2, and 3).

Allow the level of the sample solution to fall to the top of the cellulose column, and then rinse the beaker with successive 3–4-ml portions of oxidizing solvent, any aqueous phase being retained in the beaker. Keep the volume of washing solvent to a minimum. Elute the mixture by the addition of oxidizing solvent, care being taken to ensure that the level of the solvent does not fall below the top of the cellulose column.

A wide diffuse band of iridium moves rapidly down the column, its brown color masking the presence of some platinum that moves with it, sometimes a little ahead of but not detached from the iridium band. The bulk of the platinum follows the iridium as an adjacent merging band. As elution proceeds, the deep yellow platinum band moves completely away from the fairly compact orange-brown palladium band. Palladium moves away from the immobile rhodium, but at a much slower rate than does the wide platinum–iridium band.

Collect the platinum and iridium as a single fraction of about 200–250 ml in a 500-ml round-bottomed flask fitted with a ground-glass socket to facilitate the subsequent distillation (see Note 4). To avoid any loss of iridium from the platinum–iridium fraction, let the palladium band approach the bottom of the column before collecting the palladium fraction. Maintain fully oxidizing conditions to prevent any retention of iridium in the palladium band.

Collect the palladium fraction, 100–150 ml, in a similar 500-ml round-bottomed flask, discontinuing the addition of solvent to the column as soon as all of the palladium has been eluted.

Transfer any rhodium remaining in the sample beaker to the column with a jet of distilled water, and acidify the aqueous solution with a few drops of dilute hydrochloric acid (see Note 5). Collect the rhodium fraction, about 100 ml, in a 250-ml beaker, and evaporate the hexone layer with an infrared lamp.

Notes. 1.   The initial acidity of the sample solution is rather high, but it decreases as the elution proceeds.

2.   It is important to limit the initial volume of the sample solution and the solvent washings of the sample beaker. If the total volume of sample solution is too great, chromatographic separation commences before the last of the platinum-bearing solution comes into contact with the cellulose.

3.   If a few drops of aqueous phase separate from the hexone solution of the sample, this aqueous phase should be retained in the beaker by careful decantation of the hexone and the hexone washings. Some of the rhodium present in the sample will remain in the aqueous phase, and is indicated by a characteristic rose-red color. The remaining rhodium is immobile at the top of the column and is indicated by a narrow red band.

4.   Any zinc, iron, or copper present will move with the platinum and iridium, but any nickel present will be retained at the top of the column with the rhodium.

5.   Because both rhodium and nickel move rapidly down the column, the major proportion of the oxidizing solvent remaining in the column should be run either into the palladium fraction or to waste, otherwise the aqueous rhodium fraction will be accompanied by an inconveniently large volume of hexone.

The separation of platinum from iridium. Add water to the flask containing the first fraction, and fit a Claisen head and condenser. Pass a current of air through a fine capillary reaching almost to the bottom of the flask, and distill the hexone and a limited amount of water. Transfer the remaining aqueous solution to a 150-ml beaker, and remove any carbonaceous deposit (resulting from breakdown of the hexone) from the wall of the flask with hot aqua regia. Add the acid solution to the beaker, and evaporate the mixture to dryness, finishing on a steam bath. Repeatedly treat the residue with small portions of concentrated hydrochloric acid, and evaporate to ensure the complete removal of nitrate under conditions that avoid overheating the residue. Dissolve the residue in 1.0 ml of concentrated hydrochloric acid, add 0.2 g of stannous chloride, and dissolve the solid by gentle agitation.

Add 20 ml of hexone to the solution, the latter becoming highly colored (see Note 1).

Equilibrate a 15-cm cellulose column with the acid–hexone, and decant the reduced platinum–iridium mixture on to it, avoiding the transfer of any of the aqueous phase. Rinse the beaker repeatedly with 3–4 ml portions of acid–hexone and transfer the washings to the column, still retaining the aqueous phase. Collect the platinum fraction in a 500-ml round-bottomed flask, and distill off the hexone as described above, transferring the aqueous solution and acid washings of the flask to a 400-ml beaker. Elute the reduced iridium from the top of the column by adding the faintly acidified water used to rinse the sample beaker in the same way as that described for the elution of rhodium. Collect the iridium solution in a 400-ml beaker, and evaporate the solvent. At this stage, the four platinum metals are quantitatively separate, and the assay methods selected will depend on the identity and concentration of any base metals present.

Palladium is free of interference and can be precipitated by dimethylglyoxime or any other of the many suitable precipitants.

Platinum may be contaminated with copper, zinc, and tin. Evaporate the solution to a few milliliters, and add concentrated hydrochloric acid. Boil off the tin chloride, with the intermittent addition of bromine to assist the process. Evaporate to dryness, and dissolve in water containing a few drops of hydrochloric acid. Remove the base metal impurities by hydrolysis, nitrite being present to complex the platinum (Procedure 117). Evaporate the nitrate filtrate, and determine platinum by using one of the gravimetric reagents.

Iridium may be precipitated by 2-mercaptobenzothiazole (Procedure 61).

Rhodium may be precipitated by 2-mercaptobenzothiazole or hydrogen sulfide (Procedures 58 and 59).

Some Restricted Separations

The separation of iridium from rhodium. Add 0.25 g of mercuric chloride to the hydrochloric acid solution of the metal chlorides contained in a 150-ml beaker, and evaporate to dryness on a steam bath. Add a few milliliters of dilute hydrochloric acid, saturate the solution with chlorine, and again evaporate to dryness. Dissolve the residue in 1.0 ml of concentrated hydrochloric acid saturated with chlorine, and add 10 ml of generator solvent and then 10 ml of hexone. Use a 15-cm cellulose column equilibrated with solvent, and pass through a minimum of 50 ml of oxidizing solvent before decanting the sample solution on to the column. Retain any aqueous phase, and rinse the beaker with successive 3–4 ml portions of oxidizing solvent. Maintain oxidizing conditions throughout the separation to prevent any retention of trace amounts of iridium. Collect the iridium fraction in a 500-ml round-bottomed flask, and elute the rhodium with water.

The separation of rhodium from other platinum metals. Use the oxidizing solvent even in the absence of iridium, because it is desirable to repress the formation of any slower-moving reduction product of platinum. Use a 15-cm column for the separation of rhodium from platinum or palladium, but a 30-cm column for the separation of all three metals.

The separation of palladium from iridium. Use a 15-cm column, and maintain oxidizing conditions throughout to keep iridium moving away from the less mobile palladium.

Note. A separation based on the retention of reduced iridium will fail because of the marked effect of reduction on the chromatographic behavior of palladium.

The separation of platinum from iridium. This separation has already been described above.

Note. It is noteworthy that, under the reducing conditions that apply in this instance, the chromatographic behavior of platinum differs from its behavior under fully oxidizing conditions. When the platinum–iridium mixture has been reduced by stannous chloride, and the solvent solution transferred to the column, the platinum invariably appears in two forms. A wide, mobile, yellow band accounts for most of it, and a slow-moving, dull pink band represents the remainder. The two bands may separate, but both should be collected in one fraction. Sufficient solvent should be used to remove all trace of the less mobile product before commencing to elute the iridium with water. Water elutes iridium very rapidly. If a relatively large amount of platinum is being separated from a trace of iridium, it may be impossible to detect the presence of iridium at the top of the column by visual inspection. An operator unfamiliar with the characteristic colors of the various platinum metal complexes could mistake the slow-moving platinum band for iridium, and collect it as such in a separate fraction. Iridium would then be rejected with the column packing, and part of the platinum would be reported as iridium, so that the value found for platinum would be less than the true value. If sufficient iridium is present to be plainly seen at the top of the column its color is a dull grayish-green; it cannot be confused with the dull pink of the slow-moving platinum band. The separation of platinum from palladium. Use a 30-cm column and oxidizing conditions throughout the separation.

Separation of the Chloride Complexes of the Noble Elements by Partition Chromatography

Volyneto *et al.* (407, 408) developed a chromatographic method for the separation of the noble metals by separation on thin plates with a sorbent of silica gel. Standard reagents were used to develop the spots from which the quantity of separated metal was determined by direct application of reflectance densitometry to the isolated spot in question. The method is fast and reasonably accurate for microgram amounts of the noble metals. Solvents used are indicated in the following. The proportions of the solvent affects the chromatographic behavior. In general a 1-to-1 mixture of tributyl phosphate and benzene was used. This allows rhodium–palladium, iridium, platinum, and gold to be separated. By saturating the mobile solvent with 4 N hydrochloric acid and spotting some lithium chloride solution (salting out solution) on the point where the test solution is applied on the starting line the elements are separated into the following two groups: iridium, gold, and platinum at the solvent front ($R_f = 0.98$–1.0) and rhodium and palladium which remain at the start ($R_f = 0$ and 0.7, respectively).

Reversing the phases, the stationary phase in this case being tributyl phosphate and the mobile phase, 4 N hydrochloric acid, one gets a good separation of palladium and rhodium (rhodium as a narrow band on the solvent front, and palladium at the center of the plate). However, in this case, gold, platinum, and iridium remain at the start.

In some cases the noble elements affected each other mutually in their mixtures, e.g., their R_f values were different from those in the case of single elements.

PROCEDURE 112 (407, 408)

Equipment and Solutions

Solutions of noble metals.   Evaporate, dissolve the metal in aqua regia, and evaporate in the presence of hydrochloric acid and sodium chloride. Repeat several times and finally dissolve in water to which a drop or so of hydrochloric acid has been added.

Mobile solvents.   These are 1:1 mixtures of tributyl phosphate and benzene; cyclohexanone; methyl ethyl ketone; acetone. Purify by established methods.

Locating reagents.   These are 10% tin(II) chloride in 2 N hydrochloric acid for detecting platinum, palladium, rhodium on heating, and gold; 0.1% solution of pyridylazonaphthol in 95% ethanol for palladium; 0.5–1% solution of p-nitrosodimethylaniline in ethanol for palladium; 1.0% aqueous solution of bismuthol II for platinum, palladium, and gold; 0.5–1% solution of rubeanic acid for platinum, gold, nickel, cobalt, and copper; 1–1.5% solution of malachite green leuco base in concentrated acetic acid for iridium(IV), and gold; and 0.5% solution of 8-hydroxyquinoline in 60% ethanol for iron.

Sorbent and plates.   KSK grade of silica gel ground to 150–200 mesh (0.07 mm) is washed free from iron with hot hydrochloric acid (1:1) until the washings give a negative test with ammonium thiocyanate, then wash free from chloride ions, treated with silver nitrate. Dry the silica gel in air then in a drying oven at 120°C for 10–12 hr; store in a bottle with a ground glass stopper. Grind 14.25 g of silica gel with 0.75 g of rice starch, adding 29 ml of water to the mixture. Cook the mixture in a thermostat at 85°C stirring continuously. Add a further 20 ml of water and mix the whole again and apply to plates (1.3 × 18 cm). Smooth the absorbent, first with a glass rod, and shake gently in the palm of the hand. Dry the plates and sorbent overnight in air on a horizontal surface. Activate by drying for 20 min at 110°C in a drying oven.

Air tight tanks. Use battery type beakers with ground lids. Line the inside walls of the tanks with paper sheets impregnated with mobile solvent so as to saturate the tanks with the vapor of the solvent. Use atomizers for spraying the dried chromatograms and locating the elements. Use refelcting densitometer, automatic recording and integrating (Chromoscan) "Joyce" (England).

Procedure

Solutions of the chloride complexes of the platinum metals should contain 0.5–1.0 mg/ml of each element and gold. Evaporate the solution to dryness, treat the dry residue with aqua regia and evaporate to dryness in the presence of sodium chloride. Add concentrated hydrochloric acid and evaporate. Repeat this several times with intermittent additions of water. Solutions which contain hexachlororiridate are given an additional treatment with hydrogen peroxide to convert iridium to its tetravalent state in which form it is more chromatographically mobile. Iridium(III) usually remains at the start and is difficult to identify. Finally dissolve the residue in hydrochloric acid of the required concentration. Maximum differences in R_f values occur on using tributyl phosphate and benzene (1:1) as the mobile solvent for solutions which are 4 N hydrochloric acid; for cyclohexanone use 6 N hydrochloric acid; for acetone use 2 N hydrochloric acid; for methyl ethyl ketone use 2–4 N hydrochloric acid.

Apply the test solutions 1.5 cm from the edge of the plate by means of a capillary. The test solutions are applied to chromatograms in 0.001–0.003-ml portions. The space between the spots should be about 2 cm. Place the plates in a vertical, hermetically sealed tank which has been saturated beforehand with solvent vapor, and run the chromatograms by the ascending method. After the solvent front has advanced 12 cm, remove the plates from the tank, dry, and spray with an appropriate reagent. Place the plates in a drying oven at 120°C for 20 min after spraying. Separation time is determined by the mobility of the solvent used; with tributyl phosphate and benzene (1:1) 30 min is required. Calculate R_f values by means of several determinations. The variations in R_f values are seldom more than 0.02–0.03. Conditions for the separation and identification of the noble metals are given in Table 7.4. The behavior of associated base metals, iron, nickel, cobalt, and copper prepared as above is indicated in Table 7.5. The elements being located by rubeanic acid and 8-hydroxyquinoline solutions. The effect of the above base metals can be removed from the noble metals by running the chromatograms under conditions which are optimal for the separation of the noble metals. Prepare standards with known values for each metal. Prepare a calibration curve from the density of the spot as determined by reflectance densitometry; thus relating the density readings and the corresponding amount of an element in its spot. Use the least squares method. The complete analysis, including separation, can be accomplished in about 2 hr.

Concentration and Separation of Platinum Group Elements and Gold by Ion Exchange and Paper Partition Chromatography

Razina and Viklorova (409) used a cation exchange column for the separation of platinum, palladium, rhodium, and gold. Samples of ore are treated first with

TABLE 7.4
Conditions for the Separation and Identification of the Noble Metals

Mixture no.	Elements separated	HCl conc. of original soln. (M)	Mobile solvent	Location method	R_f (for multicomponent systems)	Notes
1	Au–Pt–Ir(IV)–Pd, Rh	4	TBP + benzene (1:1)	Au–SnCl$_2$, Pt–SnCl$_2$, bismuthol II Pd–SnCl$_2$, PNDA[a] Rh–SnCl$_2$ Ir(IV)–MGLB[b]	Au, 0.74 Pt, 0.43 Ir(IV), 0.37 Pd, 0.15 Rh, 0.0	Saturating the layer with 4 N HCl improves the separation of Au from Pt
2	Rh–Pd–Pt, Ir(IV), Au	4	4 N HCl	Rh, Pd, Pt, Au–bismuthol II, SnCl$_2$ Ir(IV)–MGLB	Rh, 1.0 Pd, 0.57 Pt, 0.07 Ir(IV), 0.06 Pd, 0.0	Reversed phase experiment. Layer saturated with TBP + C$_6$H$_6$ (1:1)

3	Ir(IV)–Pt–Pd, Rh	6	Cyclohexanone	The same as above	Ir(IV), 0.54; Pt, 0.33; Pd, 0.11; Rh, 0.05
4	Au–Ir(IV), Pt–Rh, Pd	2	Acetone	Au, Pt, Rh–SnCl$_2$, bismuthol II; Pd–PNDA; Ir–MGLB	Au, 0.87; Pt, 0.60; Ir(IV), 0.53; Pd, 0.01; Rh, 0.0
5	Au–Ir(IV), Pt–Pd, Rh	4	Methyl ethyl ketone	Au–SnCl$_2$, RA[c]; Ir–MGLB; Pt–SnCl$_2$, bismuthol II; Pd–PAN[d]	Au, 0.50; Ir(IV), 0.32; Pt, 0.30; Rh, 0.02; Pd, 0.01

[a] PNDA stands for p-nitrosodimethylaniline.
[b] MGLB stands for malachite green leuco base.
[c] RA stands for rubeanic acid.
[d] PAN stands for pyridylazonaphthol.

TABLE 7.5
The Behavior of Accompanying Elements

Mobile solvent	Elements separated				
	From Pt	From Pd	From Rh	Ir(IV)	From Au
TBP^a + benzene (1:1)	Ni	Fe, Co, Cu	Fe, Co, Cu	Ni	Ni
Acetone	Ni	Fe, Co, Cu	Fe, Co, Cu	Ni	
Methyl ethyl ketone	Ni	Fe, Co, Cu	Fe, Co, Cu	Ni	Ni

^{a}TBP stands for tributyl phosphate.

aqua regia, followed by a barium peroxide sintering, then a hydrochloric acid treatment and removal of barium by sulfate precipitation. The filtrates are combined and adjusted to 5 to 8% acid concentration, passed through a cation exchange. Finally the solution is adjusted to pH of 1.0–1.5 and passed again through an exchanger to remove residual amounts of base metals.

The method can be terminated quantitatively by visual comparison of the stains with those from standards or by a simple treatment with aqua regia the product of which is prepared for a spectrochemical determination. A new feature of the separation by ion exchange involves the ability to use a cation separation of noble and base metals employing an aqua regia solution for the initial ion exchange separation; the aqua regia dissolution allows direct application of a chromatographic separation and avoids the need for hydrochloric acid evaporations required to remove the nitrogen products of dissolution.

PROCEDURE 113 (409)

Apparatus and Chemicals

Tower.

Glass tank (with tight-fitting lid).

Dowex-50W-X2 (in its H form).

Chromatographic paper (265 × 210 mm).

Primary amyl alcohol.

Ethanol.

Hydrofluoric acid.

Hydrogen peroxide (30%).

Tin(II) chloride (in 20% hydrochloric acid).

Acetone.

Sulfuric acid (1 to 9 volumes).

Procedure

Treat 5–10 g of ore by heating with 50 ml of aqua regia. Evaporate to 15 ml and dilute to 100 ml with water. Filter off the insoluble and ignite it, then dissolve in 30 ml of 1-to-1 hydrochloric acid. Evaporate to 5 ml, dilute with water, and filter again. Ignite the paper and residue, weigh, and sinter with 5 times its weight of barium peroxide at 800–900°C for 1.5–2 hr. Dissolve the sintered mass in 1-to-1 hydrochloric acid. Filter again and precipitate the barium in the filtrate by adding 1-to-9 sulfuric acid at the rate of 3.5 ml of acid to 2 g of barium peroxide. Filter the barium sulfate and dilute the three filtrates with water so that the acid concentration is 5–8 vol %. Pass the solution of the noble and base metals through a column (36 × 1 in. i.d.) packed with DowexW-X2 in its hydrogen form. Wash the column with dilute hydrochloric acid. Evaporate the filtrate of noble metals to a small volume and dilute with water to pH 1.0–1.5. Pass this solution through the cation exchange column to remove residual base metals. Evaporate the filtrate to a moist salt stage and treat with aqua regia to destroy organic material and evaporate again to produce a moist salt stage. Add not more than 1 ml of aqua regia and dilute the solution to 1 to 3 ml with water and determine the platinum metals by ascending paper partition chromatography. Toward this end add to the chromatographic paper 0.01 or 0.02 ml aliquots of the above test solution and on separate strips the same amount of standard solution, containing 0.3, 0.5, 1.0, 1.5, 2.0, and 2.5 μg of platinum, palladium, rhodium, and gold. The latter, 265 × 210 mm, is divided into 25 mm strips maintaining 5 mm gaps between the strips. Dry the paper to which the solutions have been applied and place in a tank with a tightly fitting lid. Add to the tank 80 ml of a solvent consisting of primary amyl alcohol, ethanol, 10 N hydrochloric acid, 4% hydrogen fluoride and 10% sulfuric acid taken in volume proportions of 36, 21, 21, 21, 1. Locate the spots by means of tin(II) chloride. As an example of the separations which can be achieved the following R_f values have been obtained: for platinum, 0.80; palladium, 0.71; rhodium, 0.47; gold, 1.0. Gold forms a well-defined violet zone, platinum an orange one, palladium, a red one, and rhodium a yellow zone. Determine visually by comparing the color intensity with that of the standards which are applied to paper simultaneously with the test solutions. The determinations can be made by cutting up the chromatograms along the zone boundaries of the separated elements, washing the elements off the paper, and applying appropriate spectrophotometric methods or spectrochemical methods. In the latter case the colored zones are cut out and wet oxidized with aqua regia and then transferred to the crater of a carbon electrode (see Chapter 2).

In the presence of iridium and the absence of other platinum metals a zone with a R_f value of 0.79 is obtained. In the presence of other platinum metals there is a second zone for iridium with an R_f value of 0.53 with overlapping of a ruthenium zone and partially overlapping of rhodium. In the case of ruthenium there is a zone of R_f, 0.53 which is closer to that of rhodium. Here it is best to use a separation with hydrochloric acid; thiourea producing a pale blue complex. The lower limit for thus detecting ruthenium is 0.3 μg.

Selenium and tellurium interfere with their respective R_f values of 0.71 and 0.80. With tin(II) chloride the zones of the platinum, palladium, and rhodium should be developed with p-nitrosodimethylaniline. Red zones differing in shade are obtained. Palladium shows up immediately; the color then disappears and reappears once more when the chromatograms have dried out. It takes about 10–12 hr for rhodium to show up. Total identification of platinum sometimes takes as long as three days. Gold does not form a colored compound with p-nitrosodimethylaniline.

Other Chromatographic Separations

A high voltage electrophoretic separation of the platinum and other metals was investigated by Meier *et al.* (410). The action of the electrolyte, nitrilotriacetic acid, and EDTA were described in relation to various experimental factors, e.g., the former gave better separations than the latter. Mobilities of the ions were affected by pH, electrolyte concentration, and hydrolysis of the ligands in the complexes. The mobilities of the chloride complexes were higher than those of the corresponding bromide complexes. The gradation of the mobility obtained under certain conditions were employed as the basis of a procedure separating various mixtures. They decreased in the order Ir > Os Pt > Rh ≧ Pd > Ru > Rh.

TABLE 7.6
Paper Chromatographic Separations (253)

Separation	Metal solutions	Solvent solutions	R_f value
Ru Pd Pt Au	Chloro complexes	n-Butanol:3.5 N HCl	Au > Pt > Pd > Ru
Rh Pd Pt Au	in 1 M HCl	= 7:3	Au > Pt > Pd > Rh
Ir Pt		25% NaNO$_2$ solution	Ir > Pt
Os Ir Pt			Ir > Pt > Os
Ru Rh Pd	Chloro complexes (Rh, Pd), nitroso–chloro complexes (Ru) in 1 N HCl	Methyl ethyl ketone–HCl (1:19) = 7:3	Ru > Pd > Rh
Ru Rh	Nitrito complex	Ethanol:isopropanol:	Ru > Rh
Ru Rh Pd	in 3.5 N NaNO$_2$	25% NaNO$_2$ solution	Ru > Pd > Rh
Ru Rh Pt	solution	= 6:1:3	Ru > Pt > Rh
Ru Rh Pd	Chloro complexes in 1 M HCl	n-Butanol:3.5 N HCl = 7:3 (for Ru, Rh, Pd, Os, Pt, Au detection)	Au > Os = Pt Pd > Ru = Rh
Os Ir Pt Au		25% NaNO$_2$ solution for Ir detection	Ir ≈ Rh ≈ Ru > Pd > Pt > Os > Au

Chromatographic methods for the detection of the seven noble metals were recorded by Blasius and Fischer (410a). The concentrations of the metals were of the order of 20 g/l and nitrito- and nitrosochloro complexes were used as well as the cutomary chloro complexes.

Three types of solvent solutions were used. One contained by volume 70 parts of n-butanol and 30 parts of 3.5 M hydrochloric acid. A second was a 25% aqueous sodium nitrite solution, and a third, used only for small proportions of ruthenium in the presence of large proportions of rhodium, contained, in parts by volume, 60 of ethanol, 10 of isopropanol, and 30 of the 25% sodium nitrite solution. Ruthenium and osmium deposits were developed by thiourea, rhodium, palladium, platinum, and gold by a tin(II) chloride solution containing sodium iodide. Iridium collected by the nitrite solvent was treated with concentrated hydrochloric acid to produce a narrow brown zone. The use of these reagents precluded any subsequent quantitative applications. In order to detect each of the seven metals a variety of chromatograms, solvents, and feed solutions were required. Table 7.6 indicates the various separations and the conditions required. These are included to indicate potential methods for quantitative separations.

THE DETERMINATION OF OSMIUM, RUTHENIUM, PALLADIUM, PLATINUM, RHODIUM, AND IRIDIUM

Westland and Beamish (411) developed a separational scheme for the six platinum metals in solutions. The method is applicable to sample solutions obtained by dissolving small crystals of iridosmine and other platinum group minerals or insoluble residues of assay buttons and beads. The following procedure was developed specifically for work on the microgram scale. The method has been tested for both milligram and microgram amounts of osmium and ruthenium, but has been proven only for microgram amounts of palladium, platinum, rhodium, and iridium.

A summary of the method is given in Fig. 7.6.

PROCEDURE 114

Apparatus

The distillation equipment is shown in Fig. 3.

Reagents

Bromic acid. Prepared by adding a stoichiometric quantity of sulfuric acid to a hot, nearly saturated solution of barium bromate. The latter is made by dissolving 345.7 g of potassium bromate in 700 ml of hot water. To this is added a hot solution of 253 g of

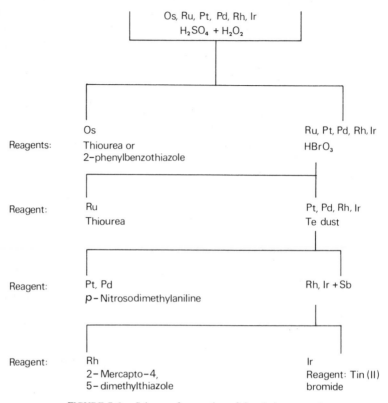

FIGURE 7.6.   Scheme of separation of the platinum metals.

barium chloride dihydrate in 400 ml of water. Cool, and decant from the crystals. Wash the latter several times with 100-ml portions of cold water. The yield should be about 10 g.

Tellurium powder.   Chips of tellurium ground to a fine powder in an agate mortar. The particles should be fine enough to form a suspension in the boiling sample.

Procedure

The separation of osmium.   Transfer the mixture of soluble platinum metals salts or the solution of these salts to the distillation flask. If the amount of osmium is sufficiently large to allow a gravimetric finish add to the three receivers respectively 15, 5, and 5 ml of 40% hydrobromic acid. If spectrophotometric determinations of the osmium are to be made, add 15, 5, and 5 ml of a 5% thiourea solution in 1:1 ethanol–hydrochloric acid to these receivers (these liquids must be chilled in ice). Add 5 ml of 70% C.P. perchloric acid to the trap and then 30 ml of 18 N sulfuric acid to the distillation flask. Heat the flask, and add dropwise a 10% hydrogen peroxide solution. When the temperature has reached

115°C continue the boiling and the dropwise addition of hydrogen peroxide solution for a further 15 min. A total of about 30 ml of the hydrogen peroxide solution is recommended. During the final 15 min reflux the perchloric acid trap solution in order to transfer the osmium to the receivers. Allow the distillation flask to cool to 30 or 40°C, and add 25 ml of water and 5 ml of a 30% hydrogen peroxide solution in this order. Boil the flask and trap solutions for 15 min.

To determine osmium gravimetrically rinse the chilled receivers, containing hydrobromic acid, with 1:4 hydrobromic acid into a 250-ml beaker. Add 0.2 g of hydroxylammonium chloride, and place the beaker on a steam bath for 30 min. Precipitate the osmium with thionalide (Procedure 66). Filter the precipitate, reduce, and weigh. For a spectrophotometric finish, transfer the chilled thiourea receiving solutions to a 50-ml volumetric flask, rinsing with 1:1 ethanol–hydrochloric acid. Make up to volume with the same solution, and filter through an 11-cm filter paper into the transmittance cell. Use the 1:1 ethanol–hydrochloric acid as a blank. Distill and prepare the osmium standards in a manner identical with the sample. The spectrophotometric method is described in Procedures 46 and 47.

The determination of ruthenium. After the distillation of osmium, allow the solution in the trap to cool, and dilute to 3 times its volume with water. Add to the trap 10 ml of 1:1 sulfuric acid, and to the three receivers 15, 5, and 5 ml of 1:1 ethanol–hydrochloric acid. Chill the receiving liquids in ice. Boil the liquid in the distillation flask to destroy the hydrogen peroxide. Add 5 ml of 5% bromic acid, boil for 15 min, then add another 5 ml of the bromic acid to the distilling liquid and also to the trap. Boil both the flask and the trap liquids for 15 min. Boil the solutions in the three receivers for a few seconds with the air stream passing through. Add 8 ml of a 5% solution of thiourea in 1:1 ethanol–hydrochloric acid to the first receiver and 2 ml to the second and to the third receiver to reduce the bromine and to develop the color of the ruthenium complex. Place a 600-ml beaker of water at 85°C around each receiver, and continue to warm the solutions for 15 min. Transfer the receiving solution to a 50-ml flask, and make up to volume with 1:1 ethanol–hydrochloric acid. Filter into a 2-cm cell, and determine the ruthenium as described in Procedure 44 or 45. For gravimetric determinations, evaporate the receiving solutions to a few milliliters, filter, and adjust the acidity and volume for a precipitation by thionalide (Procedure 65).

The separation of platinum and palladium. Transfer the liquid in the distillation flask after the removal of osmium and ruthenium to a suitable beaker, and evaporate and fume to 1 ml. Do not heat excesively at this stage, otherwise the platinum will be complexed to a degree that appreciable proportions will pass into the filtrate from the tellurium precipitation. Cool the solution, and dilute with 1 ml of water. Add 5 ml of concentrated hydrochloric acid, cover with a glass, and boil for 30 min, adding hydrochloric acid to maintain the original volume. Remove the cover glass, and again evaporate carefully to just faint fumes of sulfuric acid. Dilute to 10 ml, cool, add approximately 50 mg of tellurium powder, and boil vigorously for 10 min. Continue heating to reduce the volume to 8 ml, and add 3 ml of concentrated hydrochloric acid, a few crystals of sodium metabisulfite ($Na_2S_2O_5$) and 1 drop of 1% hydriodic acid to precipitate tellurium, platinum, and palladium (see Note 1).

Stir the mixture of platinum, palladium, and tellurium for 15 min, and add a few more crystals of sodium metabisulfite. Filter the metals by decantation through a 4-cm paper, and wash well with a 0.2% solution of sodium metabisulfite in 3 M hydrochloric acid, so as to prevent peptizing the tellurium precipitate.

Reserve the filtrate for the rhodium and iridium determinations.

Transfer the filter and residue to a 5-ml silica crucible, and char over a low flame. Then place the crucible and its contents in a suitable Vycor combustion tube, and heat over a Meker flame for 15 min in a stream of hydrogen (see Note 2). Continue heating the mixed precipitate in air. Transfer the cold crucible and its contents to a beaker, and add to the crucible a few milliliters of aqua regia. Heat to dissolve and then remove the washed crucible from the beaker with a stirring rod, taking care not to remove any of the metal solution. Transfer the beaker containing the leached solution to a steam bath, add 40 mg of sodium chloride, and evaporate. Repeat the evaporation thrice in the presence of a few milliliters of hydrochloric acid. Dissolve the dry residue in a few milliliters of water. Determine the platinum and palladium by the additative colorimetric method, using p-nitrosodimethylaniline as the color reagent. In this method the solution is adjusted to pH 2.2 and diluted to 10 ml. The method is described in Procedure 37 (see Note 2).

Notes. **1.** The latter two reagents are added to precipitate the tellurium in solution, and also the platinum and palladium, so as to avoid their interference in the subsequent determinations of rhodium and iridium. The reduction of the tellurium by the sulfite is catalyzed by the iodide; this species will interfere with the later determination of rhodium, however, if more than a trace is present. The iodide in excess would also produce hydrogen sulfide by reducing the sulfuric acid in the solution.

2. The heating period should ensure the removal of mechaically admixed tellurium, which would otherwise interfere in the determination of palladium and platinum.

3. In order to increase the sensitivity one may use an aqueous solution of p-nitrosodimethylaniline rather than the ethanolic solution. The increase in sensitivity is attained at the expense of color stability. Because of the low solubility of the reagent its solution must be prepared hot, with the resulting later deposition of a solid. Consequently one must prepare standards for each lot of reagent. For this modification see Procedure 36.

The aqueous solution is prepared by adding 165 mg of the p-nitrosodimethylaniline to 100 ml of boiling water, stirring for 30 min, and filtering.

4. The tellurium separation of platinum and palladium from rhodium and iridium may be adversely affected by high proportions of palladium. Under these conditions the palladium seems to catalyze the precipitation of iridium. The present method has been proved for amounts of palladium of the order of 20 μg. With larger amounts the separation of the palladium dimethylglyoximate can be made by extraction with chloroform, as described in Procedure 104, or by the usual precipitation, as described in Procedure 52. In these instances it would be necessary to evaporate the filtrate containing the platinum, rhodium, and iridium, and to destroy the organic matter. This can be accomplished by the conventional method with nitric acid and 30% hydrogen peroxide solution or with sulfuric acid and a few drops of nitric acid. If the latter mixture is used care must be exercised to avoid excessive fuming. Furthermore, the final sulfuric acid solution should be boiled with a few milliliters of concentrated hydrochloric acid. Note that the prior history of a platinum metals solution is of significant influence on subsequent determinations, particularly that of rhodium.

The determination of rhodium. Evaporate the filtrate from the tellurium precipitation of platinum and palladium. Continue the heating to slight fumes only, avoiding temperatures above 200°C. Cool the solution, and dilute to about 8 ml. Add 75 mg of antimony dust of such a fineness as to homogeneous mixture on boiling. Boil vigorously for 30 min, adding water to maintain the volume. Filter through a 4-cm paper, and wash with 5 ml of water. Set the filtrate aside for the determination of iridium. Transfer the paper and precipitate to the beaker used for the separation from iridium, and place it in a muffle furnace. Heat to 400–450°C for 10 hr, which period should produce complete combustion. Cool, and break up the residue with a glass rod. Add 0.5 ml of concentrated sulfuric acid, 1 ml of a 30% hydrogen peroxide solution, and 2 ml of concentrated hydrochloric acid. Cover with a ribbed cover glass, and evaporate to 1 ml (see Notes 1 and 2).

Add to this rhodium solution 10 ml of concentrated hydrochloric acid, and treat as described in Procedure 118 except add twice the recommended amount of reagent, and boil the solution for 2 hr. Filter the colored solution before making the absorbance measurements (see Note 3).

Notes. **1.** The spectrophotometric reagent 2-mercapto-4,5-dimethylthiazole (Procedure 118) is used for the determination of rhodium in preference to tin(II) chloride because with the latter there is a slight darkening due to the small amount of tellurium in the separated rhodium. If the mixture is allowed to stand the tellurium coagulates so that it can be removed by a fine texture filter paper.

2. The tin(II) chloride method (Procedure 40) can be applied providing a standard curve is prepared from samples containing the same amount of dissolved antimony as is used to effect the rhodium separation. This is necessary because the absorbance characteristics of the rhodium complex with stannous chloride are changed in the presence of antimony. If tin(II) chloride is to be used the 1 ml of evaporated solution must be treated with 1 ml of hydrochloric acid, and the solution must then be diluted to 5 ml. The dense precipitate of antimony oxychloride redissolves on heating.

3. It should be emphasized that the blank error is high if the solutions are not filtered prior to making absorbance measurements. This error should not exceed 0.4 μg.

The determination of iridium. Transfer the filtrate from the rhodium–antimony precipitation to a 250-ml flask fitted with a dropping tube and a thermometer well. Pass a current of nitrogen through the flask, and heat the contents to fuming. When the temperature reaches 160–200°C add dropwise 75 ml of concentrated hydrochloric acid, the antimony chloride then being readily volatilized (see Note 1). Fume to about 1 ml and add a few milliliters of concentrated hydrochloric acid containing a few drops of hydrogen peroxide solution. For the determination of iridium the tin(II) bromide method is recommended (Procedure 119).

Notes. **1.** Temperatures over 200°C will give low iridium recoveries.

2. An intense red color may appear during the color development with the bromide. This color constituent should disappear when the samples are diluted to the recommended 50 ml. This dilution encourages the dissociation of the bromo complexes.

3. When the quantity of iridium to be determined is large, and therefore a relatively small aliquot is required, the antimony interference is correspondingly reduced, and the distillation can be omitted.

It is emphasized that certain variations in the above procedure may be advisable depending upon the absence of some of the platinum metals. Some of these modifications will be obvious in the absence of such constituents as iridium. Furthermore, the analytical chemist may prefer the introduction of different methods at the stage of the rhodium and iridium separation. Here, particularly, the copper separation (Procedure 96) or the tin(II) bromide extraction method (Procedure 97) may be preferable.

Anyway, the complete determination of the six or seven noble metals is a lengthy procedure; its satisfactory accomplishment requires more than a casual acquaintance with the recommended analytical methods.

THE DISSOLUTION OF PLATINUM AND OTHER NOBLE METALS BY PRESSURE TECHNIQUE

Platinum may be dissolved in aqua regia in an open vessel at atmospheric pressure, but small undissolved residues containing some of the impurity elements may remain. Dissolution under pressure by a method such as that devised by Wichers *et al.* (378) has been found more satisfactory in the present authors' laboratory. This technique is described in the following procedure.

PROCEDURE 115 (378)

Transfer a 2.000-g sample of platinum metals to the reaction tube, previously chilled in a slurry of dry ice and carbon tetrachloride (see Fig. 2.5). During the cooling process introduce through a thin-stemmed funnel 15.4 ml of concentrated hydrochloric and 0.6 ml of concentrated nitric acid. Seal the end of the reaction tube with a hot oxygen–gas flame while the body of the tube remains in the cold slurry to retard the start of the reaction. Rotate the tube while heating; a slight suction in the molten glass indicates when sealing is complete. Turn off the oxygen, and anneal the heated portion of the tube with the luminous yellow gas flame until it is covered with a thick layer of soot. Allow the tube to cool at room temperature, and place it in a steel shell (see Fig. 2.8) to which has been added 30 g of calcium carbonate to react with the acid if the reaction tube should break. Place sufficient dry ice in the shell to compensate for the pressure within the reaction tube when heated, and immediately screw on the cap over an aluminum gasket. Test the shell in warm water for leakage, and place it in a tube furnace at 270 ± 5°C for a minimum of 18 hr (see Fig. 2.8). After removal from the furnace, allow the bomb to cool at room temperature for about 3 hr. Then loosen the screw cap slightly and release the carbon dioxide by moving the plug projection. Remove the screw cap, and take the reaction tube out of the shell. Place it in a dry ice–carbon tetrachloride slurry, and chill it for about 15 min. With the body of the reaction tube still in the slurry, heat the tip of the stem with a luminous gas flame until it is covered with soot, and then use a fine, hot, oxygen–gas flame until the glass softens and a small hole is blown out. Remove the tube from the slurry, and cut it just below the neck with a hot-wire cutter.

Transfer the chloroplatinic acid or other platinum metals salt to a clean unetched beaker, and wash the tube and stem with 1:10 hydrochloric acid. Fume the solution to a small volume twice with concentrated hydrochloric acid, cool to room temperature, transfer to a volumetric flask, and make up to a volume of 100.00 ml which should contain 20 mg of platinum per milliliter.

PRECIPITATION OF PLATINUM BY ZINC

PROCEDURE 116

Evaporate the hexachloroplatinic acid solution, containing approximately 10 mg of platinum, to a syrup on a steam bath; add 0.5 ml of concentrated hydrochloric acid, and dilute to 85 ml with water. Prepare an aqueous suspension of zinc containing 5.5 mg/ml by pouring water quickly over some zinc dust in a dry beaker, and stirring immediately. To the platinum solution on the steam bath add 100 mg of zinc, in 0.5 ml portions every 10–15 sec, stirring both the suspension and the reaction mixture in the interim. Add another 100 mg of zinc at approximately twice this rate. Boil the resulting mixture gently for 1 hr, filter hot, and wash the residue once by decantation with 5 ml of hot 1% hydrochloric acid, followed by a total of 60–70 ml of hot water or preferably the same volume of a hot 1% ammonium chloride solution. Cleanse the beaker wall with a small piece of filter paper moistened with 1% hydrochloric acid.

Ignite the residue and paper by raising the temperature slowly to 600°C in the muffle, and continue heating at this temperature for 1 hr. Cool, then ignite in an atmosphere of hydrogen at Meker burner temperature for 1 hr. Cool in nitrogen, then air, and weigh as metallic platinum. Subtract a blank which is determined by adding zinc to a hydrochloric acid solution, pH 1.3, heating until all visible reaction has ceased, then boiling, filtering, and igniting as previously described.

THE USE OF SODIUM NITRITE AS A SEPARATORY REAGENT
FOR GOLD AND BASE METALS FROM SOLUTIONS OF
PLATINUM METALS

PROCEDURE 117 (412)

Treat 0.2–0.4 g of the alloy with aqua regia until it is completely dissolved, or chlorinate the alloy by an appropriate method. If the former treatment accomplishes only partial dissolution, filter, burn, and chlorinate the residue. Add the resulting acid solution and 0.1 g of sodium chloride to a 400-ml beaker, and evaporate to dryness on a steam bath. Add a few milliliters of concentrated hydrochloric acid, and evaporate to dryness. Repeat this procedure to remove the nitric acid completely. Add 50 ml of water and a few drops of concentrated hydrochloric acid, and warm to dissolve the residue. Filter through a 7-cm Whatman filter paper, and wash well with water. Transfer the paper to a suitable crucible, and ignite to an ash. Add a few milliliters of aqua regia, and evaporate in the presence of a few milligrams of sodium chloride. Add a few drops of hydrochloric acid, and evaporate carefully to dryness. Repeat once or twice, then add a few milliliters of water and a drop of hydrochloric acid, filter into the original solution, and wash to make up to about 100 ml.

The precipitation of gold and base metals.   Heat the slightly acid filtrate to 60°C, and add dropwise a saturated solution of sodium nitrite until the evolution of gas ceases and the solution becomes a light yellow-green; insufficient nitrite will result in a subsequent precipitation of rhodium. Boil the mixture for 15–30 min to ensure the coagulation of the gold. At this stage precipitated copper may stain the beaker wall. Cool the mixture to 60°C, add a few drops of phenolphthalein, and neutralize to a permanent pink by adding a 10% sodium carbonate solution. Continue heating on a water bath at 60°C for 15 min, then filter through a Whatman No. 42 filter paper previously washed with a 1% solution of sodium nitrite made neutral to the blue end point of thymol blue. Wash the precipitate with 50 ml of the 1% nitrite solution. Then wash thoroughly with warm water, in order to ensure the absence of nitrite, which may dissolve some gold in the presence of acid. Retain the filtrate and washings for the determination of the platinum metals. Add to the mixed precipitate of gold and base metal hydrated oxides a solution of 0.1 N nitric acid, and allow to drain. Wash again with hot water and repeat the selective dissolution with nitric acid until the base metals are removed. The leaching acid can be treated to determine base metals by any conventional method.

Ignite the gold residue and weigh as the metal.

Note.   If it is desired, the filtration of the mixed gold and base metals can be accomplished by using porous filtering crucibles with a suitable arrangement to collect the filtrate. It should be noted that the procedure described here is based on the assumption that the nitrite precipitation effectively removes gold and separates the palladium. Gilchrist (413) stated that the partial instability of the palladium–nitrite complex at a pH of 8–9 results in the retention of some palladium by the hydrated oxides of the base metals. Furthermore it is stated that the selective dissolution of the base metals to isolate the gold from the mixed precipitate may result in some dissolution of the gold if the nitrite is not removed completely from the mixed precipitate.

In order to recover coprecipitated platinum, rhodium and palladium and the gold in the acid extract of base metal oxides the latter are reprecipitated after an evaporation of the acid extract, dilution, adjustment of the acidity to pH 1.5 (red-orange end point of thymol blue), addition of sodium nitrite, and neutralization of the boiling solution by sodium hydroxide to the blue end point of thymol blue (pH 8). The second nitrite filtrate and washings are combined with the first nitrite filtrate, and will contain the platinum metals. The second acid dissolution of the hydrated oxides of the base metals allows the recovery of the traces of gold. This recovered gold is added to the first gold precipitate. This second acid extract is treated with dimethylglyoxime to recover palladium prior to the determination of such base metals as iron, copper, zinc, and nickel.

Determination of Rhodium

PROCEDURE 118 (414)

Reagent

2-Mercapto-4,5-dimethylthiazole. First recrystallize from 25% ethanol. Dissolve 0.5 g of the recrystallized material in 100 ml of 50% ethanol. This solution is colorless and stable for several months if kept in a dark bottle. In colorless bottles, the solution will turn yellow within 2 weeks. 2-Mercapto-4,5-dimethylthiazole can be prepared by the method of Buchman *et al.* (415).

Procedure

Transfer the rhodium sample, either as a chloride or sulfate, to an Erlenmeyer flask, add 10 ml of concentrated hydrochloric acid, and dilute to about 40 ml with water. Heat the solution to boiling, add 2 ml of reagent solution per microgram of rhodium per milliliter expected, and boil the solution vigorously for 1 hr. Keep the volume approximately constant by adding distilled water. Cool the samples in running water, make up the volume to 100 ml, and measure the absorbance at 430 μm (Ilford No. 601 filter) against a reagent blank.

Determination of Iridium

PROCEDURE 119 (416)

Reagents

Tin(II) chloride solution. Dissolve 25 g of tin(II) chloride dihydrate in 75 ml of concentrated (about 40%) hydrobromic acid.

Hydrobromic acid. Redistill to remove impurities such as bromine.

Procedure

Transfer the iridium solution to a 20 ml test tube, and dilute with 5 ml of water. Add 5 ml of concentrated hydrobromic acid, and place the tube in a boiling water bath. After 10 min, add 5 ml of the tin(II) chloride solution. Remove the test tube from the water bath exactly 2 min after the addition of the reagent, and immediately cool the tube and its contents under a stream of cold water. Transfer the solution to a 25.0-ml volumetric flask, and dilute to volume with water. Measure the absorbance at 402 μm against a reagent blank.

Determination of Ruthenium

PROCEDURE 120 (417)

Reagents

Thiocyanate solution. Dissolve 81 g of sodium thiocyanate in distilled water, and make up to 1 liter.

Aluminum nitrate solution. Add 452 g of aluminum nitrate nonahydrate to water, and then add 12.6 ml of concentrated nitric acid. Dilute to 1 liter.

Standard ruthenium solution. Ruthenium nitrosohydroxide, $Ru(NO_3)(OH)_2$ is available from A.D. MacKay, Inc., New York, New York. Dissolve appropriate amounts in 1 M nitric acid, and dilute to volume. This solution may be standardized by distillation and the application of thionalide (Procedure 63) to the distillate.

Silver(II) oxide. This is available from Merck and Co., Inc.

Procedure

Transfer the sample of a sulfuric or nitric acid solution of ruthenium containing from 20 to 100 μg of metal to a separatory funnel containing about 5 ml of the aluminum nitrate solution. Add 30 mg of silver(II) oxide and then 10 ml of carbon tetrachloride. Shake for 2 min, and immediately after the separation of the layers transfer the organic layer to a second separatory funnel containing 10.0 ml of the 1 M sodium thiocyanate solution. Repeat the extraction with a further 5 ml of carbon tetrachloride, and combine the extracts. Shake for 1 min, then drain the sodium thiocyanate layer into a centrifuge cone. Centrifuge to separate any mechanically mixed residue. Allow the color to develop for 30 min, then measure the absorbance at 590 μm against a blank.

Prepare a standard curve in this way.

CUPELLATION OF THE LEAD ASSAY BUTTON AND PARTING THE SILVER BEAD

PROCEDURE 121

Remove adhering slag from the lead button by gentle tapping with a small iron rod. Transfer the clean button to a bone-ash cupel that has been preheated at 900°C for at least 10 min. Heat at about 1000°C for 5 min, and then arrange a plentiful supply of air, but avoiding an excess that would cause the button to freeze through excessive cooling. Then continue heating at about 1000°C.

Note. The lead button may contain a higher proportion of base metals than is normally found with other methods of assay. In some instances the addition of about 1 g of flour to the charge prior to fusion will assist in providing a clean fusion.

Add to a 5–20 ml porcelain crucible about 4 ml of 1:4 nitric acid solution. Heat to incipient boiling, and transfer the bead to the acid. Continue careful heating to the completion of the reaction. Decant the acid, add a few milliliters of 1:1 nitric acid, and continue heating for 5–10 min. Decant the acid, and wash with water to remove the silver nitrate. Transfer the crucible with the gold to a cold muffle, and raise the temperature to about 700°C. Cool and weigh the gold bead. The accuracy, as indicated by salted samples, including both fusion and cupellation, is about 0.2%.

The Separation of Palladium from Lead Alloys

It is a surprising fact that no proven wet procedure has been recorded for the determination of all of the platinum metals in a lead button, despite the advantage that osmium is thus collected quantitatively and that a greater accuracy can be attained for at least the more insoluble platinum metals, which do not alloy with the silver bead. The latter is a far from satisfactory collector for osmium. Even small amounts of osmium(VIII) oxide are emitted from a silver bead with explosive force, and large amounts will cause a complete and violent disintegration of the bead.

During the parting of the lead button with nitric acid most of the palladium is dissolved, as are traces of platinum. The extent of dissolution of these and other platinum metals has not been recorded, but one may expect that the proportions dissolved will vary with the parting technique, the strength of acid and the metallurgical history of each alloy. The lead assay button will generally contain small amounts of iron, copper, and nickel. Irrespective of the technique of parting, appreciable proportions of palladium are attacked by nitric acid, and the separation of palladium from large proportions of lead is always a necessity.

For amounts of palladium between 10^{-3} and 10^{-6} g one may use solvent extraction by chloroform, followed by a spectrophotometric determination. For larger amounts the selective precipitation by salicylaldoxime is applicable. With the solvent extraction method there is no interference from platinum, nickel, and copper. As little as 3 mg of iron prevents the extraction of 60% of the palladium from a sample containing 250 μg of palladium. The interference is masked by the addition of EDTA, 0.5 g of EDTA complexing 50 mg of iron.

PROCEDURE 122 (418)

Transfer the clean button to a 250-ml beaker, add 90 ml of 1:2 nitric acid, and place on a steam bath until parting is complete. Evaporate to dryness, and add 100 ml of water containing 3 or 4 drops of concentrated nitric acid. Transfer the solution to a 250 ml pear-shaped separatory funnel, and add 0.5 g of EDTA, and 3 ml of a 1% aqueous solution of sodium dimethylglyoximate. Shake well and allow to stand for 1 hr. Extract the parting solution with four portions of chloroform (50, 25, 15, and 10 ml, respectively) (see Note 1). Collect the chloroform extracts in a 125-ml conical flask, and evaporate on a steam bath.

Add 12 ml of a 1:1 solution of concentrated nitric and sulfuric acids, and evaporate on a hot plate to fumes of sulfur trioxide. Add a few milliliters of water, and evaporate; repeat this three times. Add 25 ml of water, cool, and filter into a 100-ml volumetric flask. Wash the filter paper well with water. Add the potassium iodide reagent (500 g of potassium iodide and 5 ml of concentrated ammonia per liter), 1 ml per 100 μg of palladium. Add 5 ml of a sodium sulfite solution (1.5 g of anhydrous sodium sulfite in 250 ml of water). Dilute to volume with water (see Note 2). Measure the absorbance of the solution at 408 μm, the optimum range of concentration being 1–10 ppm of palladium metal when the solution is contained in 1-cm cells (see Note 3). Prepare the standard curve from blank lead buttons or the equivalent lead nitrate solution.

Notes. **1.** Three extractions are usually sufficient but the fourth is included as a precautionary measure. With >0.25 mg of palladium a precipitate of palladium dimethylglyoximate may form, but this does not hinder the extraction.

2. The sulfite effectively prevents the usual development of iodine in acid iodide solutions. The stability of the iodide solution is thus extended to more than 18 hr.

3. In general the iodide method is not a recommended spectrophotometric procedure because a wide variety of constituents will liberate iodine. This method, however, is particularly suitable for the determination of traces of palladium in lead buttons. After solvent extraction the maximum color develops immediately for amounts of palladium up to 1 mg.

REFERENCES

1. R. C. Mallett, C. G. Pearton, E. J. Ring, and T. W. Steele, *Talanta* **19** (2) (1972) 181–195.
1a. G. G. Carlson and J. C. Van Loon, *Atomos Absorp Newsl.* **9,** No. 4 (1970) 90–91.
2. T. P. Mikhailova and V. A. Rezepina, *Analyst (London)* **95,** No. 1134 (1970) 769–775.
3. T. P. Mikhailova *et al., Izv. S. B. Otdel Akad. Nauk SSR Ser Khim Nauk* No. 2 (1970) 170–110.
4. R. G. Mallett, D.C.G. Pearton, and E. J. Ring, *Johannesburg, Natl. Inst. Metall.* Rep. No. 1086 (1970).
5. R. G. Mallett, D.C.G. Pearton, and E. J. Ring, *Johannesburg Natl. Inst. Metall.,* Rep. No. 970 (1970).
6. A. Jansen, and F. Umland, *Fresenius Z. Anal. Chem.* **251,** No. 2 (1970) 101–107.
7. A. E. Pitts, J. C. Van Loon, and F. E. Beamish, *Anal. Chim. Acta* **50,** No. 2 (1970) 195–199.
8. M. M. Schnepfe and F. S. Grimaldi, *Talanta* **16,** No. 11 (1969), 1461–1465.
9. J. C. Van Loon, *Atomos. Absorp. Newsl.* **8** (1) (1969), 6–7.
10. F. S. Grimaldi, and M. M. Schnepfe, *Talanta* **17,** No. 7 (1970) 617–621.
11. R. C. Mallett, and R. L. Breckenridge, Johannesburg, Natl. Inst. Metall., Rep. No. 1318 (1971).
12. W. B. Rowston, and J. M. Ottaway, *Anal. Lett.* **3,** No. 8 (1970) 411–417.
13. J. C. Van Loon, *J. Fresenius Anal. Chem.* **246,** (1969) 122–124.
14. C. Huffman, J. D. Mensik, and L. B. Riley, U.S. Geol. Survey, Circ. 544 (1967).

15. A. E. Pitts and F. E. Beamish, *Anal. Chim. Acta* **52**, (1970) 405–415.
16. E. W. Fowler *et al.*, Johannesburg Natl. Inst. Metall., Rep. No. 970 (1970).
17. M. A. Hildon and G. R. Sully, *Anal. Chim. Acta* **54**, (1971) 245–251.
18. M. Boncetta and J. Fritshe, Dosage Elem. Etat. Trace Roches Antres Subst. Mines. Natur. Actes. Coll. (1968) No. 26308.
19. N. Ichinose, *Bunseti Kagaku* **20**, (1971) 660–665.
20. C. E. Thompson, N. H. Nakagawa, and G. J. Vansickle, Washington, U.S. Geol. Survey, Prof. Paper 600 B, pp. 130–132 (1968).
21. T. T. Chao, *Econ. Geol.* **64**, (1969) 287–290.
22. M. P. Bratzel, C. L. Chakrabarti, R. E. Sturgeon, M. W. McIntyre, and H. Agemian, *Anal. Chem.* **44**, (1972) 372–374.
23. F.W.E. Strclow, E. C. Feast, P. M. Mathews, C.J.C. Bothma, and C. R. Van Zyl, *Anal. Chem.* **38**, (1966) 115.
24. A.E. Pitts, F. E. Beamish, and J. C. Van Loon, *Anal. Chim. Acta* **50**, (1970) 181.
25. T. W. Oslinski, and N. H. Knight, *Appl. Spectrosc.* **22**, (1968) 532.
26. W. Luecke and H. J. Zieke, *Fresenius Z. Anal. Chem.* **253**, No. 1, (1971) 20–23.
27. J. Matonsek, and V. Sychra, *Anal. Chim. Acta* **49**, (1970) 175–181.
28. M. W. Skougstad and G. F. Scarbro, *Environm. Sci. Technol.* **2**, (1968) 298.
28a. M. P. Bratzel, Jr., C. L. Chakrabarti, R. E. Sturgeon, M. W. McIntyre, and H. Agemian, *Anal. Chem.* **44**(2), (1972) 372–374.
29. J. Aggett and T. S. West, *Anal. Chim. Acta* **55**, (1971) 349–357.
30. L.R.P. Butler, A. Strasheim, and F.W.E. Strclow, *Proc. Colloq. Spectrosc. Int. Exeter, 12th* p. 289 (1965).
31. F. J. Flanagan, *Geochim. Cosmochim. Acta* **16**, (1969) 591.
32. E. E. Bugbee, "A Textbook on Fire Assaying," 3rd ed. Wiley, New York, 1940.
32a. R. C. Mallett and R. L. Breckenridge, Nat. Inst. Metall. S. Afr. Rep. NIM-1318 (1971).
32b. M. Kruger Magdalena and R.V.D. Robert, Nat. Inst. Metal., Repub. S. Afr. Rep. No. 1432 (1972).
32c. N. L. Fishkova, *Anal. Technol. Blagorod. Metal.* (1971) 248–253.
33. M. M. Kruger and R.U.D. Robért, Nat. Inst. for Metall., Rep. No. 1432 (July 1972).
34. S. J. Fothergill, D. F. Withers, and F. S. Clements, *Brit. J. Med.* **2**, (1945) 199.
35. V. P. Khrapai, *Mater. Pervoga Ural. Soveshchaniya Po Spektroskopii* (Sverdlovsk: Gosvdarst. Nauch. Tekh. Izdatel. Lit. Chernoi tsvotnoi Met.), Sbornik 1956, 90–6 (publ. 1958).
36. V. P. Khrapai, *Anal. Blagorod. Metal., Akad. Nauk S.S.S.R., Inst. Obshchei i Neorg. Khim. im. N. S. kurnakova* (1959) 128–132.
37. A. Strasheim, D. B. de Villers, and D. Brink, *J. S. Afr. Inst. Min. Met.* **62**, (1962) 728.
38. N. Tomingas and W. C. Cooper, *Appl. Spectrosc.* **11**, (1957) 164.
39. K. Naka and S. Naka, *Bunko Kenskyu* **11**, (1963) 257.
40. Yu. Yokoyama and J. P. Faris, U.S.A.E.C. Rep. ANL-6913 (1964).
41. B. M. Talalaev, *Zh. Anal. Khim.* **19**, (1964) 1163.
42. B. M. Talalaev, *Zh. Anal. Khim* **21**(12), (1966) 1443–1446.
43. V. P. Khrapai, *Anal. Blagorod. Met. Moscow* (1965) 122–123.
44. A. D. Gut'Ko, G. V. Tkacheva, N. I. Panratova, Z. N. Kozyaeva, and N. K. Vorb'ev, *Anal. Blagorod. Met. Moscow* (1965) 114–8.
45. K. E. Egizbaeva, S. K. Kalinen, and E. E. Fain, *Jh. Pripl. Spektrosk.* **7**(6), (1967) 924–927.
46. L. E. Berenshtein, D. M. Livshits, and O. B. Fal'Kova, *Tr. Tsent. Nauch.-Issled. Gornorazved. Inst.* no. 77, (1967) 239–247. From *Referat. Jh. Met.* (1968) Abstr. No. 3K15.
47. D. M. Livshits, *Tr. Tsent. Nauch.-Issled. Gornorazved. Inst.* No. 70 (1967) 176–188 (Russ.); *Referat. Zh. Met.* (1968) Abstr. No. 4K8.

48. I. Kawashima, T. Miyazaki, and I. Tanaka, *Bunko Kenkyu* **16**(2), (1967) 68–77 (Japan).
49. S. L. Terekhovich, *Zavod. Lab.* **34**(4), (1968) 426.
50. E. A. Smith, "The Sampling and Assay of the Precious Metals," 2nd ed. Griffin, London, 1947.
51. H. de Lazlo, *Ind. Eng. Chem.* **19**, (1927) 1366–1368.
52. H. Schneiderhohn, *Chem. Erde* (1929) 252.
53. Ida and W. Noddack, *Naturwissenschaften* **18**, (1930) 757.
54. H. Schneiderhohn and H. Moritz, *Siebert Festschr.* (1931) 257–285.
55. A. Iwamura, *Mem. Coll. Sci. Univ. Kyoto Ser. A* **15**, (1932) 359.
56. W. Noddack, *Metall. Erz* **29**, (1932) 67.
57. V. M. Goldschmidt and C. Peters, *Nachr. Akad. Wiss. Goettingen, II. Math.-Phys. Kl.* **2**, (1932) 377.
58. C. Peters, *Metallwirtschaft* **12**, (1933) 17.
59. V. M. Goldschmidt and C. Peters, *Nachr. Akad. Wiss. Goettingen, II. Math.-Phys. Kl.* **2**, (1933) 371.
60. V. V. Nedler, *J. Tech. Phys. (U.S.S.R.)* **6**, (1936) 1138–1143.
61. J. Seath and F. E. Beamish, *Ind. Eng. Chem., Anal. Ed.* **10**, (1938) 535–537.
62. K. Toisi, *Sci. Papers Inst. Phys. Chem. Res. (Tokyo)* **38**, (1940) 87.
63. P. Pardo, *Afinidad* **18**, (1941) 257.
64. P. Pardo, *Anal. Real Soc. Espan. Fis. Quim. (Madrid)* **37**, (1951) 321.
65. V. V. Nedler and F. M. Efendiev, *Zavodskaya Lab.* **10**, (1941) 164.
66. V. V. Nedler, *Tr. Vsesoyuz. Konferentsii Anal. Khim. Acad. Nauk SSSR* **1**, (1939) 385–391.
67. J.M.L. de Azcona and P. Pardo, *Spectrochim. Acta* **2**, (1942) 185–201.
68. A. G. Scobie, *Trans. Can. Inst. Mining Met.* **58**, (1945) 309; *Anal. Chem.* **15**, (1943) 79–80.
69. C. Minguzzi, *Atti. Soc. Toscana Sci. Nat. (Pisa)* **54**, (1947) 34 pp.
70. J. E. Hawley, C. L. Lewis, and W. J. Wark, *Econ. Geol.* **46**, (1951) 149–162.
71. J. E. Hawley, Y. Rimsaite, and T. V. Lord, *Trans. Can. Inst. Mining Met.* **66**, (1953) 19.
72. J. R. Churchill and R. G. Russell, *Ind. Eng. Chem. Anal. Ed.* **17**, (1945) 24–27.
73. J. E. Hawley and Y. Rimsaite, *Am. Mineral.* **38**, (1953) 463.
74. V. L. Ginzburg, I. N. Gramenitskii, S. E. Kashlinskaya, and D. M. Livshits, *Izv. Akad. Nauk SSSR Ser. Fiz.* **19**, (1955) 211–216.
75. I. N. Maslenitskii, *Zapiski Leningrad. Gorn. Inst.* **32**, (1956) 235–248.
76. C. L. Lewis, *Can. Mining Met. Bull.* No. 539 (1957) 163–167.
77. D. M. Livshits and S. E. Kashlinskaya, *Zh. Anal. Khim.* **12**, (1957) 714–716.
78. V. Ya. Van'kin, V. L. Ginzburg, Yu. L. Polyakin, I. N. Gramenitskii, E. E. Kashlinskaya, and D. M. Livshits, U.S.S.R. Pat 108,985 (1957).
79. O. I. Bufatin, A. N. Zaidel, and N. I. Kaliteevskii, *Zh. Anal. Khim.* **13**, (1958) 116.
80. N. F. Losev, *Sbornik Nauch. Trudov. Irkutsk. Nauch.-Issledovatel. Inst. Redkikh Metallov* No. 7 (1958) 25–28; *Referat. Zh. Khim.* (1959) Abstr. No. 45563.
81. N. I. P'yankov, *Materialy. Pervogo Ural. Soveshchaniya po Spektroskopii (Sverdlovsk: Gosudarst. Nauch.-Tekh. Izdatel. Lit. Chernoi i Tsvetnoi Met.) Sbornik* (1956) 97–101 (Publ. 1958).
82. R. R. Brooks and L. H. Ahrens, *Spectrochim. Acta* **16**, (1960) 783–788, (in English); *Chem. Abstr.* **55**, (1961) 223.
83. M. Miyamoto, *Nippon Kagaku Zasshi* **82**, (1961) 686.
84. V. L. Ginzburg, *Tr. Tsentr. Nauchn.-Issled. Goronorazved. Inst.* (57) (1963) 111–122.
85. P. R. Barnett, D. L. Skinner, and C. Huffman, U.S. Geol. Survey, Prof. Paper No. 600-C, pp. 161–163 (1968).
86. J. Haffty and L. B. Riley, *Talanta* **15**(3), (1968) 111–117.
87. A. B. Whitehead and H. H. Heady, *Appl. Spectrosc.* **24**(2), (1970) 225–228.
88. A. F. Dorrzapf, Jr., and F. W. Brown, *Appl. Spectrosc.* **24**(4), (1970) 415–418.

89. A. H. Gillieson, Dept. of Energy, Mines and Resources, Ottawa, Personal communication.
90. F. E. Beamish and W.A.E. McBryde, *Anal. Chem.* **25,** (1953) 1613–1617.
91. F. E. Beamish, Private communication.
92. W. L. Ott and C. C. Cornett, Personal communication.
93. Standard Method for spectrochemical Analysis of Ores, Minerals and Rocks by the Fire Assay Preconcentration-Spark Technique, submitted (1965) to the ASTM by C. L. Lewis, W. L. Ott, and H. R. MacMillan. Accepted 1970; E400-70.
94. I. Hoffman, A. D. Westland, C. L. Lewis, and F. E. Beamish, *Anal. Chem.* **28,** (1956) 1174
95. W. F. Meggers, C. C. Kiess, and F. J. Stimson, *Sci. Papers Bur. Std.* 1922, **18,** No. 444 (1922) 235–255.
96. W. A. Gerlach and K. E. Schweitzer, *Z. Anorg. Chem.* **181,** (1929) 103.
97. W. A. Gerlach and K. Ruthardt, *Seibert Festschr.* (1931) 51.
98. W. A. Gerlach and E. Reidl, *Z. Phys.* **34,** (1933) 516.
99. H. E. Stauss, *Proc. Am. Soc. Testing Mater.* **35,** (1935) 57–60.
100. W. Rollwagen and K. Ruthardt, *Metallwirtschaft* **15,** (1936) 187.
101. A. Hauser, *Degussa Metall-Ber.* **1,** (1941) 210.
102. A. V. Babaeva, V. I. Belova, and S. A. Borovik, *C. R. Acad. Sci. URSS* **37,** (1942) 101–103.
103. H. J. Ishitsuka, *Soc. Chem. Ind. Jpn.* (1943) **46,** 258.
104. A. R. Raper and D. F. Withers, *Collected Papers on Metallurgical Analysis by the Spectrograph,* p. 144. Brit. Non-Ferrous Metals Res. Assoc. (1945).
105. P. van der Voort, *Bull. Soc. Chim. Belges* **54,** (1945) 57.
106. A. L. Kheifits and S. M. Katchenkov, *Bull. Acad. Sci. URSS Ser. Phys.* **11,** (1947) 301–302.
107. A. V. Babaeva, V. I. Belova, and S. A. Borovik, *Izvest. Sektora Platiny i Drug. Blagorod. Metal., Inst. Obshchei i Neorg. Khim., Akad. Nauk S.S.S.R.* No. 20, (1947) 168–171.
108. A. V. Babaeva, V. I. Belova, and L. A. Nazarova, *Izvest. Sektora Platiny i Drug. Blagorod. Metal., Inst. Obshchei i Neorg. Khim., Akad Nauk S.S.S.R.* No. 20, (1947) 172–175.
109 A. V. Babaeva and E. S. Lapir, *Izvest. Sektora Platiny i Drug. Blagorod. Metal., Inst. Obshchei i Neorg. Khim., Akad. Nauk S.S.S.R.* No. 22, (1948) 145–148.
110. A. V. Babaeva and E. S. Lapir, *Izvest. Sektora Platiny i Drug. Blagorod. Metal., Inst. Obshchei i Neorg. Khim., Akad. Nauk SSSR* No. 23 (1949) 94–96.
111. W. Pasveer, *Congr. group. avance. method. anal. spectrog. produits met.* **12,** (1949) 125–127.
112. A. L. Kheifits and S. M. Katchenkov, *Izv. Akad. Nauk S.S.S.R., Ser. Fiz.* **11,** (1950) 696.
113. D. F. Withers, *in* "Metal Spectroscopy" (F. Twyman, ed.), pp. 446–57. Griffin, London, 1951.
114. J. E. Hawley, W. J. Wark, C. L. Lewis, and W. L. Ott, *Trans. Can. Inst. Mining Met.* **54,** (1951) 669.
115. H. Oberlander, *Heraeus Festschr.* (1951) 169.
116. D. M. Smith, *Met. Ital.* **43,** (1951) 121.
117. C. L. Lewis and W. L. Ott, *Trans. Can. Inst. Mining Met.* **56,** (1953) 17.
118. W. Koehler, *Spectrochim. Acta* **4,** (1950) 229–232.
119. C. L. Lewis, W. L. Ott, and J. E. Hawley, *Trans. Can. Inst. Min. Met.* **58,** (1955) 208.
120. B. Vorsatz, *Congr. Group. Avance Methodes Anal. Spectrog. Produits Met.* (1957) 155.
121. N. I. Chentsova, *Mater. Pervogo Ural. Soveshch. Spektrosk. (Sverdlovsk: Gosudarst. Nauchn.-Tekh. Izdatel. Lit. Chernoi Tsvet. Met.) Sb.* (1956) 102–104 (Publ. 1958).
122. N. I. Pankratova, *Mater. 2-go [Vtorogo] Vses Soveshchan. Spectrosk. (Sverdlovsk: Gosudarst. Nauchn.-Tekh. Izdatel. Lit. Chernoi Tsvet. Met.) Sb.* (1958) 128 (Publ. 1959).
123. A. A. Kuranov, V. D. Ponomareva, and N. I. Chentsov, *Zh. Anal. Khim.* **15,** (1960) 476–480.
124. A. J. Lincoln and J. C. Kohler, *Anal. Chem.* **34,** (1962) 1247–1251.

125. A. D. Gut'ko, I. N. Firsova, and Z. I. Kozyaeva, *Spektral Anal. Geol. Geokhim., Mater. Sib. Soveshch. Spektrosk., 2nd (Irkutsk), USSR* (1963) 243–245 (Pub. 1967).

126. A. D. Gut'ko and N. M. Pyatkova, *Anal. Blagorod. Met., Moscow* (1965) 118–122.

127. W. Diehl, *Metall.* **19**, (1965) 712–714.

128. P. Tymchuk, D. S. Russell, and S. S. Berman, *Spectrochim. Acta* **21**, (1965) 2051–2062.

129. V. M. Vukanovic, M. R. Todorovic, V. J. Vajgand, and N. M. Svilar, *Glas. Hem. Drus., Beograd* **31**(2), (1966) 83–86.

130. J. Chwastowsk, R. Dybcznski, and B. Kucharzewski, *Chem. Anal. (Warsaw)* **13**(4), (1968) 721–735.

131. J. C. Kohler and A. J. Lincoln., *Engelhard Ind. Tech. Bull.* **10**, No. 3 (1969) 92–98.

132. R. E. Michaelis, U.S. Dept. of Commerce, Nat. Bur. of Std., Washington, D.C., Personal communication.

133. Methods for Emission Spectrochemical Analysis, E-130-57T, p. 36. Am. Soc. Testing Mater. (1960).

133a. F. E. Beamish, "Analytical Chemistry of the Noble Metals." Pergamon, Oxford, 1966.

134. G. H. Ayres and E. W. Berg, *Anal. Chem.* **24**, (1952) 465–469.

135. A. Bardocz and F. Varsanyi, *Magyar Kem. Folyoirat* **60**, (1954) 292.

136. A. A. Kuranov, *Anal. Blagorodn. Metal. Akad. Nauk SSSR Inst. Obshchei Neorg. Khim N.S. Kurnakova* (1959) 143–144.

137. A. A. Kuranov, N. P. Ruksha, and M. M. Svirodova, *Anal. Blagorodn. Metal. Akad. Nauk SSR Inst. Obshchei Neorg. Khim N.S. Kurnakova* (1959) 139–142.

138. A. A. Kuranov, *Izv. Akad. Nauk SSSR Ser. Fiz.* **23**, (1959) 1140–1143.

139. E. L. Gunn, *Anal. Chem.* **28**, (1956) 1433–1436.

140. W. M. MacNevin and E. A. Hakkila, *Anal. Chem.* **29**, (1957) 1019–1022.

141. A. J. Lincoln and E. N. Davis. *Anal. Chem.* **31**, (1959) 1317–1320.

142. R. Rabillon and R. Griffoul, *Rev. Univ. Mines* **15**, (1959) 536.

143. R. Neeb, *Z. Anal. Chem.* **179**, (1961) 21–29.

143a. R. B. Diocheva, *Dokl. Bulg. Akad. Nauk* **25**(5), (1972) 649–652 (Eng.).

144. B. K. Merejkovsky, *Bull. Soc. Chim. Biol.* **15**, (1933) 1336–1338.

145. N. S. Poluektov, *Tr. Vsesovuz. Konf. Anal. Khim* **2**, (1943) 393–398.

146. E. B. Sandell, *Anal. Chem.* **20**, (1948) 253–256.

147. S. Natelson and J. L. Zuckerman, *Anal. Chem.* **23**, (1951) 653–655.

148. S. Hara, *Bunseki Kagaku* **7**, (1958) 147–151.

149. T. M. Cotton and A. A. Woolf, *Anal. Chim. Acta* **22**, (1960) 192–194.

150. E. B. Sandell, "Colorimetric Determination of Traces of Metals," 3rd ed. Wiley (Interscience), New York, 1959.

151. W. Bettel, *Min. Eng. World* **35**, (1912) 987–988.

152. C. B. Brodigan, *Met. Chem. Eng.* **12**, (1941) 460.

153. W. S. Stanbury, *Tubercle* **13**, (1932) 396–399.

154. C. G. Fink and G. L. Putnam, *Ind. Eng. Chem. Anal. Ed.* **14**, (1942) 468–470.

155. H. I. Cole, *Philippine J. Sci.* **21**, (1923) 361–364.

155a. M. E. Yaskevich, L. I. Kononova, and N. Ya. Saveleva, *Electron. Tekh. Nauch-Tekh. Sb. Poluprov Prib.* No. 1, (1970) 162–167; *Referat. Zh. Khim.* (1970) Abstr. No. 23, G175.

156. A. Chow and F. E. Beamish, *Talanta* (1963) 983.

157. W.A.E. McBryde and J. H. Yoe, *Anal. Chem.* **20**(11), (1948) 1094.

158. H. Fischer and W. Weyl, *Wiss. Veroffentlich Siemens-Werken* **14ii**, (1935) 41–53.

159. B. Bleyer, G. Nagel, and J. Schwaibold, *Sci. Pharm.* **10**, (1939) 121–124.

160. F. T. Beaumont, *Metallurgia* **29**, (1944) 217–220.

161. M. Shima, *Kagaku Kenkyusho Hokoku* **32**, (1956) 152–157.

162. L. Erdey and G. Rady, *Z. Anal. Chem.* **135**, (1952) 1–10.

163. J. Adam and R. Pribil, *Talanta* **18**(4), (1971) 405–409.

164. G. V. Knyagheva, *Izv. Sektora Platiny y Drug. Blagorodn. Metal., Inst. Obshch. Neorg. Khim. Akad. Nauk SSSR* No. 22 (1948) 129–135.

165. G. H. Ayres and A. S. Meyer, Jr., *Anal. Chem.* **23**, (1951) 299–304.

166. G. H. Ayres and J. H. Alsop, *Anal. Chem.* **31**, (1959) 1135–1138.

167. F. Pantani and G. Piccardi. *Anal. Chim. Acta* **22**, (1960) 231–236.

168. L. G. Overholser and J. H. Yoe, *J. Am. Chem. Soc.* **63**, (1941) 3224–3229.

169. D. E. Ryan, *Analyst* **76**, (1951) 167–171.

170. J. H. Yoe and L. G. Overholser, *J. Am. Chem. Soc.* **61**, (1939) 2058–2063.

171. E. S. Przheval'skit, V. I. Shlenskaya, and I. S. Razina, *Vestnik Moskow Univ. Ses. Mat. Mekh. Astron. Fiz. Khim* **12**, No. 1 (1957) 111–116.

172. R. Gilchrist and E. Wichers, *J. Am. Chem. Soc.* **57**, (1935) 2565–2573.

173. J. H. Yoe and J. J. Kirkland, *Anal. Chem.* **26**, (1954) 1335–1340.

174. C. W. Davis, U.S. Bureau of Mines Tech. Paper 1921–270.

175. It. Wölbling, *Ber.* **67**, (1934) 773–776.

176. G. H. Ayres, *Anal. Chem.* **25**, (1953) 1622–1627.

177. A. S. Meyer, Jr. and G. H. Ayres, *J. Am. Chem. Soc.* **77**, (1955) 2671–2675.

178. O. I. Milner and G. F. Shipman, *Anal. Chem.* **27**, (1955) 1476–1478.

179. S. S. Berman and E. C. Goodhue, *Can. J. Chem.* **37**, (1959) 370–374.

180. G. H. Faye and W. R. Inman, *Anal. Chem.* **33**, (1961) 278–283.

181. H. G. Coburn, F. E. Beamish, and C. L. Lewis, *Anal. Chem.* **28**, (1956) 1297–1300.

182. E. Wagner, *Z. Anal. Chem.* **147**, (1955) 10–20.

183. Ya E. Shmulyakovskii, *Khim Tekhnol. Topliv Masel* **3**, No. 12 (1958) 56–58.

184. M. Struszyński and J. Chwastowska, *Chem. Anal.* **3**, (1958) 949–954.

185. N. A. Figurovskii, *Ann. Sect. Platine, Inst. Chim. Gen. USSR* (1938) 129–135.

186. N. S. Poluektov and F. G. Spivak, *Zavodskaya Lab.* **11**, (1945) 398–402.

187. R. S. Young, *Analyst* **76**, (1951) 49–52.

188. W. B. Pollard, *Analyst* **67**, (1942) 184–186.

189. M.E.V. Plummer and F. E. Beamish, *Anal. Chem.* (1959) 31, 1141–3.

190. R. Gilchrist, *J. Res. Nat. Bur. Std.* **20**, (1938) 745–771.

191. M.E.V. Plummer, C. L. Lewis, and F. E. Beamish, *Anal. Chem.* **31**, (1959) 254–258.

191a. B. S. Rabinovich, G. M. Baevskaya, and E. E. Rakovskii, *Inst. Tsvet. Redk. Blagorod Metal.* No. 97 (1971) 178–182; *Referat. Zh. Khim* No. 20 (1971) G81.

192. J. J. Kirkland and J. H. Yoe, *Anal. Chem.* **26**, (1954) 1340–1344.

193. A. I. Conrad and J. K. Evans, *Anal. Chem.* **32**, (1960) 46–49.

194. A. D. Westland and F. E. Beamish, *Am. Mineralog.* **43**, (1958) 503–516.

195. L. D. Johnson and G. H. Ayres, *Anal. Chem.* **38**, (1966) 1218.

196. R. Keil, *Fresenius Z. Anal. Chem.* **254**(3), (1971) 191–192.

197. G. H. Ayres and F. L. Johnson, Jr., *Anal. Chim. Acta* **23**, (1960) 448–457.

198. G. H. Ayres, B. L. Tuffly, and J. S. Forrester, *Anal. Chem* **27**, (1955) 230–238.

199. A. D. Maynes and W. A. E. McBryde, *Analyst* **79**, (1954) 230–238.

199a. M. E. Smith, *Anal. Chem.* **30**, (1958) 912–913.

200. R. D. Gardner and D. Hues, *Anal. Chem.* **31**, (1959) 1488–1489.

201. J. O. Karttunen and H. B. Evans, *Anal. Chem.* **32**, (1960) 917–920.

202. R. B. Cheneley, R. G. Osmond, and S. G. Perry, U. K. At. Energy Res. Estab 1956, C/S 1870.

203. J. J. Markham, Univ. Microfilms, Ann Arbor, Michigan. U.S.A., L.C. Card No. Mic 58-7013; *Diss. Abstr.* **19**, (1958) 1198–1199.

204. S. K. Kalinin and G. A. Yakovleva, *Zh. Anal. Khim.* **25**(2), (1970) 312–314.

205. S. S. Berman and R. Ironside, *Can. J. Chem.* **36**, (1958) 1151–1155.

206. G. H. Ayres and W. T. Bolleter, *Anal. Chem.* **29**, (1957) 72–75.

207. M. Ewen and E.B.T. Cook., Nat. Inst. Met. Republic S. Afr. Rep. No. 1179 (1971).

208. R. P. Yaffe and A. F. Voigt, *J. Am. Chem. Soc.* **74**, (1952) 3163–3165.

209. Z. Kolarik and C. Konecny, *Coll. Czech. Chem. Comm.* **25**, (1960) 1775–1779.

210. H. Wölbling and B. Steiger, *Mikrochemie* **15**, (1934) 295–301; H. Wölbling, *Ber.* **67B**, (1934) 773.

211. G. H. Ayres and F. Young, *Anal. Chem.* **22**, (1950) 1281–1283.

212. W. D. Jacobs and J. H. Yoe, *Talanta* **2**, (1959) 270–274.

213. V. I. Shlenskaya, *Vesnik Moskov. Univ., Khim. Ser. II* **15**, No. 2 (1960) 69–72.

214. B. Steiger, *Mikrochemie* **16**, (1934) 193–202.

215. R. P. Yaffe and A. F. Volgt, *J. Am. Chem. Soc.* **74**, (1952) 2503–2506.

216. A. T. Pilipenko and I. P. Sereda, *Ukr. Khim. Zh.* **27**, (1961) 257–60.

217. D. D. DeFord, Chemistry of Ruthenium, Oak Ridge, U.S. At. Energy Commission Document NP-1104 (Dec. 1949).

218. K. S. Bergstresser, U.S. At. Energy Comm. LA-2025. (1956)

219. G. H. Ayres and F. Young, *Anal. Chem.* **22**, (1950) 1277–1280.

220. A. D. Westland and F. G. Beamish, *Anal. Chem.* **26**, (1954) 739–741.

221. L. Tschugaeff, *C. R. Acad. Sci. Paris* **167**, (1918) 235.

222. E. B. Sandell, *Ind. Eng. Chem. Anal. Ed.* **16**, (1944) 242–243.

223. W. J. Allan and F. E. Beamish, *Anal. Chem.* **24**, (1952) 1608.

224. F. P. Dwyer and N. A. Gibson, *Analyst* **76**, (1951) 104–106.

225. G. H. Ayres and W. N. Wells, *Anal. Chem.* **22**, (1950) 317–320.

226. R. D. Sauerbrunn and E. B. Sandell, *Anal. Chem. Acta* **9**, (1953) 86–90.

227. W. Geilmann and R. Neeb, *J. Anal. Chem.* **152**, (1956) 96–107.

228. B. Steiger, *Mikrochemie* **16**, (1934) 193–202.

229. Z. Bardodej, *Chem. Listy* **48**, (1954) 1870–1871.

230. J. H. Yoe and L. G. Overholser, *Ind. Eng. Chem. Anal. Ed.* **14**, (1942) 435–437.

231. E. L. Steele and J. H. Yoe, *Anal. Chim. Acta* **20**, (1959) 205–211; *Anal. Chem.* **29**, (1957) 1622–1624.

232. K. Ishii, *J. Chem. Soc. Jpn.* **52**, (1931) 167–171.

233. P. Champ, P. Fauconnier, and C. Duval, *Anal. Chim. Acta* **5**, (1951) 277–81.

234. T. Kiba and T. Ikeda, *J. Chem. Soc. Jpn.* (1939) 60.

235. W. B. Pollard, *Analyst* **62**, (1937) 597–603.

236. G. Kemp, *Rep. Für. Pharm. (Buchner)* **24**, (1841) 235.

237. V. A. Magdalena, *Anales Assoc. Quim, Farm. Uruguay* **50**, (1948) 27–28.

238. L. S. Malowan, *Mikrochemie* **35**, (1950) 104–106.

239. L. Vanino and L. Seemann, *Ber.* **32**, (1899) 1968–1972.

240. F. E. Beamish, J. J. Russell, and J. Seath, *Ind. Eng. Chem. Anal. Ed.* **9**, (1937) 174–176.

241. J. Seath and F. E. Beamish, *Ind. Eng. Chem. Anal. Ed.* **9**, (1937) 373–374.

242. F. Hecht and G. Lamac-Brunner, *Mikrochemie* **35**, (1950) 390–399.

243. L. Duparc, *C. R. Soc. Phys. List. Nat. Genève* **29**, (1912) 20.

244. M. Wunder and V. Thüringer, *Ann. Chim. Anal.* **17**, (1912) 201.

245. M. Wunder and V. Thüringer, *Z. Anal. Chem.* **52**, (1913) 660–664.

246. A. Gutbier and C. Fellner, *Z. Anal. Chem.* **54**, (1915) 205–208.

247. C. W. Davis, *U.S. Bur. Mines. Rep. Invest.* (1922) 2351.

248. Z. E. Gol'braikh, *J. Gen. Chem. USSR* **14**, (1944) 810–811.

249. H. E. Zschiegner, *Ind. Eng. Chem.* **17**, (1925) 294.

250. H. Holzer, *Z. Anal. Chem.* **95**, (1933) 392–400.

251. V. M. Peshkova and V. I. Shlenskaya, *Vestn. Moskov. Univ.* **8**, No. 3; *Ser. Fiz.-Mat. Estestven. Nauk* No. 2, (1953) 129–133.

252. Yu. V. Morachevskii, I. A. Tserovnitskaya, and Z. G. Golubtsova, *Zh. Obshch. Khim.* **29**, (1959) 1405–1408.

253. P. Filott, R. K. Vitek, and K. L. Cheng, *Anal. Chim. Acta* **19**, (1958) 323–327.

254. A. G. Sharpe and D. B. Wakefield, *J. Chem. Soc.* (1957) 3323.

255. P. Champ, P. Fauconnier, and C. Duval, *Anal. Chim. Acta* **6,** (1952) 250–258.

256. M. Tashiro, *J. Chem. Soc. Jpn.* **52,** (1931) 232.

257. G. H. Ayres and E. W. Berg, *Anal. Chem.* **25,** (1953) 980–981.

258. L. J. Kannen, E. D. Salesin, and L. Gordon, *Talanta* **7,** (1961) 288–290.

259. F. E. Beamish and J. J. Russell, *Ind. Eng. Chem. Anal. Ed.* **8,** (1936) 141–144.

260. S. O. Thomson, F. E. Beamish, and M. Scott, *Ind. Eng. Chem. Anal. Ed.* **9,** (1937) 420–422.

261. F. E. Beamish and M. Scott, *Ind. Eng. Chem. Anal. Ed.* **9,** (1937) 460–463.

262. R. Gilchrist, *Bur. Std. J. Res.* **12,** (1934) 291–303.

242a. T. Fedorova, *Zh. Anal. Khim.* **27**(7), (1972) 1340–1343.

263. R. C. Voter, C. V. Banks, and H. Diehl, *Anal. Chem.* **20,** (1948) 652–654.

264. J. J. Berzellius, *Lehrb. Chem. (Dresden)* **21,** (1826), 180.

265. U. Antony and A. Lucchesi, *Gazz.* **261,** (1896) 211–218.

266. E. von Meyer, *J. Prakt. Chem. (2),* **16,** (1877), 1–22.

267. D. S. Jackson and F. E. Beamish, *Anal. Chem.* **22,** (1950) 813–817.

268. P. Champ, P. Fauconnier, and C. Duval, *Anal. Chim. Acta* **10,** (1954) 443–7.

269. F. Feigl, *Z. Anal. Chem.* **65,** (1924) 25–46; *Z. Anorg. Chem.* **157,** (1926), 269.

270. I. M. Kolthoff and E. A. Pearson, *J. Phys. Chem.* **36,** (1932) 549–566.

271. W. F. Hillerbrand, G.E.F. Lundell, H. A. Bright, and J. I. Hoffman, "Applied Inorganic Analysis," 2nd ed., p. 343 Wiley, New York. 1953.

272. F. P. Treadwell and W. T. Hall, *Anal. Chem.* **11,** (1942) 141.

273. R. Gaze, *Chem. Zentralbl.* **1,** (1913) 464.

274. V. N. Ivanov, *J. Russ. Phys. Chem. Soc.* **48,** (1916) 527.

275. R. Doht, *Z. Anal. Chem.* **64,** (1924) 37.

276. D. E. Ryan, *Can. J. Chem.* **34,** (1956) 1683–1686.

277. A. D. Westland and L. Westland, *Talanta* **3,** (1960) 364–369.

277a. D. E. Ryan, *Can J. Chem.* **40,** (1962) 570.

278. S. C. Surk and K. Shome, *Anal. Chim. Acta* **57**(1), (1971) 201–208.

279. W. W. Scott, "Standard Methods of Chemical Analysis," Vol. 1, p. 746. Van Nostrand-Reinhold, Princeton, New Jersey, 1946.

280. W. J. Allan and F. E. Beamish, *Anal. Chem.* **22,** (1950) 451–454.

281. I. K. Taimni and G.B.S. Salaria, *Anal. Chim. Acta* **11,** (1954) 329–338.

282. R. L. Haines and D. E. Ryan, *Can. J. Res.* **27B** (1949) 72–75.

283. L. Moser and H. Graber, *Monatsh* **59,** (1932) 61–72.

284. F. Mylius and A. Mazzucc Helli, *Z. Anorg. Chem.* **89** (1914) 1–38.

285. E. Wichers, *J. Am. Chem. Soc.* **46** (1924) 1818–1833.

286. R. Gilchrist, *Bur. Std. J. Res.* **9,** (1932) 547–556.

287. R. R. Barefoot, W. J. McDonnell, and F. E. Beamish, *Anal. Chem.* **23,** (1951) 514–516.

288. W. D. Jacobs and J. H. Yoe, *Talanta* **2,** (1959) 270–274.

289. J. F. Flagg, "Organic Reagents," Wiley (Interscience), New York, 1948.

290. W. J. Rogers, F. E. Beamish, and D. S. Russell, *Ind. Eng. Chem. Anal. Ed.* **12,** (1940) 561–563.

291. Q. E. Zvyagintsev, *Issled. V. Oblas. Geol. Khim. Met. Sb.* (1955), 216–218; *Zh. Khim* (1956), Abstr. No. 39314.

292. R. Gilchrist, *Bur. Std. J. Res.* **3,** (1929) 993, 1004.

293. R. Gilchrist, *Bur. Std. J. Res.* **6,** (1931) 421–448.

294. W. Geilmann and R. Neeb, *Z. Anal. Chem.* **156,** (1957) 411–420.

295. F. J. Welcher, "Organic Analytical Reagents," p. 289. Van Nostrand-Reinhold, Princeton, New Jersey, 1948.

296. I. Hoffman, J. E. Schweitzer, D. E. Ryan, and F. E. Beamish, *Anal. Chem.* **25,** (1953) 1091–1094.

297. H. Peterson, Z. Anorg. Chem. **19**, (1898) 59–66.
298. F. A. Gooch and F. H. Morley, Am. J. Sci. **8**, (1899) 261–266.
299. E. Rupp, Ber. **35**, (1902) 2011–2015.
300. V. E. Herschlag, Ind. Eng. Chem. Anal. Ed. **13**, (1941) 561–563.
301. V. Lenher, J. Am. Chem. Soc. **35**, (1913) 733–736.
302. L. Vanino and F. Hartwagner, Z. Anal. Chem. **55**, (1916) 377–388.
303. L. Brüll and F. Griffl, Ann. Chim. Appl. **28**, (1938) 536–544.
304. L. Brüll and F. Griffi, Ric. Sci. **9**, (1938) 11, 622–623.
305. A. Chow, Private communication.
306. C.B.F. Young, Prod. Finish. **7**, (1943) 20, 22–24, 28, 30, 32.
307. E. A. Parker, Mon. Rev. Am. Electroplaters Soc. **34**, (1947) 33–40.
308. E. M. Sudilovskaya, Zavod. Lab. (1950) 1312–1316.
309. A. R. Jamieson and R. S. Watson, Analyst (1938) 702–704.
310. O. E. Zvagintsev, Zolotaya Prom. **3**, (1939) 36.
311. O. E. Zvagintsev, S. K. Shabarin, V. A. Yorob'eva, and A. P. Bockkareva, Khim. Referat. Zh. **2**, (1940) 63–64.
312. Z. P. Zharkova and E. I. Zhacheva, Nauchn.-Issl. Tr. Tsentral. Nauchno.-Issled. Inst. Vspomogat. Izdel. Zapas. Detaleik Tekstel. Oborud No. 4, (1956) 44–47; Referet. Zh. Met. (1957) Abstr. No. 11190.
313. R. Belcher and A. J. Nutten, J. Chem. Soc. (1951) 550–551.
314. I. A. Peshkov, Nauchn. Tr. Tul. Gorn. Inst. Sb. **1**, (1958) 229–238.
315. G. Milazzo, Rend. Ist. Super. Sanita (Italian Ed.) **11**, (1948) 801–816.
316. G. G. Tertipis and F. E. Beamish, Anal. Chem. **32**, (1960) 486–489.
317. E. I. Ryabchikov and G. V. Knyazheva, C. R. Acad. Sci. URSS **25**, (1939) 601–604.
318. A. Czaplinski and J. Trokowicz, Chem. Anal. (Warsaw) **4**, (1959) 463–469.
319. A. F. Bukharin, Sb. Tr. Vses. Nauchno. Issled. Gorno-Met. Inst. Tsvet. Met. **5**, (1959) 209–220.
320. A. W. Titley, Analyst **87**, (1962) 349.
321. R. Kersting, Ann. **87**, (1853) 25.
322. N. K. Pshenitsyn and S. I. Ginzburg, Izv. Sektora Plating i Drug Blagorod Metal., Inst. Obshcheu i Neorg. Khim, Acad. Nauk S.S.S.R. **25**, (1950) 192–199.
323. N. K. Pshenitsyn and S. I. Ginzburg, Sekt. Plat. Drug. Blagorodn. Met. Inst. Obshch Neorg Khim. Acad. Nauk SSSR **32**, (1955) 31–37.
324. R. H. Atkinson, Analyst **79**, (1954) 368–370.
325. R. H. Atkinson, R. N. Rhoda, and L. G. Lomell, Analyst **80**, (1955) 838–839.
326. R. N. Rhoda and R. H. Atkinson, Anal. Chem. **28**, (1956) 535–537.
327. A. I. Busev and M. I. Ivanyutin, Zh. Anal. **13**, (1958) 18–30.
328. S. Barabas and J. Uinaric, Anal. Chem. **36**, (1964) 2365.
329. W. B. Pollard, Trans. Inst. Min. Met. **47**, (1937–1938) 331–346.
330. W. B. Pollard, Trans. Inst. Min. Met. **48**, (1938–1939) 65.
331. G. H. Ayres, and P. W. Glass, Anal. Chim. Acta **60(2)** (1972) 357–365.
332. D. I. Ryabchikov, Izv. Sekt. Platiny Drugikh Blagorodn. Met. Inst. Obshch. Neorg. Khim. Akad. Nauk SSSR No. 22 (1948) 35–42.
333. D. I. Ryabchikov, Zh. Anal. Khim. **1**, (1946) 47–56.
334. G. Milazzo and L. Paoloni, Rend. Ist. Super Sanita (Italian Ed.) **12**, 693–704.
335. N. K. Pshenitsyn and I. V. Prokof'eva, Akad. Nauk SSSR **28**, (1954) 229–234.
336. S. G. Bogdanov and S. E. Krasidov, Izv. Sekt. Platiny Drugikh Blagorodn. Met. Inst. Obshch. Neorg. Khim. Akad. Nauk SSSR No. 16, (1939) 77–80.
337. W. B. Pollard, Bull. Inst. Min. Met. No. 497 (1948) 9–17.
338. V. S. Syrokomskii, C. R. Acad. Sci. URSS **46**, (1945) 280–282.
339. L. Meites and R. E. Cover, Anal. Chem. **40**, (1968) 209.

340. J. J. Lingane and R. L. Pecsok, *Anal. Chem.* **20,** (1948) 425.

340a. V. V. Lichadeev, A. K. Dement'eva, M. P. Lukicheva, and L. G. Krasnova, *Nauch. Tr. Sib. Nauch. Issled. Proekt. Inst. Tsvet. Met.* No. 4(2), (1971) 255–336.

341. A. Chow and E. E. Beamish, *Talanta* **14,** (1967) 222.

342. A. Strasheim and J. Van Wamelen, *J. S. Afr. Chem. Inst.* **15,** (1962) 60.

343. C. H. Fulton and W. J. Sharwood, "Manual of Fire Assaying." McGraw-Hill, New York, 1929.

343a. G. T. Georgiev and D. Apostolov, *Zh. Anal. Khim.* **27**(3), (1972) 506–511.

344. A. Chow and F. E. Beamish, *Talanta* **14,** (1967) 219–231.

345. J. M. Kavanagh and F. E. Beamish, *Anal. Chem.* **32,** (1960) 490–491.

346. J. C. Van Loon, *Anal. Chem.* **36,** (1964) 892; **37,** (1965) 113.

347. F. E. Beamish and J. C. Van Loon, "Recent Advances in the Analytical Chemistry of the Noble Metals." Pergamon, Oxford, 1972.

348. J. Seath and F. E. Beamish, *Ind. Eng. Chem.* (1940) 12.

349. J. G. Fraser and F. E. Beamish, *Anal. Chem.* **26,** (1954) 1474–1477.

350. I. Hoffman and F. E. Beamish, *Anal. Chem.* **28,** (1956) 1188–1193.

351. L. M. Banburg and F. E. Beamish, *Anal. Chem.* 211, Band 3 Heft 178–187.

351a. V. V. Lichadeev, R. M. Nazarenko, and M. P. Lukicheva, *Nauch. Tr. Sib. Nauch. Issled. Proekt. Inst. Tsvet. Met.* No. 4(2), (1971) 57–68.

352. M.E.V. Plummer, J. M. Kavanagh, J. C. Hole, and F. E. Beamish, *Met Soc. AIME* **221,** (1961) 145–151.

353. R.V.D. Robert, E. VanWyk, and R. Palmer, Natl. Inst. Metal. Republick S. Africa, Rep. 1971, No. 1341, 1487.

354. E. A. Jones, M. M. Kruger, and A. Wilson, Natl. Inst. Metal. Johannesburg South Africa, Rep. 1971 No. 1232.

355. G. H. Faye and P. E. Moloughney, *Tantala* **23,** (1976) 377.

356. G. H. Faye, Mineral Sciences Division p. 17 (April 1965).

357. J. C. Sen Gupta and F. E. Beamish, *Anal. Chem.* **34,** (1962) 1761.

358. A. S. Lee, F. E. Beamish, and M. G. Bapat, *Mikrochim. Acta* (1969) 329–344.

359. D. Fridman and N. N. Popova, *Tr. Tsentr. Nauchno.-Issled. Gorno-Razvedoch Inst.* **26,** (1958) 113–115.

360. J. B. Kushner, *Ind. Eng. Chem. Anal. Ed.* **11,** (1939) 223.

361. C. T. Creed and C. F. Clayton-Boxall, *J. Chem. Met. Min. Soc. S. Afr.* **33,** (1932) 190–192, 331, 398.

362. J. T. King and S. E. Wolfe, *Can. Min. J.* **59,** (1938) 6–8.

363. W. Bettel, *Min. Eng. World* **36,** 774, Abstracted in *Chem. Abstr.* **6,** (1912) 1414.

364. A. Woqrinz, *J. Anal. Chem.* **108,** (1937) 266–267.

365. C. E. Roodhouse, *Ind. Eng. Chem. Anal. Ed.* **10,** (1938) 641–642.

366. J. B. Kushner, *Ind. Eng. Chem. Anal. Ed.* **10,** (1938) 641–642.

367. C. Crichton, *J. Chem. Met. Min. Soc. S. Afr.* **12,** (1911–1912) 90.

368. A. Fraser, *Chem. Eng. Min. Rev.* **28,** (1936) 312.

369. E. A. Marenkov, *Zolotaya Prom.* **11,** No. 7 (1939) 30–31.

370. B. Shah, *Current Sci.* **9,** (1940) 73.

371. R. J. Rochat, *Plating* **36,** (1949) 817.

372. I. I. Kalinichenko and P. A. Sokolova, *Zh. Khim.* (1955) No. 9670; *Metody Anal. Chem. Tsvet. Metall.* (1953) 105–111.

373. L. C. Wilson, *Metal. Ind.* **14,** (1961) 378–379.

374. W. E. Caldwell and L. E. Smith, *Ind. Eng. Chem. Anal. Ed.* **10,** (1938) 318–319.

375. A. Chiddy, *Eng. Min. J.* **75,** (1903) 473.

376. R. V. Lundquist, *Eng. Min. J.* **141,** (1940) 51–52.

377. M. A. Hill and F. E. Beamish, *Anal. Chem.* **22**, (1950) 590–594.

378. E. Wichers, W. G. Schlecht, and C. L. Gordon, *J. Res. Nat. Bur. Std.* **33**, (1944) 364, 459, 461.

379. L. Ubaldini, *Proc. Int. Congr. Pure Appl. Chem. 11th, London* **1**, (1947) 293–295.

380. W. Gibbs, *Chem. News* **7**, (1863) 61, 73, 97.

381. N. K. Pshentsyn, I. A. Federov, and P. V. Simanovskii, *Izv. Lektora. Plat. Drug Blagorod. Met. Inst. Obshchei Neorg. Khim. Nauk SSSR* **22**, (1948) 22–27.

382. V. M. Mukhachev, *Javodskaya Lab.* **12**, (1946) 927–929.

383. S. Aoyama and K. Watanabe, *J. Chem. Soc. Jpn. Pure Chem. Sect.* **75**, (1954) 20–23; **76**, (1955) 597–602.

384. G. G. Tertipis and F. E. Beamish, *Anal. Chem.* (1962) 623–625.

385. R. Gilchrist, *Sci. Papers Bur. Std.* **19**, No. 483 (1924) 325–345.

386. R. Thiers, W. Graydon, and F. E. Beamish, *Anal. Chem.* **20**, (1948) 831.

387. R. Gilchrist, *Am. Chem. Soc. J.* **57**, (1931) 2565–2573.

388. R. Gilchrist, *Bur. Std. J. Res.* **12**, (1934) 283; R.P. 654.

389. S. M. Anisimov, P. G. Shulakov, V. N. Alyanchikova, V. M. Klypenkov, and P. A. Gurin, *Anal. Blagorod. Metal. Akad. Nauk SSSR Inst. Obshch. Neorg. Khim. I.M. N.S. Kurnakova* (1959) 88–102.

390. N.K. Pshenitsyn, K. A. Gladyshevaskaya, and L. M. Ryakhava, *Zh. Neorg. Khim.* **2**, (1957) 1057–1068.

391. K. Kimura, N. Ikeda, and K. Yoshihara, *Bull. Electrotech. Lab. Tokyo* **19**, (1955) 913–917.

392. M. V. Suvic, *Bull. Inst. Nucl. Sci. "Boris Kidrich"* **7**, (1957) 39–41.

393. F. E. Beamish, "The Analytical Chemistry of the Noble Metals." Pergamon, Oxford, 1966.

394. A. G. Marks and F. E. Beamish, *Anal. Chem.* **30**, (1958) 1464–1466.

395. H. Zachariasen and F. E. Beamish, *Anal. Chem.* **34**, (1962) 964–966.

396. A. E. Pitts and F. E. Beamish, *Anal. Chem.* **41**, (1969) 1107–1109.

396. I. Palmer, G. Streichert, and A. Wilson, Nat. Inst. for Metallurgy, Johannesburg, South Africa, No. 1218, Rep. No. 8 (1971).

397. P. Palmer and G. Streichert, Nat. Inst. for Metallurgy, Johannesburg, South Africa, No. 1273, Rep. No. 13 (1971).

398. J. Seath and F. E. Beamish, *Ind. Eng. Chem. Anal. Ed.* **10**, (1938) 639–641.

399. G. S. James, *S. Afr. Ind. Chem.* **15**, (1961) 62–68.

400. R. R. Barefoot and F. E. Beamish, *Anal. Chim. Acta* **9**, (1953) 49–58.

401. S. T. Payne, *Analyst* **85**, (1960) 698–714.

402. N. F. Kember and R. A. Wells, *Analyst* **80**, (1955) 735–751.

403. O. Konig and W. R. Crowel, *Mikrochemie* **33**, (1947) 300–302.

404. M. Lederer, *Nature (London)* **162**, (1948) 776–777.

405. D. B. Ress-Evans, W. Ryan, and R. A. Wells, *Analyst* **83**, (1958) 356–361.

406. F. H. Burstall and R. A. Wells, *Analyst* **76**, (1951) 396.

407. M. P. Volynets, A. N. Ermakov, L. P. Nikitina, and N. A. Ezerskaya, *Zh. Anal. Khim.* **25(4)**, (1970) 759.

408. M. P. Volynets, A. N. Ermakov, and L. P. Nikitina, *Zh. Anal. Khim* **25(2)**, (1970) 294.

409. I. S. Razina, and M. E. Vikorova, *Zh. Anal. Khim* **25(6)**, (1970) 1160–1165.

410. H. Meier *et al.*, *Mikrochim. Acta* (3) (1970) 553–563.

410a. D. Blasius and M. Fischer, *Z. Anal. Chem.* **177**, (1960) 412–420.

411. A. D. Westland and F. E. Beamish, *Mikrochim. Acta* **5**, (1957) 625–639.

411a. A. P. Blackmore, M. A. Marks, R. R. Barefoot, and F. E. Beamish, *Anal. Chem.* **24**, (1952) 1815–1819.

412. R. Gilchrist, *J. Res. Nat. Bur. Std.* **30**, (1943) 89; *Chem. Rev.* **32**, (1943) 277–372.

413. R. Gilchrist, *J. Res. Nat. Bur. Std.* **20**, (1938) 745–771.

414.   D. E. Ryan, *Analyst* **75**, (1950) 557–561.
415.   E. R. Buchman, A. O. Reims, and H. Sargent, *J. Org. Chem.* **6**, (1941) 764–773.
416.   S. S. Berman and W. A. E. McBoyde, *Analyst* **81**, (1956) 516–570.
417.   W. L. Belew, G. R. Wilson, and L. T. Corbin, *Anal. Chem.* **33**, (1961) 886–888.
418.   J. G. Fraser, F. E. Beamish, and W. A. E. McBoyde, *Anal. Chem.* **26**, (1954) 495–498.

INDEX